Springer Series in
Surface Sciences

37

Springer
Berlin
Heidelberg
New York
Barcelona
Hong Kong
London
Milan
Paris
Singapore
Tokyo

Springer Series in Surface Sciences

37 **Magnetic Multilayers and Giant Magnetoresistance**
Fundamentals and Industrial Applications
Editor: U. Hartmann

Volume 1–36 are listed at the end of the book

Uwe Hartmann (Ed.)

Magnetic Multilayers and Giant Magnetoresistance

Fundamentals and Industrial Applications

With Contributions by
R. Coehoorn
M.A.M. Gijs
P. Grünberg
T. Rasing
K. Röll
H.A.M. van den Berg

With 151 Figures

Springer

Professor Dr. Uwe Hartmann

Institute of Experimental Physics
University of Saarbrücken
66041 Saarbrücken, Germany

Series Editors

Professor Dr. Gerhard Ertl

Fritz-Haber-Institut der Max-Planck-Gesellschaft, Faradayweg 4–6,
14195 Berlin, Germany

Professor Robert Gomer, Ph.D.

The James Franck Institute, The University of Chicago, 5640 Ellis Avenue,
Chicago, IL 60637, USA

Professor Dr. Hans Lüth

Institut für Schicht- und Ionentechnik
Forschungszentrum Jülich GmbH,
52425 Jülich, Germany

Professor Douglas L. Mills, Ph.D.

Department of Physics, University of California,
Irvine, CA 92717, USA

Library of Congress Cataloging-in-Publication Data.

Die Deutsche Bibliothek - CIP-Einheitsaufnahme
Magnetic multilayers and giant magnetoresistance: fundamentals and industrial applications /
Uwe Hartmann (ed.). With contributions by R. Coehoorn... – Berlin; Heidelberg; New York; Barcelona;
Hong Kong; London; Milan; Paris; Singapore; Tokio: Springer, 1999
(Springer series in surface science; 37)
ISBN 3-540-65568-9

ISBN 3-540-65568-9 Springer-Verlag Berlin Heidelberg New York

Typesetting: Camera ready copy from the authors
Printed on acid-free paper SPIN: 10701894 57/3144 - 5 4 3 2 1 0

Preface

Up to now, electronic devices have relied on the precise control of electronic charge. Yet in all microelectronic devices used today, we neglect the fact that nature has given electrons a spin. On the other hand, ferromagnetic phenomena are, of course, ultimately the result of just this electron spin. For a long period of time spin-dependent electronic properties of ferromagnetic materials and micromagnetic phenomena have, even in research, been considered in a completely independent way. In the past, a large amount of effort was concentrated on the understanding of micromagnetic phenomena, but research on the specific electronic properties of ferromagnetic materials was far less intensive. The main reason for this is that industrial applications of magnetic materials have so far been based on their induced or remanent magnetization rather than on their electronic transport properties. However, today we are confronted with the fact that the electron spin as the relevant physical phenomenon explicitely enters industrial applications.

The new area in magnetic research and in industrial application of magnetic materials started with the discovery of exchange coupling in periodic arrangements of thin sheets of ferromagnetic metal with spacers made of normal metal. Under suitable circumstances, these exchange-coupled multilayers exhibit a dramatic change in their electrical resistance under the influence of varying external fields. Since this is quite surprising for a metallic system, the effect was named the giant magnetoresistance (GMR) effect.

About ten years after the discovery of the basic phenomena, the first GMR-based devices are starting to enter the market, in particular as read heads in magnetic recording systems. Peculiar micromagnetic and magnetooptic properties which can be tailored with exchange-coupled multilayers make these devices extremely promising, in particular for high-density magneto-optical recording. Furthermore, the detailed interplay between the spin-polarized electronic transport and the micromagnetic behavior is the basis of new magnetoelectronic devices, like spin valves, spin transistors or magnetic random access memories. For some of these devices, prototypes are already available, and the technology is maturing. The increasing technological interest considerably stimulates today's research on magnetic multilayers and, as a consequence, the number of dedicated scientific papers is already vast.

I felt that at this point in time it is worthwhile to pause for a moment, reflect on past accomplishments and think about future developments. The purpose of this book is to present a unified view of the rapidly growing field of magnetic multilayer research, while especially emphasizing the most important particular industrial applications. It is not the purpose of this book to just present an exhaustive list of all scientific contributions which have been published so far. Apart from a general introduction, six chapters describe a wide range of physical aspects, experimental and theoretical methods, and present as well as future applications of magnetic multilayers. The detailed discussions are based on suitable amounts of key references.

I have been fortunate in being able to enlist excellent scientists, actively working in the field of magnetic multilayers. This leads, in particular, to a well-balanced representation of the industrial point of view. Apart from the basic physics underlying the interlayer exchange coupling and the GMR effect, the applications of the exchange-coupled multilayers in magnetooptical recording and in read heads for longitudinal recording are discussed in detail. The discussion is completed by a survey of the interesting nonlinear magnetooptical phenomena which can be observed on multilayers.

The field of magnetic multilayers and their applications is a multidisciplinary area. The book is intended for scientists from the various specialized fields involved, such as magnetism, materials science, surface science, and physics and chemistry in general. It will certainly be of some value to graduate students who are interested in becoming familiar with the exciting new field of research and development that has strongly increasing industrial importance. Even undergraduate students or readers with as yet little knowledge of magnetism will benefit from an impression of the basic physics underlying numerous devices with which we are already familiar in technical applications or which will become widely used in the near future.

I would like to express my thanks to all contributors for being willing to share their experience and expertise with the interested reader. The efficient collaboration with Springer-Verlag is gratefully acknowledged. It is hoped that it will be a pleasure for the reader to use this book and that the presentation will lead to further stimulate basic and applied research in the fascinating field of magnetic multilayers.

Saarbrücken, *U. Hartmann*
October 1999

Contents

List of Contributors

R. Coehoorn
Philips Research Laboratories
Prof. Holstlaan 4,
NL-5656 AA Eindhoven
The Netherlands

M.A.M. Gijs
Institute of Microsystems
Swiss Federal Institute of Technology
Lausanne (EPFL)
CH-1015 Lausanne
Switzerland

P. Grünberg
Institute for Solid State Research
Jülich Research Center
D-52425 Jülich
Germany

U. Hartmann
Institute of Experimental Physics
University of Saarbrücken
D-66041 Saarbrücken
Germany

Th. Rasing
Research Institute for Materials
University of Nijmegen
NL-6225 ED Nijmegen
The Netherlands

K. Röll
Physics Department
University of Kassel
D-34109 Kassel
Germany

H.A.M. van den Berg
Siemens Corporate Research and Development
D-91052 Erlangen
Germany

1 Introduction

U. Hartmann

Magnetoelectronics is certainly one of the most rapidly expending fields in basic research as well as in industrial application. Magnetic multilayers are the key configurations in this field which allow the utilization of unique micromagnetic, magnetooptic, and magnetoelectronic phenomena which cannot be realized on the basis of conventional materials. The improved understanding of interface-dominated magnetic phenomena results to a considerable amount from the development of new and powerful analysis and imaging methods. The progress in this field as well as the main trends in magnetic thin film research and the main applications of magnetic multilayers are briefly reviewed in the following.

1.1 Recent Trends in Magnetic Thin Film and Multilayer Research

Traditionally, the community of researchers being interested in ferromagnetic phenomena was subdivided into two categories. By far most of the experimental and theoretical work during the past more than fifty years was concentrated on the macroscopic and microscopic behavior of ferromagnetic materials, where the resulting overall properties can be classified as hard or soft magnetic. The subject of this category of research is generally the magnetization reversal process of the ferromagnetic materials and its microscopic origin. Due to both a continuous basic interest and still strongly increasing industrial applications, our knowledge about domain configurations, interdomain boundaries and, in general, about micromagnetics, i.e., the detailed energy balance in a ferromagnetic material, can be considered as fairly profound [1]. This solid knowledge about micromagnetic phenomena enabled us to obtain some continuous progress in tailoring a variety of particular materials exhibiting, for example, extremely hard or soft overall properties [2]. Materials for special applications involve today nanocristalline and amorphous configurations.

For quite some time it has been noticed that magnetic properties at surfaces and interfaces deviate from those of bulk materials. Due to the considerable progress made in the development of techniques suitable for the high-quality deposition of magnetic thin films, the surface- and interface-determined properties of magnetic materials have attracted much interest.

The reason for this is twofold: On the one hand, it is possible to use magnetic thin film systems to tailor special properties which cannot be achieved solely by employing bulk materials. On the other hand, in particular in the magnetic recording industry, we have, as in the whole information technology, a rigorous trend to further miniaturize basic components. This requires thin film technology as the main basis for device fabrication. It is thus not surprising that current research in magnetism is concentrated on ultrathin magnetic films to a considerable amount [3].

The second of the aforementioned categories of research on ferromagnetic phenomena can be attributed to the field of electronic transport properties. Compared to the considerable effort which for many years has been put into the understanding of the micromagnetics, progress in the description of the spin-dependent band structure of ferromagnetic materials and the related electronic transport phenomena was far less pronounced for quite some time. A comparably small amount of research has been devoted to this "exotic" field of physical phenomena. Apart from the complexity of the research subject, clearly the direct industrial applications of spin transport were missing.

The less pronounced interest in magnetoelectronic phenomena abruptly changed when the possibility of an exchange interaction between ferromagnetic films across a nonmagnetic metallic interlayer was clearly experimentally identified in Fe/Cr structures and rare earth-based multilayers in 1986 [4–6]. While exchange coupling is certainly a phenomenon which cannot be treated in terms of purely micromagnetic considerations but only in terms of band-structure-related properties, its really stimulating influence on the renascent area of magnetoelectronics became definitely clear with the discovery of the "Giant Magnetoresistance" (GMR) in antiferromagnetically coupled Fe/Cr systems [7,8]. Today, of course, it is well known that the antiferromagnetic coupling is not a necessary prerequisite for the GMR effect which can be obtained in spin-valve structures if an antialignment of individual ferromagnetic layers is obtained by other means [9].

Actually since more than 140 years another galvano-magnetic mechanism, namely the Anisotropic Magnetoresistive effect (AMR) [10], is known. It relies in its technical realization on a detailed interplay between the micromagnetic configuration and the electronic transport properties. The AMR effect found a considerable range of applications over the past twenty years. Its ultimate origin is the spin-orbit coupling and thus the GMR effect and the AMR effect have a somehow close resemblance from a basic physical point of view. However, from the technical optimization point of view the GMR effect, which is realized in magnetic thin film and multilayer systems, opens a completely new and very wide area of research because, in contrast to AMR configurations, the boundary conditions, involving the system design and the processing parameters, are much more manifold. Additionally, already existing and potential future industrial applications, which benefit from the fact that the GMR effect can by far exceed the AMR effect, further push to a considerable amount the currently performed basic research. While it is still questionable if another spin-transport phenomenon, the Colossal Magnetoresistance (CMR) effect [11] found in a completely different class of materials, will ever gain the status of an industrially widely used phenomenon, the strongly growing applicational relevance of the GMR effect and its related physics can clearly be emphasized for the years to come.

In conclusion, recent trends in magnetic thinfilm and multilayer research are concentrating on the interesting phenomena which occur (i) if the properties of magnetic layers are determined by surface and interface effects, (ii) if the layers are so thin and eventually additionally structured so that low-dimensional properties play a role, (iii) if successive magnetic layers are electronically coupled through nonmagnetic interlayers, or (iv) if the electronic transport through multilayer structures is considered. It is clear, that, apart from the purely electronic properties of multilayer systems, still the micromagnetic behavior is of basic importance. While this is evident for multilayer configurations used in magnetooptical recording in the absence of any electronic transport current, it also holds for devices based on magnetoelectronic transport. The magnetization of a multilayer system of some finite size is, of course, generally determined by the actual domain configuration. Additionally, if we are dealing with a magnetoresistive device, a central experimental parameter is given by an external magnetic field which affects the electronic resistance of the device via changing its magnetization and thus its domain configuration. One of the main characteristics in modern research on magnetism is certainly that electronic and micromagnetic properties of ferromagnetic materials, which have been considered almost independently in the past, are now viewed as interplaying properties of a system, where the predominant goal is to optimize this interplay. Such an optimization requires, apart from a profound understanding of the micromagnetic and electronic phenomena, systematic materials research.

1.2 Progress in High-Resolution Imaging

Progress in materials science and also in the understanding of micromagnetic phenomena and magnetization reversal processes was always closely coupled to the progress in the development of tools being suitable to directly observe phenomena in real space at high resolution. While for really basic research it is only important that suitable methods somewhere exist, it is of particular importance for concentrated applied research that methods exist which allow a wide employment locally at many places. Thus, it should not be underestimated that the introduction of the so-called scanning probe methods [12] particularly supported research on magnetic thin films and multilayers, in addition to the whole variety of other well-established and recently developed dedicated methods [1]. It is important to emphasize that not only direct imaging of micromagnetic configurations is of interest but that the reproducible

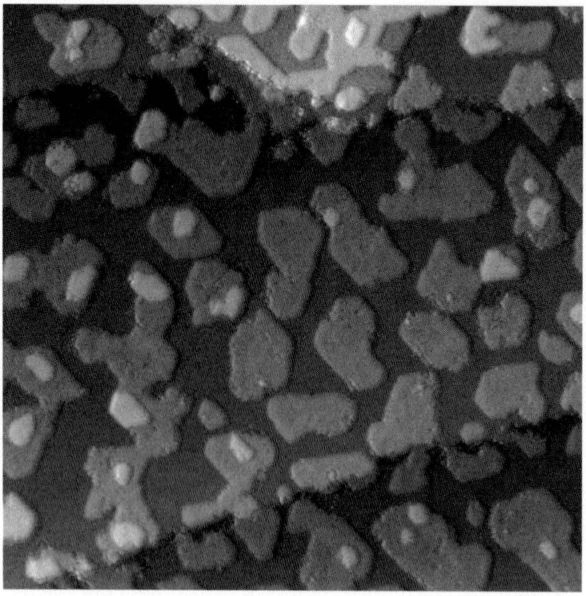

Fig. 1.1. STM image of Fe islands growing on an Au(111)-substrate. The scanned area is 50 nm × 50 nm [13]

fabrication of optimized thinfilm and multilayer systems largely relies on a detailed characterization of the morphology during processing. In this context the Scanning Tunneling Microscope (STM) has become an important standard tool which allows the monitoring of thin film growth even under in situ conditions, e.g., in Ultra-High Vacuum (UHV). Figure 1.1 shows an example of an STM analysis of the growth of a particular multilayer system [13].

For GMR devices as well as for material configurations used in magnetic recording as media in the absence of any transport current it is, of course, of particular interest to monitor the domain configuration in the magnetic films. Magnetic Force Microscopy (MFM) is dedicated to this purpose. Using this method it is possible to monitor under in situ conditions the domain configuration of thin films at high resolution. In this way, complex domain structures extended over large areas can be interpreted in detail, as shown in Fig. 1.2. Additionally, not only the detailed magnetization rotation in individual domain boundaries (Fig. 1.3) but also, for example, the occurence of a ripple structure (Fig. 1.4) can today be verified in a routine way by MFM.

It should be kept in mind that, apart from magnetoelectronic phenomena, also magnetooptic properties of thinfilm and multilayer systems are of basic as well as of a strong applied interest. The industrial interest mainly results from the fact that, on the basis of coupled multilayers, magnetooptical properties can be engineered in much the same way as the magnetoelectronic properties, based on a parameter space more extended than given if only bulk

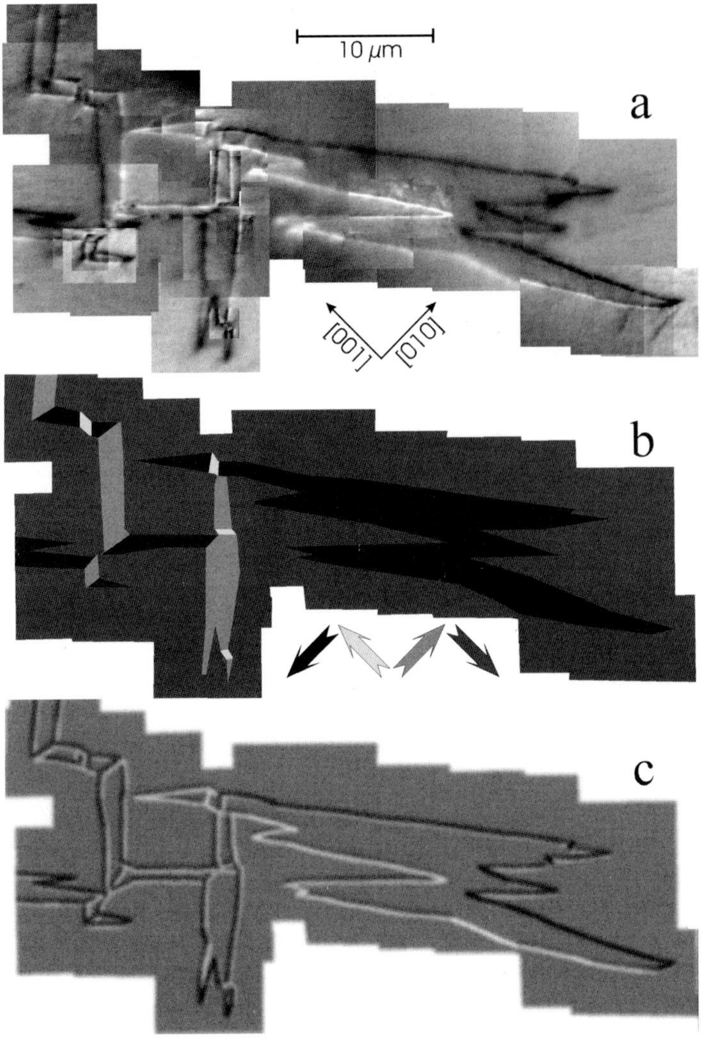

Fig. 1.2. a–c Large-area MFM image of a complex domain arrangement in a Fe thin film, in situ observed under UHV conditions. (**a**) shows the complete domain arrangement which is composed of numerous individual images, (**b**) gives the derived magnetization orientations, and (**c**) yields details of the local variation of the wall contrast, as derived from the variations in wall magnetic charging [14]

properties of materials are employed. Thus, it is not surprising that presently new concepts in magnetooptical recording are directly derived from the peculiar properties of multilayer systems. With respect to materials analysis it is certainly of major interest in magnetooptical recording to be able to lacally verify the magnetooptical behavior of a multilayer configuration. For this

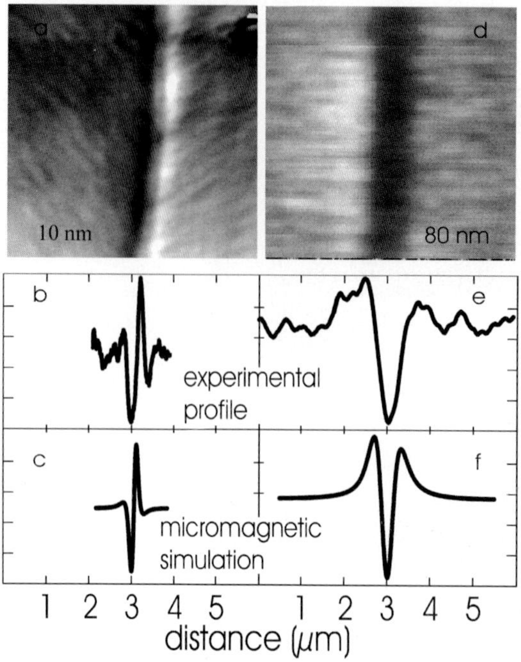

Fig. 1.3. Experimentally recorded cross-sectional profiles of interdomain boundaries in Fe thin films in comparison with theoretical results. The left part of the image shows a 90° Néel-type boundary in a 10 nm thick film while the right part shows a 90° Bloch-type wall in a 80 nm thick film. The MFM data have been acquired under UHV conditions [14]

purpose another variant of the family of scanning probe methods is well suitable, namely Scanning Near-field Optical Microscopy (SNOM). This method allows realization of the Kerr or Faraday effect at sub-wavelength resolution. Figure 1.5 exhibits a representative example for sub-wavelength magnetooptical imaging in the Faraday mode on a particular multilayer system [16].

Already today, scanning probe methods in combination with a large variety of other dedicated methods are an extremely important support for the development of optimized magnetic thin film and multilayer systems. Apart from being powerful analysis tools, the most important advantage of the scanning probe methods in comparison to other sophisticated analysis instruments is that, due to their moderate costs, they are widely distributed giving a large number of researchers access to them.

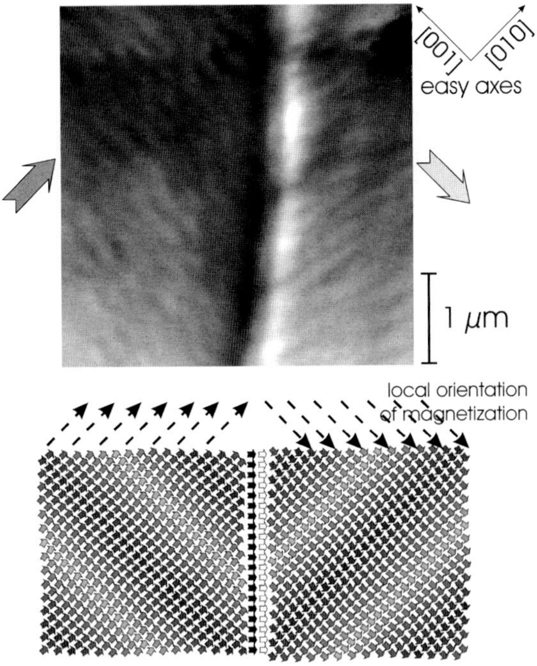

Fig. 1.4. UHV-MFM image of the magnetic ripple structure, detected in a 10 nm thick Fe film in the vicinity of a 90° Néel-type wall [15]

1.3 Present and Future Applications of Magnetic Multilayers and Related Devices

Apart from the ever-existing basic interest the present and potential future applications of magnetic multilayers are a very strong pushing force to the current research in this area. Among all fields of possible application the magnetic recording industry certainly extended the strongest momentum.

In analog as well as digital magnetic recording there is a strong and continuous effort to permanently decrease the characteristic size of a magnetic bit, i.e., the size of the magnetized area of the recording medium that represents the information unit. In standard tape systems as well as in hard-disk drives the inductive write head magnetizes the medium within the surface plane such that adjacent domains have a head-to-head or a tail-to-tail configuration. In such a longitudinal recording set-up the recording medium consists of a multilayer configuration which involves, however, in general only one magnetically active thin film sandwiched between carrier and protective layers. Much effort is presently concentrated towards optimizing the magnetic thin films to be capable of carrying increased bit densities. Since, however, longitudinal recording media are not composed of multilayers as discussed here, their future development will not be discussed further in the present

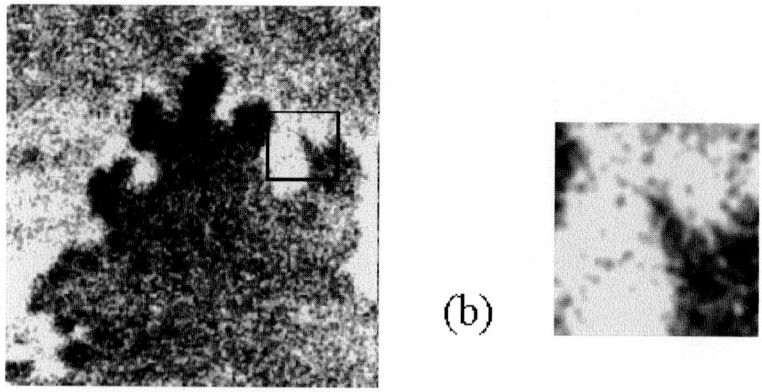

Fig. 1.5. a, b Near-field magnetooptical Faraday image of domains in a Co/Pt multilayer. While (**a**) shows the general appearance of the domains the high-resolution image (**b**) clearly confirms the obtained sub-wavelength lateral resolution [16]

context. However, another issue concerning longitudinal recording is particularly important: The areal bit density which can be achieved in longitudinal, and vertical recording as well, obviously depends on the bit size produced by the inductive write head as well as on the ability of the read head to detect those tiny bits. Actually, it turns out that the latter aspect is the crucial one. Earlier read heads were based on magnetic induction as are the write heads. The introduction of AMR read heads, since 1980 in tape systems and in the beginning of the nineties also in hard-disk systems, has considerably increased the yearly growth of the areal bit density in longitudinal recording: For hard-disk systems from 30% per year for inductive thinfilm heads to 60% per year for AMR heads. GMR recording heads which are presently introduced into the marked for hard disk drives are expected to sustain or even further increase this annual growth. It is evident that the extreme competition in magnetic-recording technology exerts considerable pressure on further developments in GMR-based magnetic multilayer systems.

Dealing with magnetic recording another issue is important in the context of applications of magnetic multilayers. There is presently a strong competition between longitudinal recording approaches, as mentioned above, and magnetooptical recording which is a completely different approach and which is based on a vertical magnetization of the recording medium, thus a priori involving some inherent advantages. The bits, thermomagnetically written into the medium, are detected via the magnetooptical Faraday effect. A major limit to the achievable areal bit density is thus the optical diffraction limit. In order to further shift the latter towards smaller bit sizes it is promising to use green and blue laser light for optical detection. Optimum performance of the storage medium is expected for sophisticated multilayer structures tailored to carry, on the one hand, ultrahigh bit densities introduced by the

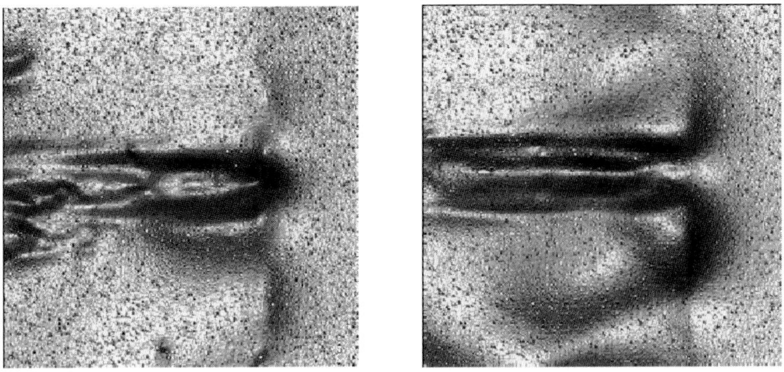

Fig. 1.6. Magnetoresistively recorded images of a weld. The difference in quality of the welded joint becomes clearly evident from the spatially resolved detection of the near-surface stray-field variations [17]. The image size is 5 cm × 5 cm

thermomagnetic writing process and exhibiting, on the other hand, a maximum Faraday effect in the reading process. As industrial applications of magnetoelectronic phenomena also the applications of magnetooptics stimulate research on magnetic multilayers considerably.

Apart from applications in magnetic data storage other present applications of thinfilm and multilayer devices mainly cover the sensitive detection of magnetic fields in position and angle detection systems (automotive industry), in magnetometers, in compass systems, and for measuring electrical currents. So far these applications are predominantly based on AMR sensors, however, the future use of GMR-based multilayer devices is considered as very promising.

Another rapidly growing area of application of magnetoresistive devices is the nondestructive testing and evaluation of metallic materials and devices. The magnetoresistive element is moved in close proximity across the surface of the sample and thus detects a variety of defects, like cracks, inclusions or other inhomogeneties, in terms of local variations of the magnetic stray field. An obvious example showing the possibility of magnetoresistively detecting defects in this way is provided by Fig. 1.6. If the sample is nonmagnetic, the magnetoresistive detection is combined with the excitation of eddy currents in the metallic material, where the magnetic field produced by the eddy currents is, as the remanent magnetic field of ferromagnetic samples, affected by the presence of defects and inhomogeneties. Nondestructive testing which is an application of magnetoresistive sensors of strongly growing importance will much benefit from the availability of devices with increased field sensitivity and minimized sensor dimensions. Thus the development of new GMR-multilayer devices will have a strong influence on the future development in this field.

In general, it is assumed that magnetoelectronics will be of increasing importance in future. Promising approaches involve, in particular, spin-valve arrangements and spin-polarized tunneling devices. Concepts exist to set up, for instance, spin-valve transistors or even complete magnetic random access memory chips.

1.4 Survey of This Book

The following chapters provide a broad overview over the recent developments in the field of magnetic multilayers involving a detailed discussion of the underlying fundamental physics, of experimental and theoretical possibilities to study the specific physical phenomena, and of present and future applications.

Chapter 2 presents a survey of the recent progress made in magnetooptical data storage. After an introduction to the field and a discussion of the basic fundamentals of magnetooptical recording, the basic requirements to thinfilm and multilayer arrangements used as recording media are discussed in detail. These include, in particular, the use of materials which are suitable for the application of green and blue laser radiation. Of central importance for the engineering of optimized magnetooptical systems is the application of exchange-coupled layers.

In Chap. 3,a detailed introduction to the physics of interlayer exchange-coupled systems and to the striking phenomenon of the giant magnetoresistance (GMR) is given. The different types of coupling are analyzed, and relevant coupling parameters are defined. The related phenomenon of multiperiodic oscillations is analyzed in terms of microscopic models. The GMR effect will be phenomenologically discribed, and the underlying microscopic mechanisms are summarized.

Chapter 4 concentrates on GMR in exchange-biased spin-valve layered structures. These structures consist of a magnetically soft ferromagnetic layer separated by a nonferromagnetic layer from a second ferromagnetic layer which has its magnetization pinned by the exchange-biasing interaction with a third magnetic (usually antiferromagnetic or ferrimagnetic) layer. For suitable layer compositions and nanometer-scale layer thicknesses, these materials combine the GMR effect with a very small field interval close to zero field in which the resistance change takes place. The resulting high sensitivity makes the spin-valve structures potentially attractive for sensor applications and other magnetoelectronic devices. The chapter reviews experimental results on GMR in the involved materials, on their theoretical interpretation, and on the magnetic interactions which determine the magnetization reversal process. Trends in magnetic recording and the potential benefit of exchange-biased spin valves in yoke-type read heads for high-density digital magnetic recording are additionally discussed.

Chapter 5 deals with experiments on the perpendicular GMR in magnetic multilayers. An overview is given on the various experimental investigations of the GMR effect in magnetic multilayers measured with the current perpendicular to the multilayer plane. Experimental results on the perpendicular magnetoresistance effect are relatively scarce. The reason is that they are much harder to obtain than those from the more conventional current-in-plane magnetoresistance measurements: The small, perpendicular resistance of magnetic multilayers requires the use of specialized ultralow-resistance measurement techniques, or of micro- and nanofabrication methods to realize samples with a higher perpendicular resistance which can be measured using a more conventional technique. The attractiveness of the perpendicular magnetoresistance comes, from a fundamental physical view, from the high symmetry of the problem enabling an easier understanding of physical processes, and, from an application point of view, from the fact that the perpendicular effect typically is a factor of 2–5 larger than the corresponding current-in-plane magnetoresistance effect. The physics of and the methods for studying metallic multilayers with interlayer exchange coupling and GMR response are obviously quite important.

Chapter 6 is devoted to the afore-mentioned subject. It involves an overview of the developments in the field of magnetoresistive sensors and discusses in its central part in much detail the physics of the interlayer coupling and presents a considerable amount of recent experimental results. The physics of the GMR effect is analyzed from a fundamental point of view on the basis of a rigorous comparison between the experimental results and the recently developed elaborate theoretical models. A discussion of prospects for the future applications of GMR sensors completes this part.

Chapter 7 finally introduces the reader to the nonlinear magnetooptics exhibited by magnetic thinfilm and multilayer configurations. Due to its extremely high surface and interface sensitivity second-harmonic generation and the resulting nonlinear magnetooptical effects turn out to be extremely promising phenomena for the analysis of thinfilm and multilayer configurations. The presentation gives a detailed introduction to the physics underlying the magnetooptical phenomena and, in particular, the magnetization-induced second-harmonic generation. The second-harmonic generation for multilayers, particular experimental considerations, and some representative experimental results on magnetic thinfilm structures and systems involving quantum-well states are discussed in detail.

References

1. A. Hubert, R. Schäfer: *Magnetic Domains - The Analysis of Magnetic Microstructures* (Springer, Berlin Heidelberg 1998)
2. For a general overview, see, e.g., J. Evetts (ed.): *Concise Encyclopedia of Magnetic and Superconducting Materials* (Pergamon, Oxford, 1992), pp. 216
3. J. A. C. Bland, B. Heinrich (eds.): *Ultrathin Magnetic Structures: An Introduction to the Electronic, Magnetic and Structural Properties* Vols. I and II (Springer, Berlin Heidelberg 1994)
4. P. Grünberg, R. Schreiber, Y. Pang, M. B. Brodsly, H. Sowers: Phys. Rev. Lett. **57**, 2442 (1986)
5. M. B. Salamon, S. Sinha, J. J. Phyne, J. E. Cunningham, R. W. Erwin, J. Borchers, C. P. Flynn: Phys. Rev. Lett. **56**, 259 (1986)
6. C. F. Majkrzak, J. W. Cable, J. Kwo, M. Hong, D. B. McWhan, Y. Yatet, J. V. Waszc-zak, C. Vettier: Phys. Rev. Lett. **56**, 2700 (1986)
7. M. N. Baibich, J. M. Broto, A. Fert, F. Nguyen-Van-Dan, F. Petroff, P. Etienne, G. Creuzet, A. Friedrich, J. Chazelas: Phys. Rev. Lett. **61**, 2472 (1988)
8. G. Binasch, P. Grünberg, F. Saurenbach, W. Zinn: Phys. Rev. B **39**, 4828 (1989)
9. J. Barnas, A. Fuss, R. E. Camley, P. Grünberg, W. Zinn: Phys. Rev. B **42**, 8110 (1990)
10. W. Thompson: Proc. R. Soc. (London) A **8**, 546 (1857)
11. See, e.g., A. P. Ramirez: J. Phys. Condens. Matter **9**, 8171 (1997)
12. R. Wiesendanger, H.-J. Güntherodt (eds.): *Scanning Tunneling Microscopy I–III*, 2nd edn., Springer Ser. in Surf. Sci. Vols. **20**, **28**, **29** (Springer, Berlin Heidelberg 1992–95)
13. University of Saarbrücken, unpublished data (1998)
14. U. Memmert, P. Leinenbach, J. Lösch, U. Hartmann: J. Magn. Magn. Mat. **190**, 124 (1998)
15. P. Leinenbach, J. Lösch, U. Memmert, U. Hartmann: Appl. Phys. A **66**, 1191 (1998)
16. University of Saarbrücken, unpublished data (1997)
17. University of Saarbrücken, unpublished data (1997)

2 Progress in Magnetooptical Data Storage

K. Röll

Magnetic recording is widely used in many present applications, for instance, in tapes for audio and video recording or in floppy and hard disks for data storage [1–5]. Optical recording, on the other hand, is still an emerging technique [5,6]. In a consumer product it has started with the Compact audio disk and is now developing in many varieties, for instance, Compact Disk-Read Only Memories (CD-ROMs), Write-Once-Read-Many media (WORMs) or magnetooptical minidisks. For comparison of magnetic and optical recording the basic principles and the physical limitations of both techniques will be discussed in the following.

2.1 Magnetic and Optical Recording

Metallic thin films are most often applied as magnetic recording media in hard disks. An inductive magnetic head slides at a small distance ($\approx 0.1\,\mu m$) on an air bearing above the rotating disk. It writes a binary information into the film by means of oppositely directed magnetic domains. Due to the shape anisotropy of the film the magnetization of the domains is parallel to the film plane in the direction of the moving track (longitudinal recording).

The stability of the written domains is limited by demagnetizing fields, which have their origin in the magnetic discontinuities at the domain walls. In order to get a stable written information, the coercivity must be larger than the demagnetizing field. The latter increases with increasing recording density. At present, the domain length is limited to about $0.3\,\mu m$ which corresponds to 83,000 flux changes per inch (83 kfci). To stabilize such an information density along the track, a coercivity of the order of 300 kA/m is required. This is achieved by the use of double layers consisting, for instance, of Cr and CoNi and/or by additional elements such as Cr, Pt, W in the magnetic material [7]. If the coercivity would be further increased, the magnetic field of the head may no longer be able to magnetize the medium for writing the domains.

In principle, this problem could be overcome by decreasing the distance between head and medium. This distance is even more critical for the read-out signal, which drops exponentially with distance and the recording density. Further reduction of the distance is limited by mechanical problems, for instance, surface roughness or mechanical instability of the sliding head, and by a protective layer which is required to reduce friction and stiction between head and medium surface. The present flying hight of 0.1 μm is almost equivalent to the mean free path of the atoms at atmospheric pressure. The thickness of the protective layer, which consists of amorphous carbon, is normally 0.01 μm. Therefore, the effective distance between head and medium is 0.11 μm.

This is important, in particular, for future perpendicular recording, where the domains will be magnetized normal to the film plane. This reduces the demagnetizing limit for the domain size and allows a density along the track as high as 700 kfci [8]. The problem to read such a high-density information, however, has not yet been solved in a technical product. Therefore, the linear density along the track is mainly limited by the head-medium distance.

A limit for the area density is given by the width of the track. Significant improvement has been achieved by the use of magnetoresistive readout heads which have a better signal-to-noise ratio than the previously employed inductive readout heads. An inductive write head is most often combined with a magnetoresistive readout head, both prepared from thinfilm stacks by a sequence of deposition, photolithography and etching processes. The track width required for readout is about 3 μm, but the written track is slightly larger to improve positioning of the head along the track. A distance of about 2 μm between the tracks leads to a typical density of 5,000 tracks per inch (tpi). As the information density can be increased by a factor of 1.5 by appropriate coding, the areal density at present is about 600 MBit/inch^2.

The sensitive conditions for the interaction between head and medium require an extremely clean air surrounding. Therefore, the disks are encapsulated in a housing, where the air is contineously filtered to particles of the order of 0.1 μm.

A completely different approach is performed in the various types of optical disks, indexoptical disks where the information is read by the reflection of a focussed laser beam. In the Compact Disk (CD), which is commonly used for audio recording or as a Read-Only-Memory (CD-ROM), the information is coded in small pits, which are printed on the disk surface during the production process (Fig. 2.1). The surface of the disk is coated by an aluminum layer with high reflectivity and is protected against damage by a thick lacquer layer. The optical contrast for readout is achieved by the interference of the laser beams reflected from the bottom and from the border of the pit. The pits are arranged in tracks 1.6 μm apart, which corresponds to a track density of 16,000 tpi. Taking, on average, a bit length of 1.5 μm, an area density of 400 MBit/inch^2 is achieved. The grooved structure along the track is used for positioning and focussing of the optical head.

In contrast to magnetic recording the distance between the information layer and the optical readout head is of the order of 1 mm. Because the readout procedure is performed through the transparent substrate, the diameter of the laser beam is $\approx$ 1 mm at the substrate surface. Scratches, dust or other

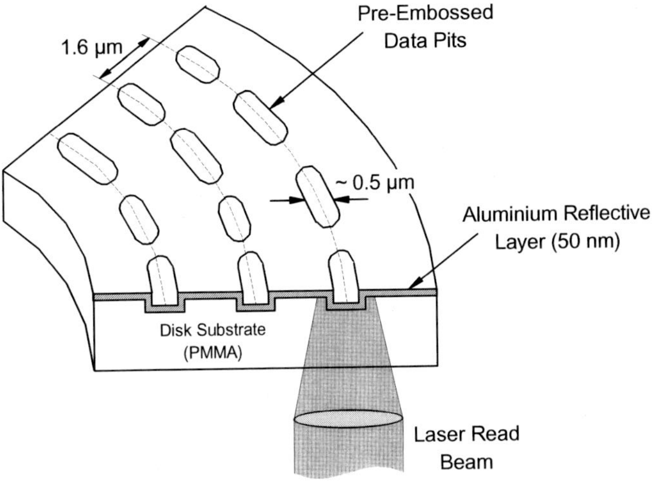

Fig. 2.1. Compact disk as an example for optical recording

damages do practically not disturb the optical signal. Therefore, handling of the disk is very easy and does not require any substantial care.

The areal density is principally limited by diffraction of the laser light ($\lambda \approx 800\,\mathrm{nm}$). At present the dimensions of tracks and pits are much larger than this limit of about $(0.8\,\mu\mathrm{m})^2$, which corresponds to an areal density of $1.5\,\mathrm{GBit/inch}^2$. The areal density, therefore, could be increased by a factor of 3.5 by changing the dimensions without a substantial change of the technology.

A concept similar to that of the compact disk is used in all optical recording techniques, which differ mainly in the way how the information is written and detected [5,6]. Media for "Write Once, Read Many" (WORM), for instance, consist of a sensitive layer, which is locally destroyed by an intense laser pulse. This can be achieved by burning holes in a low melting tellure layer or by forming small bubbles in an AuCr film by local evaporation of a polymer substrate layer. The local variations of the structure lead to an optical contrast for the reading laser. In phase-change media a reversible transition from an amorphous to a crystalline state of GeTeSb layers is used for writing, reading and erasing [9]. A similar principle is applied in magnetooptical recording [10–14], where the information is written and erased thermomagnetically and is read by the magnetooptical Kerr or Faraday effects. The reversible erasing process allows an almost unlimited number of writing cycles.

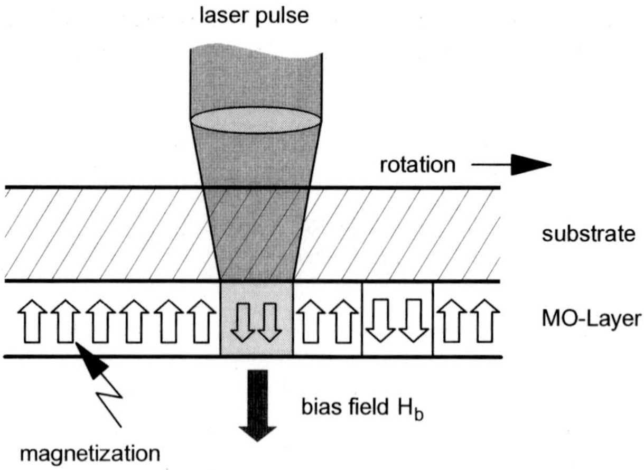

Fig. 2.2. Principle of magnetooptical recording

2.2 Magnetooptical Recording

In the thermomagnetic writing process a homogeneously magnetized medium, which may represent the information "1" (Fig. 2.2), is heated locally by a short laser pulse to a temperature near the Curie temperature. If an oppositely oriented bias field H_b is present during cooling a magnetic domain is formed, which may represent the information "0" [1]. The size of the domain is determined by the temperature profile $T(r)$ and by the profile of the temperature dependent coercivity H_c. The stability of the domain is influenced by the wall energy, the magnetization and the anisotropy energy (Sect. 2.2.1). It should be noted, that the size of the written domain can be smaller than the diffraction limit of the laser beam (Sect. 2.3.6).

The reading process is performed by employing the magnetooptical Kerr effect at typical wavelengths of 780 or 820 nm. The linear polarization of a low-power reading laser is transformed to an elliptical polarization and tilted by a small angle Θ_K when the light is reflected from a magnetized medium. For normal incidence, the Kerr rotation Θ_K is proportional to the magnetization component parallel to the incident laser beam (polar Kerr effect). To obtain maximum signal, the medium should be prepared with a magnetization perpendicular to the film plane (perpendicular recording).

The Kerr rotation and ellipticity which represent the orientation of the written domains, are detected by a polarizer which transforms the change of polarization into a modulation of the light intensity [12]. Most often the

[1] For the sake of simplicity, the magnetization directions up (↑) and down (↓) are *arbitrarily* attributed to the binary data "1" and "0", respectively, throughout the present chapter. In technical applications, however, the binary coding is derived from the transitions between the states by complex algorithms (see, for instance, [2,12])

reflected light is split into two beams with orthogonal polarizations, which include an angle of 45° with the polarization of the laser beam. A normalized signal,

$$S = \frac{(I_1 - I_2)}{(I_1 + I_2)}, \tag{2.1}$$

which is proportional to the Kerr rotation, is formed electronically from the intensities I_1 and I_2 of the two separated beams. For focussing and tracking the disk is pregrooved during fabrication with a groove distance of 1.4 µm. The same principles as in the CD technique allow focussing and tracking by means of a 4-quadrant-detector [12].

For writing new information a high-intensity laser is applied to the whole track in the presence of a bias field, to magnetize the film homogeneously along the track. This erasing requires a full rotation of the disk and limits the data access time. Present research is therefore concentrated on the development of methods avoiding the procedure by utilizing aproppriate layer systems which allow Direct Over-Writing (DOW) (Sect. 2.3).

2.2.1 Basic Material Properties

The performance quality of the writing, reading and erasing process as well as the stability of the domains depends strongly on the material properties of the magneooptic films. These properties have been extensively investigated in the past decade. They have thoroughly been reviewed by many researchers [12–21]. The main aspects will be discussed in the following.

A perpendicular magnetic anisotropy is required to keep the magnetization M_s of the domains normal to the film plane. The energy of the demagnetizing field, therefore, must be compensated by a perpendicular anisotropy energy K_u:

$$K_u \geq (1/2)\mu_0 M_s^2 . \tag{2.2}$$

This equation can be fulfilled when the magnetization is sufficiently small. This is achieved by amorphous thin films consisting of Rare Earth metals (RE) and Transition Metals (TM) , which have an antiferromagnetic coupling [17–21]. The film thickness is of the order of 50 to 100 nm. A typical composition is $Tb_{25}Fe_{75}$. In closely related compounds Tb and Fe are partially substituted by other RE metals (Dy, Gd, ...) and TM metals (Co, Ni), respectively. Because of the different temperature dependences of the magnetic TM and RE subnetworks the films have normally a compensation temperature T_{comp}, where the magnetizations of RE and TM cancel each other, and the net magnetization is zero (Fig. 2.3). By appropriate film composition the compensation temperature can be chosen close to room temperature. As a consequence of zero magnetization the coercivity becomes very large at the compensation temperature, since an infinite field would be required to

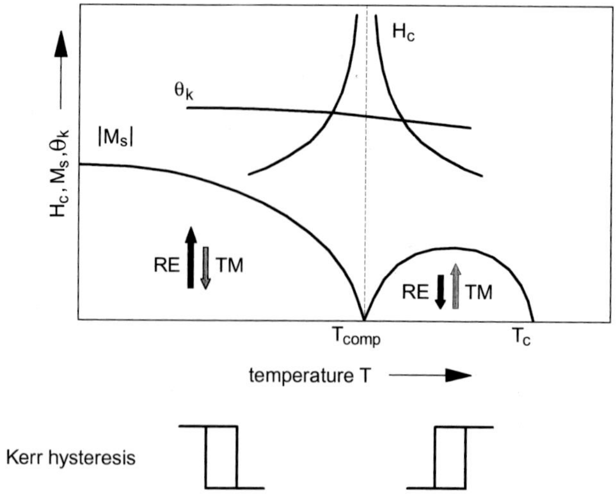

Fig. 2.3. Magnetooptical properties of amorphous rare earth (RE) transistion metal (TM) thin films. (M_s: saturation magnetization, Θ_K: Kerr rotation, H_c: coercivity). In the RE dominated range the hysteresis loop measured by the magnetooptical Kerr effect is reversed with respect to the sign, because the Kerr rotation mainly results from the TM component

change the magnetic moments. Therefore, a domain structure written at a Curie temperature of about 250°C is very stable against external fields at room temperature.

The Kerr effect required for reading is not influenced by the compensation effects. It is related to interband and intraband transitions of the RE/TM alloy and thus to the subnetwork magnetizations rather than to the net magnetization. The Kerr rotation Θ_K and the Kerr ellipticity ϵ_K can be calculated from the diagonal and off-diagonal elements of the complex conductivity, which are related to the optical constants and which depend strongly on the electronic band structure [22,23]. For infrared laser light ($\lambda \approx 800\,\mu m$), the main contribution comes from the 3d electrons of the transition metals, whereas the 4f electrons of the RE metals have less influence. The magnitude of the Kerr rotation, therefore, can be adjusted by the composition [17–21].

The films are prepared by evaporation or sputtering and grow in an amorphous state which is necessary for a low medium noise. Evaporation or rf diode sputtering is preferred for basic investigations because of the greater process variability. In disk production, however, magnetron sputtering is usually applied to obtain high deposition rates at a sufficiently low thermal load of the disk during preparation (Sect. 2.2.3).

In spite of the amorphous structure the films have a strong anisotropy perpendicular to the film plane, the origin of which is not yet clear [24]. Contributions may come from a fibrouse structure, where the atoms are ran-

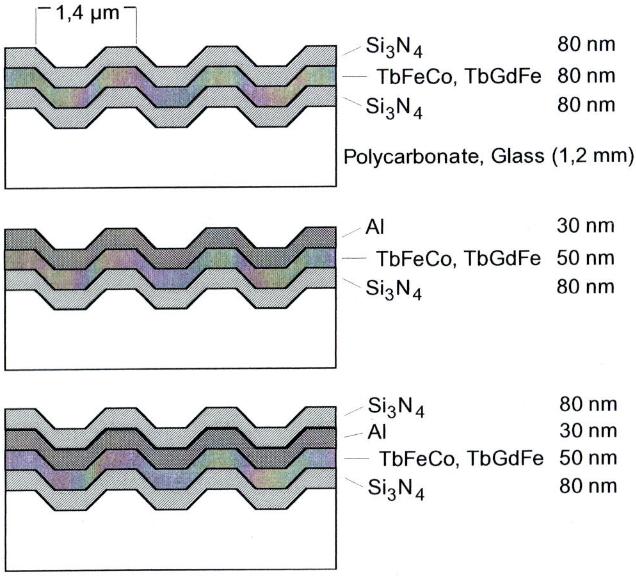

Fig. 2.4. Typical layer systems for magnetooptical disks. A grooved surface structure is necessary for focusing and tracking. All layer systems are additionally covered by a protective lacquer layer

domly arranged along fibers with a preferred direction normal to the film plane. Other contributions may be induced by single-ion anisotropy of the RE elements, by pair ordering due to different resputtering rates of the RE and TM atoms or by magnetostriction effects due to internal stresses in the films. The anisotropy stabilizes the orientation of the domains normal to the film plane, and influences the minimum size of the domains by means of the wall energy $\sigma_{\mathrm{w}} = 4(AK_{\mathrm{u}})^{1/2}$, where A is the exchange stiffness constant.

The formation of a stable domain is determined by the laser-induced temperature profile, the radial and time dependence of the magnetic properties, the domain nucleation, and the domain wall motion [25–27]. In case of nucleation-controlled domain formation the shape of the domains is irregular, which leads to a high write noise. Therefore, formation controlled by domain-wall motion is preferred, where homogeneously magnetized domains are obtained. The situation can be described by a force that acts on the domain wall and determines the final position [25]. It depends on the radial distributions of magnetization, coercivity, wall energy and demagnetizing field, which are induced by the temperature profile of the laser heating pulse. The minimum bit size can be roughly approximated [11] by

$$r \geq (\sigma_{\mathrm{w}}/M_{\mathrm{s}}H_{\mathrm{c}})\,. \tag{2.3}$$

To obtain the minimum recorded bit size, the domain-wall energy should be as small as possible while still maintaining the conditions of (2.2).

All magnetic properties, in particular the compensation temperature and the coercivity at room temperature, depend strongly on the composition which can be changed significantly by the preparation conditions. In sputtering, for instance, the composition of the film may deviate from the composition of the target due to scattering of the sputtered atoms in the sputtering atmosphere and due to preferential sputtering from the growing film induced by a substrate bias voltage. Because the various magnetic properties can not be adjusted independently of each other, an optimum compromise has to be found for contradictory demands [28]. Kryder et al. [29] have proposed a strategy to control the compensation temperature, Curie temperature, coercivity at room temperature, anisotropy and the Kerr rotation of GdTbFeCo films systematically by the composition of the target, the substrate bias voltage and the sputtering pressure.

In addition to the magnetooptical layer, the recording disk contains a film stack with three or more different films (Fig. 2.4). Most important is an antireflecting coating, which reduces the polarization component of the directly reflected light by interference, but does not influence the component induced by the Kerr rotation. Therefore, the Kerr rotation is increased, but the total reflectivity R is reduced [23,30,31]. The figure of merit is

$$(S/N)_{\mathrm{shot}} = [R({\Theta_{\mathrm{K}}}^2 + {\epsilon_{\mathrm{K}}}^2)]^{1/2}\,, \tag{2.4}$$

which corresponds to the normalized signal-to-noise ratio, when only shot noise is considered. It can be increased by multilayer structures, for instance by a magnetooptical double layer ($Tb_{11}Fe_{64}Co_{25}/Tb_{24}Fe_{65}Co_{11}$) [32] or by trilayers or quadrilayers with an additional reflecting layer [33]. For the dielectric film commonly Si_3N_4 is used, which has an optical refraction index of $n \approx 2.0$. Other materials such as AlN and Al_2O_3 have also been taken into account (Sect. 2.2.3).

There are several other sources of noise, the most important of which is the write noise due to irregularities of the written domains. As already explained, it depends on the temperature profile of the heating pulse. Therefore, the optical absorption, which controls the heating process, as well as the thermal conductivity of the complete layer system have also to be taken into account [34]. An important criterion for the disk performance is the Carrier-to-Noise Ratio (CNR) defined as

$$\mathrm{CNR} = 10 \log \left(S^2/\langle N^2\rangle_{\mathrm{b}}\right), \tag{2.5}$$

where S and N are the detector currents equivalent to signal and noise, respectively, measured at a certain bandwidth b [11,16]. A standard test for the disk quality is the measurement of CNR as a function of the write power and of the data density [32–35].

In summary, many physical properties have to be taken into consideration when the material for the magnetooptical disk is chosen. Further technical properties such as corrosion, long-term stability, defects, homogeneity of the film, availability and costs of materials are also crucial for the practical performance of the disk. An example was given by Raasch [35] who compared the recording characteristics of Dy/FeCo-based magnetooptical disks with more conventional disks using Tb-FeCo and GdTb-Fe materials (typical data for currently used disks are discussed in Sect. 2.4).

2.2.2 Materials for Green and Blue Laser Radiation

Many alternatives to the standard amorphous RE-TM films, typically TbFeCo or GdTbFe, have been investigated to solve the problems discussed in Sect.2.2.1 [36]. One of the most serious problems of the RE-TM alloys is the sensitivity against oxidation and corrosion. Attempts have been made to solve this problem by magnetic oxides such as Co ferrites (cubic), Ba ferrites (hexagonal) or garnets [36,37]. These materials are polycrystalline with crystallites of the order of 50 nm to 0.5 μm, which lead to a relatively poor CNR. Because of the transparency of the oxides the Faraday rotation, which is typically 1°–10° per μm thickness, is employed instead of the magnetooptical Kerr effect. The most interesting material is a (Bi, Ga)-substituted Dy-Fe garnet which has a good CNR at a short wavelength (400 nm) [36,37].

Research on materials for shorter wavelengths is a very important issue, since the ultimate bit density is limited by the optical diffraction in the conventional reading process. At present semiconductor lasers with a wavelength of ≈ 800 nm are employed, which are suitable for the amorphous RE-TM alloys. It is expected, that in the near future also semiconductor lasers with shorter wavelengths (green ≈ 500 nm, blue ≈ 400 nm) will be available. In this range, however, the figure of merit obtainable in the RE-TM alloys is poor. Approaches have been made to improve the reading quality by a systematical adjustment of optical constants [22,23]. One solution is the optical interaction in a magnetic-nonmagnetic bilayer film. A relatively large Kerr rotation together with a reasonable reflectivity has been demonstrated for a wavelength of 330 nm by using a very thin (11 nm) GdTbFeCo film on top of bulk silver [38]. Kerr rotations up to 0.75° in the wavelength range from 330 m to 700 nm have been obtained in superlattices consisting of multilayered (8.2 nm Ag + 9.6 nm $Fe_{50}Co_{50}$) films [39].

The most promising candidates for shorter wavelengths are Co/Pt or Co/Pd layered structures with a typical composition $25 \times (0.4$ nm Co + 1.9 nm Pt) [40–42]. The figure of merit for such films is comparable with that of GdTbFe at 820 nm and increases towards shorter wavelengths. At 458 nm the CNR of Co/Pt is 3 dB higher due to the higher signal level [43]. Although the Co/Pt multilayers are crystalline (grain size: 10 – 20 nm), no extra noise is present at short wavelengths. Domains as small as 0.25 μm could be written and read in Co/Pt with a modest writing laser power (< 3 mW) [43]. Therefore, Co/Pt is a very promising candidate for the blue laser radiation.

For green light (532 nm) the recording densities of Co/Pt disks where found to be similar to those of GdFeCo/TbFeCo double layers [44]. CNR of

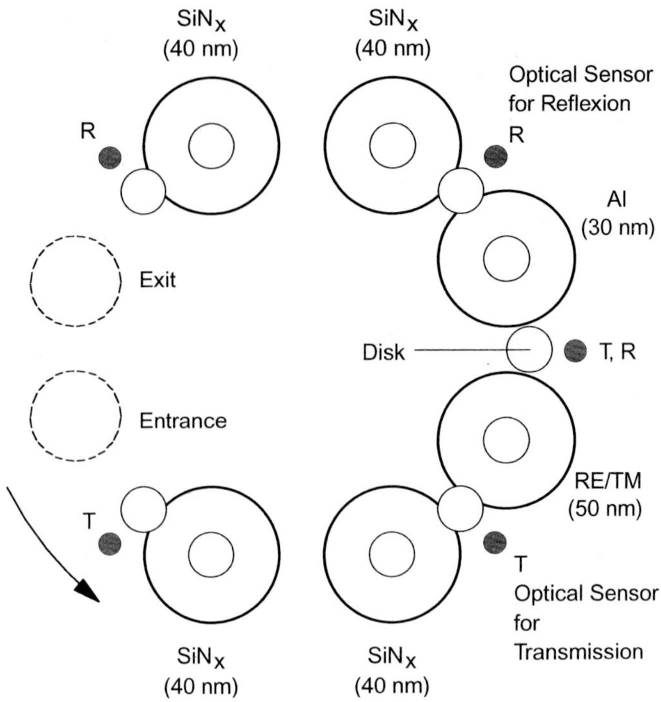

Fig. 2.5. Schematical diagram of a single-disk production system for magnetooptical quadrilayers (SiN_x, RE − TM, Al, SiN_x). The system consists of 6 sputtering chambers for static sputter deposition and 5 positions for in situ transmission and reflection measurements [47]

45 dB were obtained at a mark length of 0.4 μm for a track pitch of 0.9 μm. This permits a recording density almost three times higher than the current MO disks which operate at a minimum mark length of 0.65 μm and a track pitch of 1.4 μm using red light (780 nm). Similar results for green radiation have also been obtained in other, more complicated structures. A CNR of 46.7 dB was measured, for instance, at a domain length of 0.5 μm in a layer system constisting of a dielectric layer, an exchange-coupled (NdDyTbFeCo/NdTbFeCo/NdDyTbFeCo) triple layer, a second dielectric layer and a reflective layer [45].

2.2.3 Production of Magnetooptical Disks

The three or more different layers of the recording thin film stack are prepared sequentially in one run on the surface of a transparent substrate (polymer or glass) [46–50]. In the dynamical sputtering process [46,49] the substrates are positioned on a pallet, which is moved through several vacuum chambers for loading, degassing, surface conditioning (etching), sputtering and unloading.

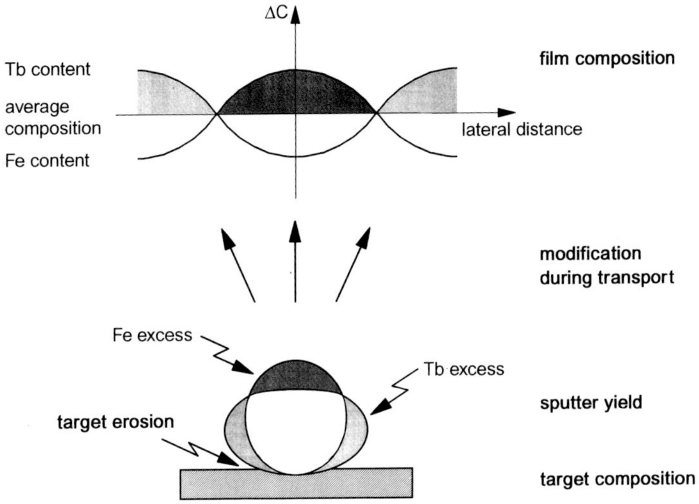

Fig. 2.6. Lateral distribution of the composition as a consequence of different sputtering-yield distributions of Tb and Fe from a TbFe target

In the sputtering chamber the pallet passes two or more rectangular cathodes, which allow the deposition of a layer stack in one run.

A different concept is the single disk coating [46–50] by static sputtering, where each layer is deposited in a seperate chamber from a circular cathode. The disk is transported through the various chambers on a circular path in order to obtain the complete layer stack (Fig. 2.5). After each process the optical properties of the layer stack are determined from transmission and reflection spectra by an online spectral photometer. Immediately after leaving the vacuum chamber the coercivity of the stack is measured by an automated Kerr-loop tracer. This concept allows control of the individual processes and improves the yield, but interferes with the requirement of a high production rate. To typically obtain 150 disks per hour, each process must take not more than 24 seconds, including the transport from one station to the other. To keep within this cycle two chambers are provided for the deposition of the SiN_x layer which is the slowest process.

The sputtering process for the magnetooptical layer is very critical, as the magnetic properties, in particular the compensation temperature and the coercivity at room temperature, are sensitive functions of the composition. In magnetron sputtering, the target erosion is a localized process, where the distribution of the sputtered TM atoms is different from that of the RE atoms (Fig. 2.6). In addition, the distributions are modified by scattering in the sputtering atmosphere. Therefore, it is very difficult, to obtain a homogeneous radial distribution of the film composition. It can be adjusted by the geometrical arrangement of substrate and cathode, by special masks, by the radial distribution of the gas flow, by the sputtering power and by the design of the erosion zone on the target [46–48]. Often, rotating magnets in the cathode are used to obtain a homogeneous erosion which also improves the target yield [47].

Even the structure of the target is important, as it influences the distribution of the sputtered atoms. Therefore, the target preparation is a crucial and expensive part of the disk production. Most often Hot Isostatically Pressed (HIP) targets are used, which should contain at least 30% of intermetallic phases to obtain a homogeneous distribution of film composition [46].

The dielectric layers improve the optical figure of merit, but also serve as protective layers and as diffusion barriers for water vapour, outgassing from the polymer substrate. They are prepared by reactive sputtering from Al or Si targets in a nitrogen atmosphere to obtain AlN or Si_3N_4 [49,50]. The gas distribution is a crucial factor for the homogeneity of the optical refraction index which should vary less than $\pm 2\%$ across the whole disk area. Oxides, for instance Al_2O_3 or SiO_2, are conventionally not applied since the oxygen in the sputtering gas may react with the RE-TM film.

In all reactive processes the stability of the glow discharge is a serious problem, because isolating films are formed not only on the substrate but also on the target, and lead to instabilities and arcing [49,50]. This is a main source for particles, which may cause defects on the film surface. In case of single disk deposition the size of the sputtering chamber is only of the order of the disk diameter. Therefore, the gas distribution can be optimized easier than in dynamic sputtering, where large sputtering chambers are required. The cathode, however, has to be switched on and off for each individual sputtering process, which increases the problem of arcing. In any case, the deposition of particles on the disk surface has to be minimized by constructive arrangements, for instance by an upright position of the disk during all preparation and transport cycles. Rotation of the disk, which would improve the circumferential homogeneity, is avoided because of the particle problem.

The sputtering procedure is only one part of the whole production process. The objective is to perform as many steps as possible immediately after each other in a fully automated production line [51,52]. The first step is the fabrication of the magnetooptical substrate exhibiting a grooved structure. Glass disks are covered by a polymer layer, into which the grooves are pressed by a stamper. Polymer substrates are formed by a high-pressure injection of molten PC resin. Then, the films are deposited by sputtering, followed by a lacquering in a spin coater. After fixing a central hub, two disks are glued together with the layers on top of each other to form a double-sided disk. Finally the disks are packed in a cartridge for transport. All steps outside the vacuum should be performed by roboters in a Class 10 clean room. After critical process steps, for instance after the injection molding and the lacquering, the quality is controlled by statistical tests.

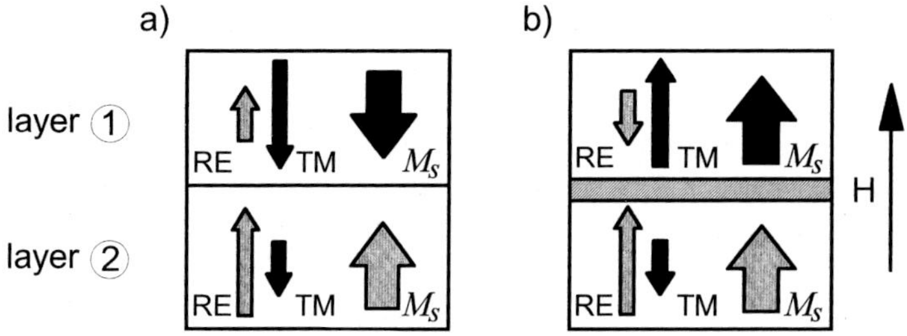

Fig. 2.7. a, b Subnetwork magnetizations of the RE and TM components and the net magnetization M_s of Exchange-Coupled Double Layers (ECDLs) with antiparallel coupling. (**a**) Magnetic configuration in the ground state without external field and (**b**) magnetic configuration in a saturation field

2.3 Application of Exchange-Coupled Double Layers

In standard magnetooptical recording the data access time is restricted by erasing a whole track before writing of new information. Research is presently concentrated on the incorporation of the erasing process, which can be considered as an initialization procedure performed by applying a constant field, into a Direct Over-Write (DOW) cycle [53,54]. One solution is the modulation of the bias field to directly obtain domains with magnetization up and down [55]. This technique is applied in the Sony Mini Disk for audio recording, where the reversal frequency is typically 44 kHz. In computer data recording, however, modulation frequencies in the range of several MHz would be necessary. Therefore, solutions are preferred, where direct overwrite is obtained by dual laser beam-techniques [56], by demagnetizing effects [57] or by laser modulation techniques [58].

A solution proposed by Shieh and Kryder [57] achieves domain nucleation and erasure in single layers by thermomagnetic processes with two different laser-pulse lengths. The field to reverse the magnetization is given by the demagnetizing field of the magnetic film at elevated temperatures.

Another solution first proposed by Saito et al. [58] applies Exchange-Coupled Double Layers (ECDLs) as storage media and a laser modulation technique with three power levels: a very low reading power and two different laser powers for writing at high or low temperatures.

2.3.1 DOW Using ECDLs

As an example, Fig. 2.7 illustrates magnetization configurations of ECDL having antiparallel coupling [59,60]. It consists of a layer 1, which is dominated by the subnetwork magnetization of the TM component and a layer 2, which is dominated by the RE component. Due to exchange coupling, the subnetwork magnetizations of RE and TM are continuous, but the net magnetizations of layer 1 and layer 2 are opposite to each other (antiparallel coupling). Because

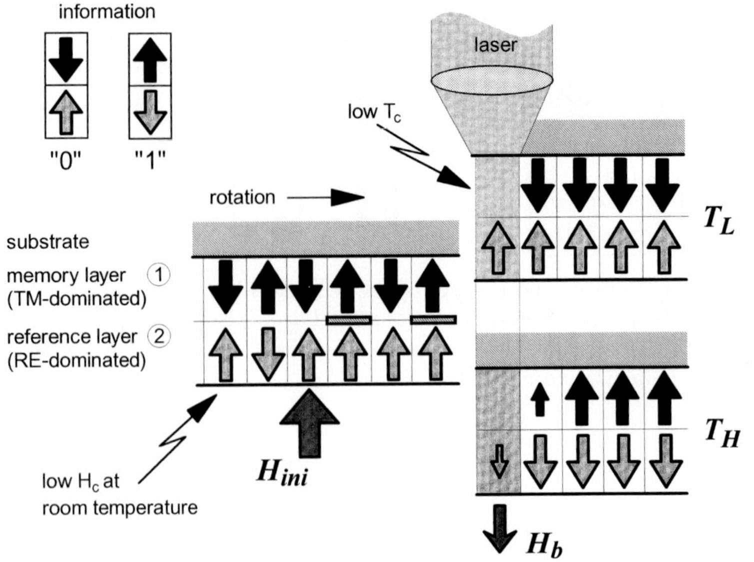

Fig. 2.8. Principle of direct overwrite in ECDL with antiparallel coupling. The film is initialized by a field H_{ini} at room temperature and is then exposed to a low- or high-power laser pulse in the presence of a bias field H_{b}. The cycle at low temperature T_{L} leads to a configuration representing "0", whereas the cycle at high temperature T_{H} leads to "1" [58]

of the high exchange coupling energy this is a stable configuration in spite of the negative dipole field. In a saturation field both net magnetizations will be turned into the direction of the field. As a consequence, the subnetwork magnetizations of layer 1 are opposite to those of layer 2, and an interface wall is formed, which contains the wall energy σ_{w} per unit area. If the coercivity of the layers is sufficiently high, this is a stable configuration. A switching field may be required, to return to a state with opposite net magnetizations and continuous subnetwork magnetizations. In this case, the wall is annihilated and the wall energy σ_{w} is released. Therefore, the switching energy per unit area for a magnetic reversal is given by [59–63]

$$E_{\mathrm{s}} = 2H_{\mathrm{c}}M_{\mathrm{s}}t \pm \sigma_{\mathrm{w}} \,. \tag{2.6}$$

The first term describes the energy required to reverse an uncoupled single layer of the thickness t, coercivity H_{c} and magnetization M_{s}. The second term represents the coupling by the wall energy. Switching takes place at the switching field [63] $H_{\mathrm{s}}^{\pm} = E_{\mathrm{s}}/2M_{\mathrm{s}}t$. Thus

$$H_{\mathrm{s}}^{\pm} = H_{\mathrm{c}} \pm \frac{\sigma_{\mathrm{w}}}{2M_{\mathrm{s}}t} \,. \tag{2.7}$$

The sign (+) is related to the creation of a wall, and the sign (−) to wall annihilation.

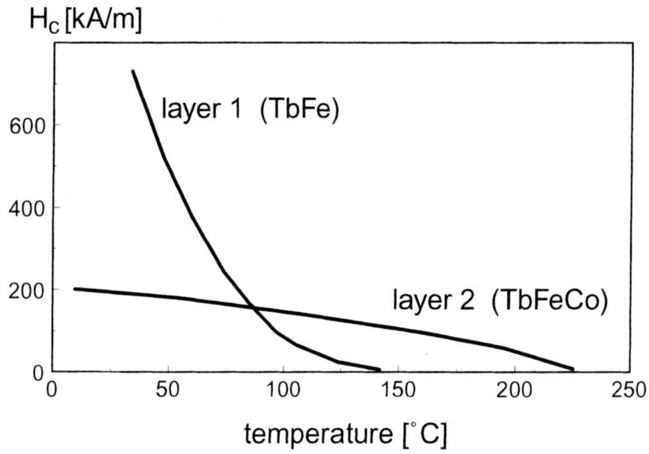

Fig. 2.9. Coercivity H_c as a function of temperature for a TM-dominated layer 1 and a RE-dominated layer 2

Figure 2.8 exhibits how ECDLs with antiparallel coupling can be applied to a direct overwrite process [58]. The medium consists of a memory layer 1 dominated by the TM component and a reference layer 2 dominated by the RE component. Since the coupling is antiparallel, the net magnetizations of layer 1 and layer 2 are opposite to each other. The information is read only from the memory layer, whereas the reference layer is used to control the writing process. The basic idea is that an initialization performed at room temperature should be followed immediately by a writing process at an elevated temperature T_L or T_H. During operation, therefore, the recording disk first runs through a region, where a permanent initialization field H_{ini} is applied and then through a region where the writing takes place by low- or high-power laser pulses in the presence of an opposite permanent bias field H_b.

The initialization field H_{ini} turns the magnetization of the reference layer 2 homogeneously into the field direction without changing the magnetization of the memory layer 1. This requires a low coercivity H_{c2} of the reference layer and a high coercivity H_{c1} of the memory layer at room temperature.

A second condition refers to the Curie temperature of the reference layer, which should be higher than that of the memory layer. A low-power laser pulse may heat the memory layer 1 to a temperature T_L close to its Curie temperature, but does not significantly influence the reference layer. During cooling a domain is formed in the memory layer, which is opposite to that of the reference layer due to the exchange energy. As a consequence, the information "0" is written into the memory layer independently of the former information.

A high-power laser pulse is applied to heat the layer system up to a temperature T_H close to the Curie temperature of the reference layer 2. Then the bias field H_b first switches the magnetization of layer 2. During cooling a domain is also formed in the memory layer 1, the magnetization of which is opposite to that of the reference layer (and opposite to the bias field) due to the exchange coupling. Eventually a magnetization configuration results, which represents the information "1".

To obtain a reliable initialization process at room temperature and the appropriate magnetic configurations after applying the low- and high-temperature cycles the coercivities of the layers 1 and 2 should have a temperature dependence as shown in Fig. 2.9. The composition of the memory layer 1 should be chosen in such a way that the compensation temperature is just below room temperature. Then, the film is TM dominated with a very high coercivity at room temperature and a relatively low Curie temperature [19]. A typical composition is $Tb_{22}Fe_{78}$. The reference layer 2, on the other side, should be coupled in an antiparallel way and should have a coercivity which does not change dramatically across a wide temperature range. This is achieved by RE-dominated films, having a composition without a compensation temperature. The high Curie temperature required for layer 2 is achieved by an additional Co concentration which partially substitutes the Fe content [33]. A typical composition, therefore, is $Tb_{30}(Fe_{90}Co_{10})_{70}$.

2.3.2 Magnetic Reversal Processes in ECDLs

Many researchers tried to determine the switching conditions for DOW only from the coercivity fields $H_c(T)$ (Fig. 2.9). This figure, however, represents only the behaviour of single layers without coupling. During a DOW cycle, interface walls are created and annihilated, and the related energy has be taken into account for a more detailed discussion. The diagram is therefore modified significantly, when exchange-coupled double layers are formed by a combination of single layers [63].

As an example, Fig. 2.10 depicts the hysteresis loops of single layers in comparison with those of a double-layer film [63]. For this experiment, single layers of TbFe (TM dominated) and TbFeCo (RE dominated) and a double layer of TbFe/TbFeCo were prepared sequentially by sputtering in one run without breaking the vacuum. The hysteresis loops were measured by the magnetooptical Kerr effect. In the double-layer stack the hysteresis loops of each of the layers were obtained separately by measurements from both sides. The switching fields of the layers in the stack were found to be significantly different from the coercivities of the single layers (the slope in the hysteresis loop of Fig. 2.10 d is caused by the Faraday effect of the substrate). Due to different growth conditions, the coercivities in a double-layer stack may not be the same as in a single layer, even if the preparation conditions were the same [63–65]. The main difference, however, results from the coupling energy σ_w.

The switching fields of the films in the stack can be determined as a function of temperature, when a negative saturation field is applied first and thereafter the field is increased to a positive saturation field (Fig. 2.11). For

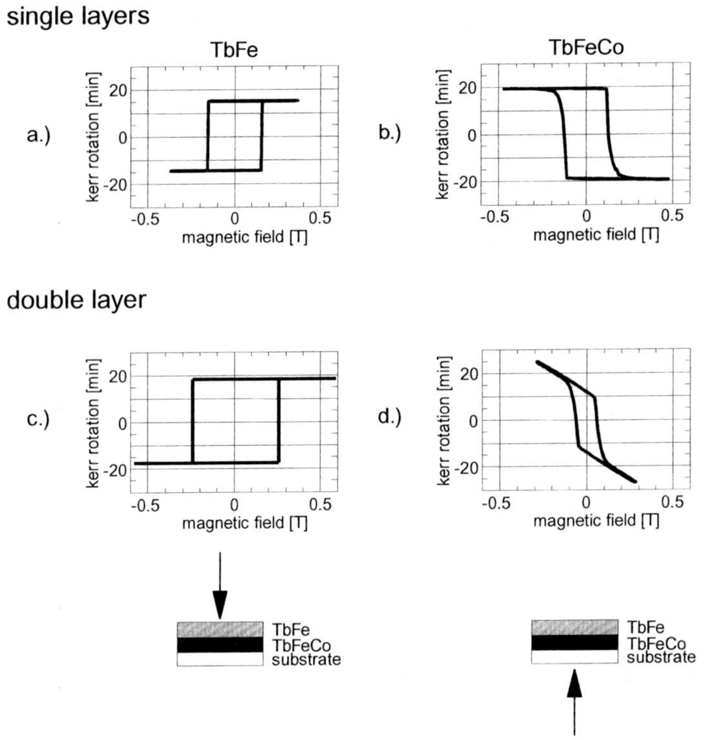

Fig. 2.10. a–d Single-layer TbFe (TM-dominated) and TbFeCo (RE-dominated) films in comparison with layers of the same composition in a double-layer stack. The hysteresis loops were measured by a Kerr magnetometer. The slope in (**d**) results from the Faraday effect of the glass substrate [63]

uncoupled layers a diagram similar to Fig. 2.9 would be expected. The behaviour of the coupled double layers is significantly different due to the exchange coupling. Below a critical temperature $T_\mathrm{s} \approx 55°\mathrm{C}$ the reference layer 2 is switched first and then the memory layer 1 is switched. Above the critical temperature the sequence is reversed: The memory layer first switches and the reference layer switches subsequently. Note that in a certain temperature range from 70 to 130°C the switching field of the memory layer is negative, i.e., the switching occurs already before the field is reversed. This phenomenon is caused by the wall energy which allows switching against the external field [63,66,67].

The switching behaviour in Fig. 2.11 can be understood when the switching fields are discussed in terms of (2.7). Figure 2.12 provides a graphical representation of (2.7) [66,67]. Without the coupling energy the switching curves would be given by the coercivities $H_{\mathrm{c}1}$ and $H_{\mathrm{c}2}$. Due to the coupling energy each of the curves splits into two branches $H_\mathrm{s}^{\pm}$, which correspond to the additional effect of creation or annihilation of an interface wall. Above

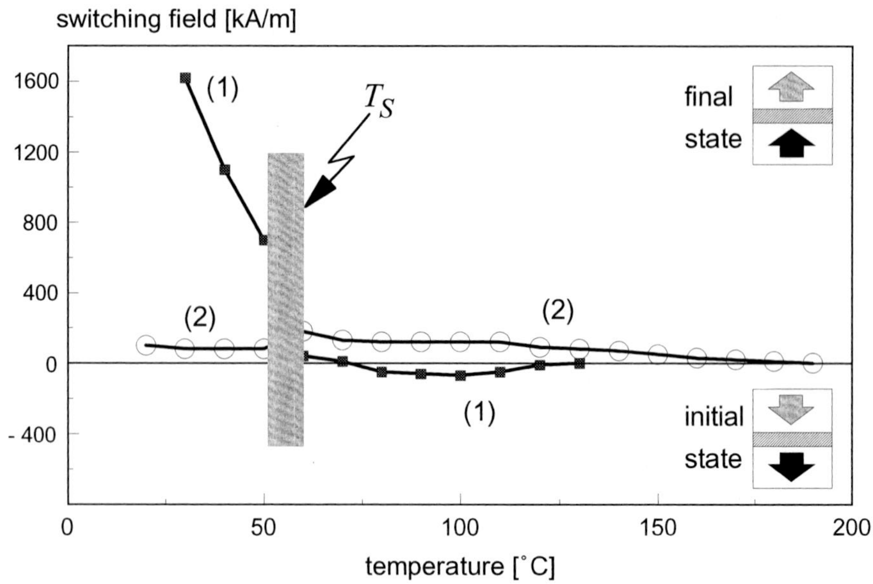

Fig. 2.11. Switching fields of ECDL with antiparallel coupling as a function of temperature. Layer 1 is TM dominated and layer 2 RE dominated. The data were measured by changing the external field from negative saturation (initial state) to positive saturation (final state) at constant temperature

the Curie temperature of the memory layer, T_{C1}, no coupling and, hence, no splitting of the curves appears.

Only a few of the various branches are observed during a complete magnetic reversal [66,67]. Let us consider a situation below the critical temperature ($T < T_s$). In a negative saturation field an interface wall exists, which is annihilated during the first switching process. It occurs at the lowest switching field available, which is H_{s2}^-. Then, the magnetization of the layers 2 and 1 are up and down, respectively, and no interface wall exists. Next switching in a positive field takes place with layer 1. It requires the formation of the interface wall again. Switching, therefore, occurs at the switching field H_{s1}^+. For temperatures above the critical temperature T_s a similar discussion explains the fact that now layer 1 switches first by annihilation and then layer 2 by creation of an interface wall. Here, the switching fields are H_{s1}^- and H_{s2}^+, respectively. In a small temperature range near T_s a more complicated switching behaviour takes place [61–63,66,67].

If the coupling energy is strong enough, a temperature range exists, where the switching field H_{s1}^- becomes negative. This is an important prerequisit for a direct overwrite (Sect. 2.3.3) [66,67]. Within this range the magnetization of layer 1 switches from the negative to the positive direction in spite of the fact that the external field is still negative. The energy to switch against the external field results from the wall energy released by the switching procedure.

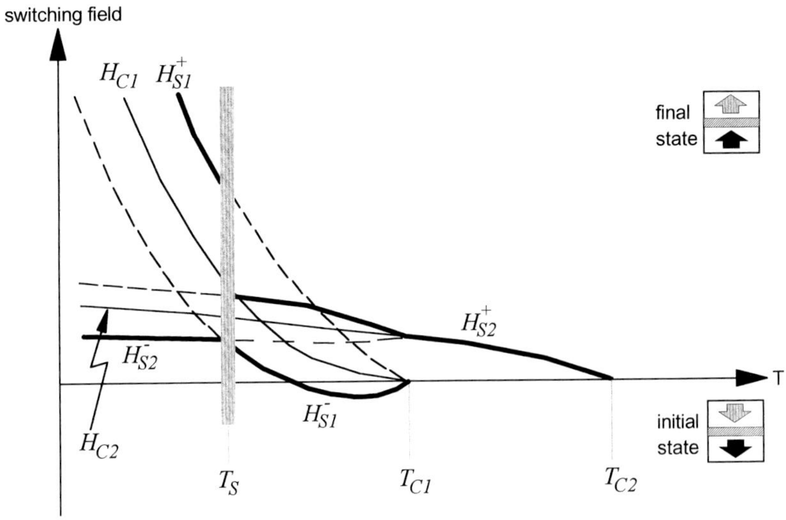

Fig. 2.12. Switching-field diagram for ECDL with antiparallel coupling. The diagram is a graphical interpretation of (2.7) and explains the experimental data of Fig. 2.11

2.3.3 Application of the Switching-Field Diagram

The switching diagram can be used to determine quantitatively the temperatures T_L and T_H and the fields H_b and H_{ini} required for a reliable DOW process. The diagram can be interpreted as a phase diagram, where the switching-field lines separate regions of stable magnetization up (↑) and down (↓) [66,67]. An original magnetization down (↓) switches up (↑), when an external field H is applied which exceeds the relevant switching lines H_s^+ or H_s^-. When the original magnetization is up (↑), the branches of the switching diagram (Fig. 2.12) have to be inverted with respect to the sign. Then, switching occurs if the external field is below the appropriate switching-field line.

As an example, the various procedures of a DOW cycle will be discussed for ECDLs with antiparallel coupling [66,67]. Let us first consider the initialization of the original information "1" (Fig. 2.13, upper diagram). Because switching of one layer would create (+) an interface wall, the relevant branches of the phase diagram are H_{s1}^+ and H_{s2}^+. As the original magnetization of layer 1 is up (↑), the line H_{s1}^+ has to be reversed with respect to the sign. An initialization field H_{ini}, as shown in the figure, is above H_{s2}^+ and, therefore, switches layer 2. It is also above the (inverted) switching field H_{s1}^+ and does, therefore, *not* influence the magnetization of layer 1. The criteria, therefore, are $0 < H_{s2}^+ < H_{ini}$ and $H_{s1}^- < H_{ini}$.

The final configuration in Fig. 2.13 (upper diagram), however, may not be stable, since the coupling energy stored in the domain wall may switch one of the layers. The stability, therefore, has to be discussed separately

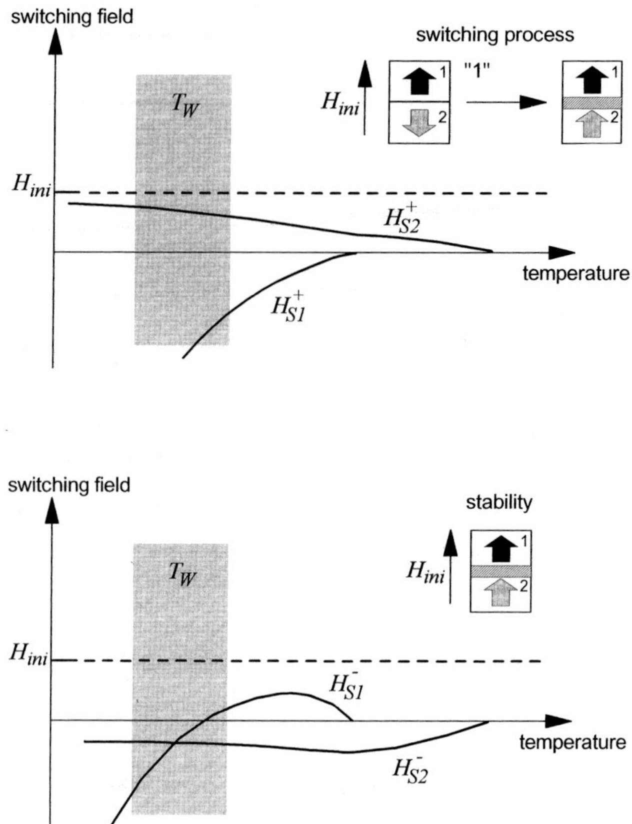

Fig. 2.13. Application of the switching field-diagram to the initialization process of the information "1". The upper diagram shows the switching process: Layer 2 is switched and layer 1 is not affected by the initialization field $H_{\rm ini}$. The lower diagram involves a discussion of the stability: The magnetic configuration remains stable when $H_{\rm ini}$ is above $H^-_{\rm s1}$

(Fig. 2.13, lower diagram) [66,67]. As an interface wall will be annihilated (−), the relevant switching curves are $H^-_{\rm s1}$ and $H^-_{\rm s2}$. Both are inverted with respect to the sign, since the initial magnetizations are up (↑). The critical curve is now $H^-_{\rm s1}$ which is the inverted curve in Fig. 2.12. When the initialization field $H_{\rm ini}$ is above this curve, as shown in the figure, the situation is stable. It would not be stable, if the magnetic field $H_{\rm ini}$ is lower than the maximum of the curve $H^-_{\rm s1}$ and the field line $H_{\rm ini}$ crosses the curve.

We now concentrate on the low-temperature procedure for the situation sketched in Fig. 2.13 (lower diagram). It should be transformed to the information "0" when a temperature $T_{\rm L}$ and a (negative) bias field $H_{\rm b}$ are applied. Under these conditions layer 1 should switch but *not* layer 2. The diagram

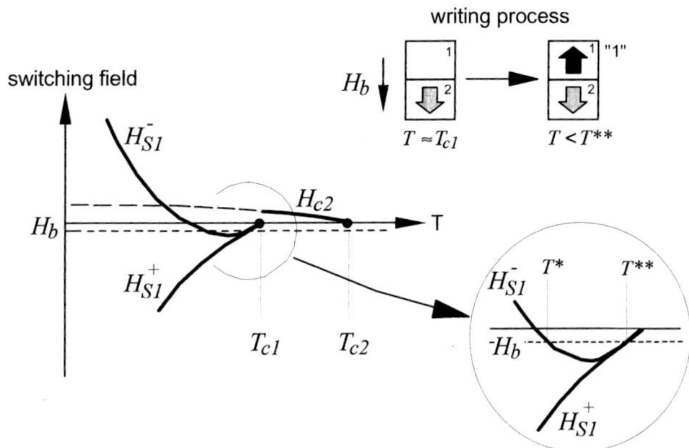

Fig. 2.14. Application of the switching-field diagram to the high-temperature writing cycle. A domain is formed in layer 2 at the high temperature $T_{\mathrm{H}} \approx T_{\mathrm{c2}}$. After cooling to T_{c1} a domain representing the information "1" has to be formed by exchange coupling. This must be stable in presence of the negative bias field H_{b} for $T < T^{**}$

in Fig. 2.13, therefore, must be discussed for a temperature range T_{L} near T_{c1}, and a negative value of the external field H_{b}. The appropriate switching is obtained for $H_{\mathrm{s2}}^{-} < H_{\mathrm{b}} < 0$ and $H_{\mathrm{b}} < H_{\mathrm{s1}}^{-}$ in the temperature range T_{L} [66,67].

When the original information is already "0" both the initialization and the low-temperature process are less critical. As the magnetic configuration for "0" is opposite to that for "1", the relevant switching field diagram is the reversed version of Fig. 2.13 (upper part). First, it has to be discussed for a positive initialization field H_{ini} at the working temperature T_{w} and then for a negative bias field H_{b} in the temperature range T_{L}. It can be concluded [66,67] that the criteria $0 < H_{\mathrm{ini}} < H_{\mathrm{s1}}^{+}$ and $H_{\mathrm{s2}}^{+} < H_{\mathrm{b}} < 0$ have to be fulfilled for a successful initialization and for low-temperature writing processes, respectively [66,67].

A critical situation is encountered during the high-temperature cycle (Fig. 2.14). Near the Curie temperature T_{C2} a domain is formed in layer 2 by the bias field H_{b}. During cooling, eventually the Curie temperature T_{C1} of the memory layer is reached. Then, a domain has to be formed in the memory layer by the exchange coupling in a direction opposite to the (negative) orientation of H_{b}. This can only occur when a temperature range exists, where H_{b} is above the curve H_{s1}^{-}, i.e., in the temperature range between T^{**} and T^{*} [66,67]. Therefore, the negative part of the switching field already shown in Fig. 2.11 is essential for the high-temperature writing process. Even the dynamical behaviour of the process can be analyzed if the fact is taken into account, that the bias field H_{b} decreases with time, because of the rotation of the disk [66].

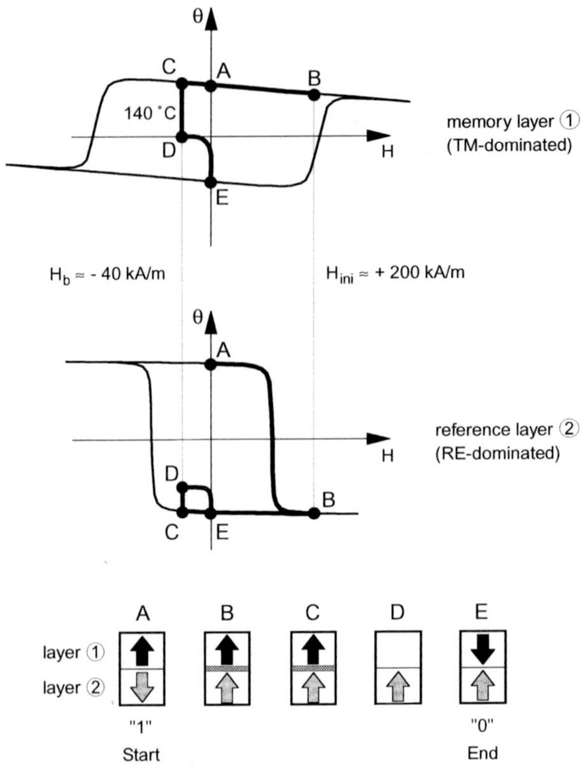

Fig. 2.15. Macroscopic simulation of a low-temperature writing cycle at $T_{\mathrm{L}} = 140°\mathrm{C}$. The magnetic configurations of the memory layer 1 and the reference layer 2 were determined by Kerr measurements at the various steps A–E

An important case is reading without writing a new information. Since the fields H_{ini} and H_{b} cannot be switched off during disk rotation, the written information must pass the field regions several times without being destroyed. This requires further criteria for the magnitude of the initialization and the bias fields, and for the maximum power of the reading laser. This can also be deduced from the switching diagram [66].

The parameters to fulfill the criteria can be adjusted according to (2.7). The magnetic moment $M_{\mathrm{s}} \cdot t$ per unit area can be changed by composition, which influences M_{s}, and by the thickness t. It is difficult, however, to independently adjust other magnetic parameters because of the mutual correlations provided by the sputtering process. Increasing the thickness, for instance, requires a higher sputtering power or a longer sputtering time which influences the temperature and, as a consequence, the wall energy [69].

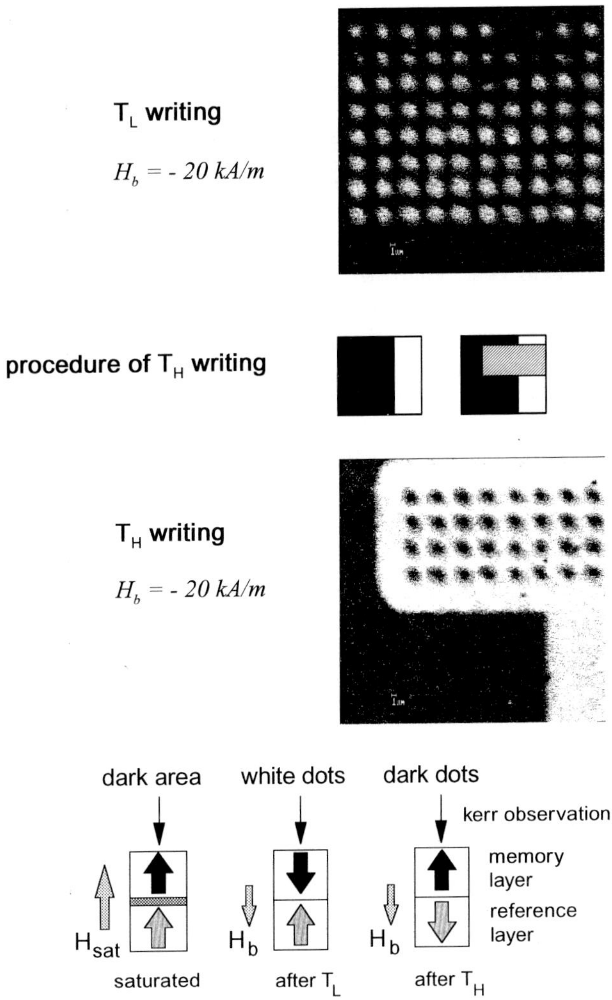

Fig. 2.16. Demonstration of the T_{L} and T_{H} writing cycles for ECDLs with antiparallel coupling. After saturation (dark area) white dots (information "0") were written by the low-temperature cycle at T_{L} and dark dots (information "1") by the high-temperature writing cycle at T_{H} [72]

2.3.4 Simulation of DOW

The various steps leading to a DOW process can be simulated by experiments on a macroscopic scale, where the double layer is exposed to appropriate fields and temperatures as well as the magnetization of the layers 1 and 2 are determined by Kerr measurements [71,72]. Figure 2.15 exhibits the results of a low-temperature cycle. First, the hysteresis loops of the memory layer 1 and of the reference layer 2 are determined at room temperature. Let us start with an information “1” (configuration A). An initialization field of 2 kG saturates the reference layer, but does not significantly change the memory layer (configuration B). A small negative bias field H_b applied at room temperature (C) does not change the magnetization configuration. If the temperature is increased up to $T_L = 140°C$, the magnetization of the reference layer 2 is slightly reduced (D), but the memory layer 1 is heated to the Curie temperature, where the magnetization is zero. Cooling in a negative field induces a reversed magnetization in layer 1 because of exchange coupling. The final situation is situation (E), where the configuration corresponds to the information “0”. Experiments like this one allow to follow macroscopically the various steps of the switching process and provide quantitative numbers for the required fields and temperatures.

The experiments give guidelines to perform microscopic studies with pulsed and focused laser beams [71,72]. Figure 2.16 shows the results of such an experiment using ECDL with antiparallel coupling. The domain observations were performed with a magnetooptical microscope. Before writing the whole sample was homogeneously magnetized by a saturating field, giving a dark area in the memory layer, which represents the information “1”. Then white dots, representing the information “0”, were written in a low-temperature cycle using laser pulses of 4 μs at various power levels and an additional, permanent, negative bias field (Fig. 2.16, upper diagram).

For the high-temperature cycle, the whole area was again saturated. Thereafter, a continuous low-power laser was applied to reverse the magnetization in the right part of Fig. 2.16 (central diagram) by the low-temperature cycle (white area). The goal was, to write the information “1” (dark dots) by the high-temperature cycle into both the white and dark areas (i.e., into the shaded area in Fig. 2.16, central diagram). The result of an experiment using maximum laser power (20 mW) and a pulse length of 200 μs is shown in Fig. 2.16 (lower diagram). In the right part of the diagram the dark dots in the white background are clearly seen. In the left part the dark dots are observed in a white surrounding. This situation comes from the temperature profile of the high-power laser pulse: it writes in the center the information “0” (dark dots) and leaves in the surrounding a low-temperature process which gives the white area. It should be noted that the magnetization in the dark dots is opposite to the direction of the bias field due to exchange coupling.

2.3.5 Layer Systems for DOW

In the previous Sections 2.3.1–2.3.4 only double layers with antiparallel coupling were discussed. Similarly, also parallel coupling in ECDLs can be applied [73]. In this case, both layers are dominated either by the RE components or by the TM components. Therefore the subnetwork magnetizations as well as the net magnetizations are parallel. An interface wall is created when the magnetization of one of the layers is switched to the opposite direction.

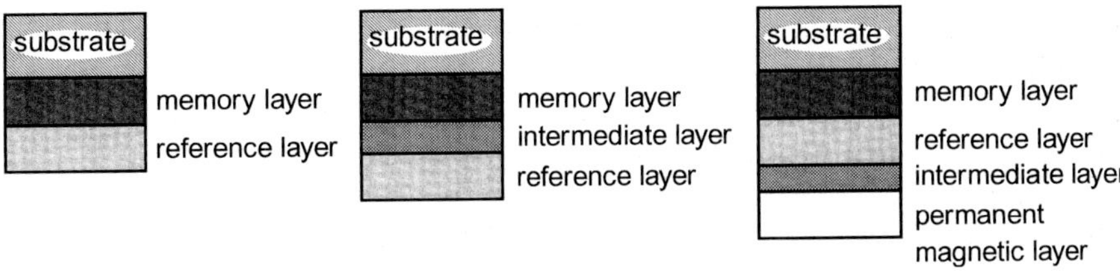

Fig. 2.17. 3-layer and 4-layer systems for direct overwrite using ECDLs

Switching diagrams similar to those for antiparallel coupling can be derived for the various steps of the DOW processes [66,67].

ECDLs discussed before have always the same kind of coupling over the whole temperature range (parallel or antiparallel). ECDLs which have a compensation temperature in layer 1 or (and) in layer 2 may also be used [73]. In this case, similar considerations as for antiparallel or parallel coupling can be applied. ECDL, for instance, having a RE-dominated reference layer and a compensation temperature T_{comp} in the memory layer behaves as a parallel coupled for $T < T_{\mathrm{comp}}$ and as an antiparallel coupled bilayer for $T > T_{\mathrm{comp}}$.

Various systems were proposed which contain three or more layers (see Fig. 2.17). In all cases the parallel or antiparallel coupling between ECDLs is the essential mechanism. The additional layers are needed to fulfill special requirements. For instance, an intermediate layer may be placed between ECDLs to adjust the magnitude of the coupling energy to obtain improved data stability [74,75]. This procedure is only reasonable in cases, where the coupling energy is sufficiently high.

A permanent magnetic layer and an additional intermediate layer with a low Curie temperature can be used to replace the initialization magnet (4-layer system [76,77]). The initialization process is achieved at room temperature by the magnetic coupling between the permanent magnetic layer, the intermediate layer, and the reference layer. At the writing temperature, the intermediate layer becomes nonmagnetic and ECDL is decoupled from the permanent magnetic layer.

Multilayer systems with a gradient of the Curie temperature were proposed, too. Most often, however, multilayer structures are used to adjust the magnetic properties. ECDLs consisting of periodic structures with typically $27\times(0.85\,\mathrm{nm\,Tb} + 1.0\,\mathrm{nm\,Fe})$ for the memory layer and $32\times(1.5\,\mathrm{nm\,Tb}+1.0\,\mathrm{nm\,Fe})$ for the reference layer have been proven to be suitable for DOW [78]. If the modulation wavelength is small enough (typically $\leq 3\,\mathrm{nm}$) the magnetic behavior is similar to that of amorphous alloys [79,80] but the multilayer shows an increased perpendicular anisotropy [81]. In all cases it is important to determine K_{u} in the surface region. This can be achieved by Kerr-torque magnetometry [82,83] which permits to detect, for instance, small changes of the anisotropy induced by diffusion of nitrogen from the SiN_x coverlayer [84].

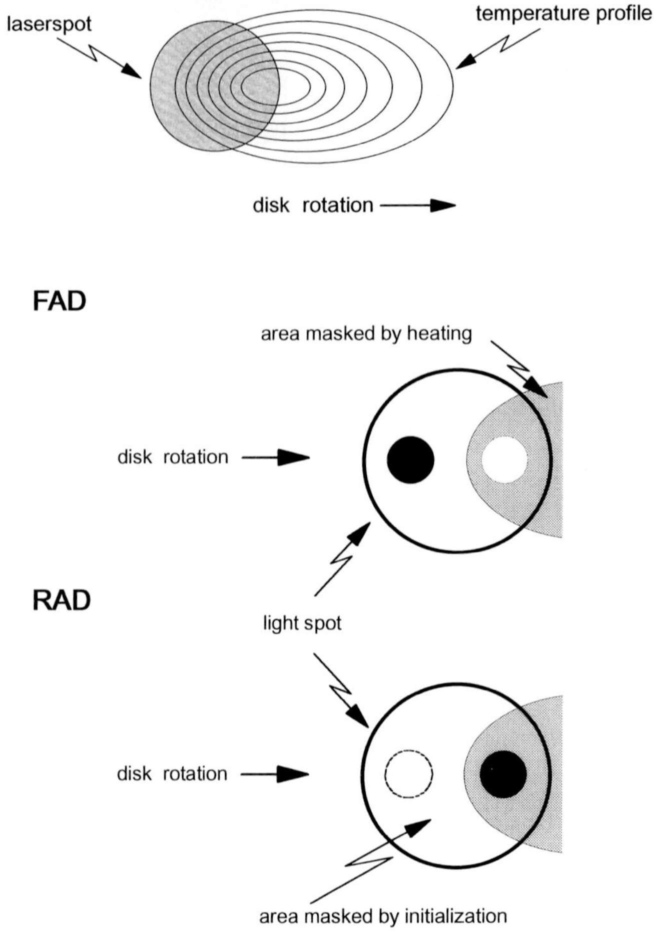

Fig. 2.18. Principle of Magnetic Super-Resolution (MSR) [89]. The upper diagram shows the temperature profile caused by the laser spot on a moving disk. The central diagram represents Front Aperture Detection (FAD) by masking a mark at the front of the laser spot. The lower diagram represents Rear Aperture Detection (RAD) by masking the marks in the rear part of the laser spot

Various other layer systems do not use the exchange coupling but are based on demagnetizing effects of a leakage field [85]. In all cases, additional optical layers are applied to improve the signal-to-noise ratio by interference effects. At the moment, not any of these devices is available as a technical product. A certain tendency can be seen for the 4-layer system, where the initialization magnet can be omitted. Then, the disk will be compatible with the standard magnetooptical drive having only one magnet. The preparation of systems with more than two layers, however, is more complex due to the increased number of interfaces.

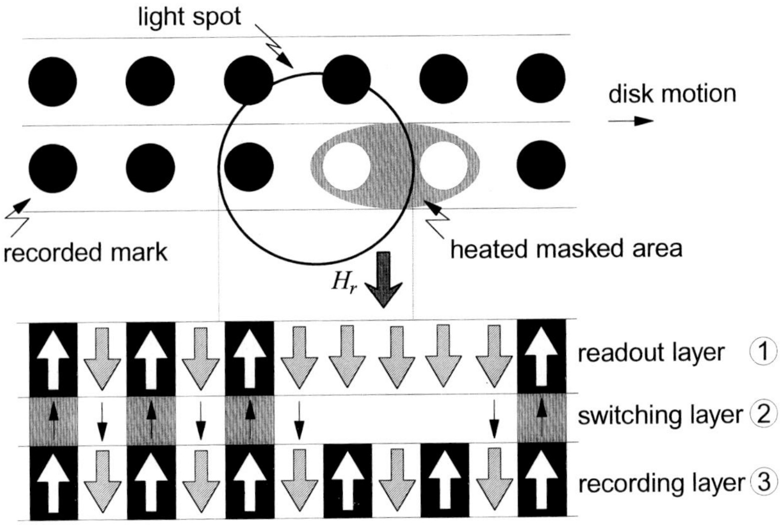

Fig. 2.19. Front Aperture Detection (FAD) in a triple-layer system with parallel exchange coupling. Masking in the readout layer 1 is achieved by laser-pulse heating in presence of a reading field [89]

2.3.6 Magnetically Induced Super-Resolution

Optical recording is usually limited by the spot size of the focused laser beam. According to optical diffraction theory the diameter is given by $d = 1.22\,(\lambda/A)$, A being the numerical aperture of the optical system. In magnetooptical recording the size of the written domain can be smaller than d, because it is determined by the temperature profile induced by the laser pulse. The reading process, however, is still limited by diffraction, because the maximum of a diffraction spot should fall at least into the minimum of the neighboring spot.

Domains with a smaller distance can be read, when part of the information is optically masked (Magnetically induced Super-Resolution, MSR) (Fig. 2.18). The effective aperture is smaller than that given by optical theory, and the information can be extracted from the intensity modulation induced by the rotating disk [86–90]. Masking can be obtained, in principle, by the temperature profile of the reading laser [89]. Due to disk rotation, the effective temperature profile in the magnetooptical layer is elliptical with a maximum near the front edge of the beam spot (Fig. 2.18, upper diagram). If the maximum temperature is sufficient for erasing, only the mark in the low-temperature crescent remains (Fig. 2.18, central diagram) and contributes to the signal (Front Aperture Detection, FAD). Exchange-coupled layer systems can be prepared which allow this type of reading in one readout layer without loosing the information stored in a second recording layer. Alternatively, masking can be obtained in the crescent at the rear part of the laser spot

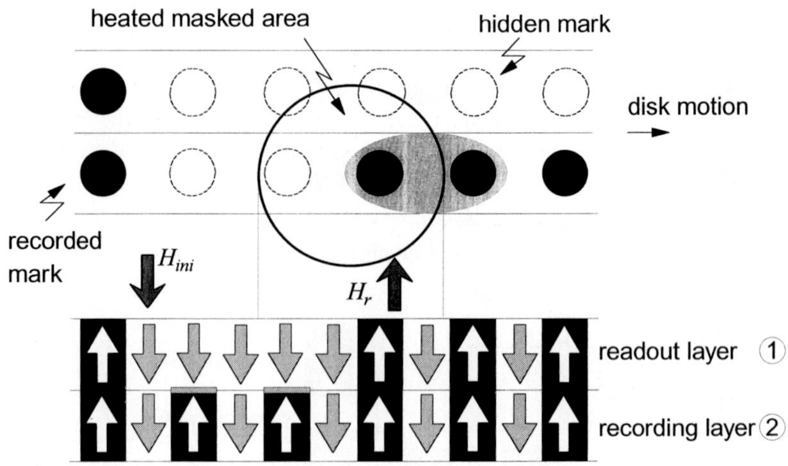

Fig. 2.20. Rear Aperture Detection (RAD) in ECDL with parallel coupling. Masking is achieved by an initialization field H_{ini} [89]

(Fig. 2.18, lower diagram). Then the information is read only from the mark at the leading edge (Rear Aperture Detection, RAD).

MSR by FAD is achieved by an exchange-coupled triple layer having parallel coupling (Fig. 2.19). It consists of a readout layer 1, a switching layer 2, and a recording layer 3. Writing is performed by a standard thermomagnetical process with a laser pulse in the presence of a bias field. The power of the laser pulse is adjusted to a level, where domains smaller than the spot size are formed in the readout layer 1, which is the layer with the maximum Curie temperature. During cooling the information is copied through the whole layer system by the exchange-coupling energy [91].

For reading, the laser is adjusted to a power level which heats the switching layer 2 just above the Curie temperature. In the high-temperature part of the spot (leading edge) the readout layer and the recording layer are decoupled, whereas the coupling is still present in the colder part of the spot. A reading field $\mathrm{H_r}$ switches the readout layer in the high-temperature area homogeneously into field direction. Therefore, the front part of the aperture is masked and the information is read only from the rear crescent. During cooling the switching layer becomes again magnetic and copies the information back from the recording layer into the whole layer system by exchange coupling.

Several conditions have to be fulfilled to obtain a reliable writing and reading process. The Curie temperatures of the three layers should be adjusted to $T_{\mathrm{C1}} > T_{\mathrm{C3}} > T_{\mathrm{C2}}$, which is achieved, for instance, by a GdFeCo readout layer ($T_{\mathrm{C1}} = 300°\mathrm{C}$), a TbFeCoAl switching layer ($T_{\mathrm{C2}} = 140°\mathrm{C}$) and a TbFeCo recording layer ($T_{\mathrm{c3}} = 250°\mathrm{C}$) [89]. To obtain a stable magnetic configuration in the recording layer the coercivities must be $H_{c1} < H_r$ and $H_{c3} > H_r$ at the readout temperature.

Exchange-coupled double layers with parallel coupling can be applied to achieve MSR by RAD (Fig. 2.20). The writing process again occurs by conventional thermomagnetic recording using a laser power appropriate for the small domain sizes. In an initialization process the readout layer is homogeneously magnetized by an initialization field H_{ini} at first. Since domain walls are created in part of the interfaces the conditions

$$H_{s1}^{+} < H_{ini} \quad \text{and} \quad H_{s2}^{-} > H_{ini} \tag{2.8}$$

have to be fulfilled at room temperature to obtain switching of layer 1 and a stable configuration in layer 2 [66]. After initialization the marks in the film are hidden and are addressed locally only for reading. This is obtained by a laser temperature profile, which copies back the information from the recording layer into the readout layer by exchange coupling in the presence of the reading field H_r. By an appropriate laser power the copying takes place only in the high-temperature part T_r of the laser spot, i.e., near the leading edge. Therefore, the rear part remains masked and the information is obtained only from the modulation in the front part.

Contrary to FAD, also the information in the neighboring tracks is masked in RAD. Due to the large-area initialization field, the marks are hidden and do not interfere with the information to be read. This avoids the problem of crosstalk . The tracks, therefore, can be arranged in a closer distance, determined by the temperature profile rather than by the optical diffraction limit.

Switching-field diagrams similar to those of DOW can be derived for MSR to obtain the conditions for a reliable reading process. In RAD, for instance, the initialization at ambient temperature is limited by $H_{s2}^{-} > H_{ini} > H_{s1}^{+}$. In the hot part of the laser spot the conditions are $H_{s1}^{-} < H_r < H_{s1}^{+}$ and $H_r < H_{sz}^{+}$, whereas in the cold part the conditions are $H_r < H_{s1}^{-}$, $H_r < H_{s1}^{+}$, and $H_{r2} < H_{s2}^{+}$, respectively [66]. In case of a 3-layer system the coupling energies for the interfaces 1/2 and 2/3 have to be taken into account. From the resulting diagrams, it can be shown, that FAD is also possible in double-layer structures, and both FAD and RAD can be achieved in ECDLs with antiparallel coupling as well [66].

Further improvements in RAD are expected from a 4-layer structure, consisting of a readout layer (GdFeCo), a subsidiary layer (TbFeCoAl), an intermediate layer (GdFeCo), and a recording layer (TbFeCo) [89]. Similarly as in 3-layer systems for DOW, the intermediate layer is used to adjust the wall energy σ_w. The readout layer and the subsidiary layer perform the masking process. The readout layer has a low coercivity at rom temperature and a high Curie Temperature. It can be switched, therefore, by the initialization field, but retains its magnetization up to high temperatures. The subsidiary layer, on the other hand, has a high coercivity at room temperature and a low Curie temperature. Data are copied from the recording layer to the readout layer in the area where the temperature is below the Curie temperature of the subsidiary layer.

Another solution is Center Aperture Detection (CAD) [92], where the information is copied from a recording layer into a readout layer during the reading process. The readout layer switches from an in-plane anisotropy to a perpendicular anisotropy at a certain temperature TR, which is achieved only in the central part of the laser pulse. In this part, the information is copied by exchange coupling from the recording layer into the readout layer, whereas the marks remain hidden in the peripheral parts of the temperature profile. By using a red laser beam (680 nm) a CNR of 45 dB has been obtained at a 0.45 μm mark length and 0.7 μm track pitch. This corresponds to a recording density of over 2 Gbit/inch^2 [92].

2.4 Outlook

The present available 5.25″ magnetooptical disks have a capacity of 650 MB per side (2X capacity), which corresponds to the CD-ROM standard. Since a disk can be used on both sides, the total capacity is 1.3 GB. The data are recorded in marks of 0.65 μm diameter at a mutual distance of 1.29 μm along the track and at a track pitch of 1.4 μm [93,94]. Detection is performed with a 780 nm laser by using the Pulse Position Modulation (PPM) coded to obtain an area density of 410 Mbit/inch^2 on average [94].

There are various possibilities to increase the area density. The 3X capacity, i.e., 1 GB per side and an areal density of 620 Mbit/inch^2, will be obtained by a track pitch of 1.34 μm and by mark-length recording [93–95]. Marks of different lengths, Pulse Width Modulation (PWM), and an appropriate coding [96] can provide a factor of 1.5 to 2 in capacity [94]. Transition to a shorter wavelength of 670 nm allows an increase by a factor of 1.3 to 1.5 [94]. Inland and -groove recording marks [97] will be written both into the groove areas and into the land areas between the grooves. This gives an additional factor of 2 [94,97]. A combination of these factors may allow an increase in capacity up to 8X without a significant change of materials [94,97,98].

Lasers with shorter wavelengths can lead to a further increase in capacity. Compared with the present standard, the blue laser radiation (428 nm) will provide, for instance, a factor of 3.8 by a linear reduction of the dimensions [94]. In this case, however, new materials, as explained in Sect. 2.2.3 are required.

Exchange-coupled layer systems provide a great variety of solutions to obtain direct overwrite and magnetic super-resolution. Although none of these techniques is applied in a present technical product, some of the various approaches will certainly be incorporated into further developments. A lot of attempts, for instance, are performed to combine the possibility of DOW with high-density recording by using more complicated layer systems [99–102]. Quadrilayers with an additional readout layer or an intermediate layer, for instance, have been studied with respect to a higher storage capacity [102]. A CNR of more than 46 dB was obtained at a 0.66 μm mark length and a 1.2 μm track pitch using a 680 nm diode laser. This would allow a capacity of 2.5 to 3.5 GB for a direct overwriteable double-sided 130 mm disk [102].

Magnetooptical recording will be in competition with other recording techniques, such as phase-change recording or magnetic recording. An area density of 2 Gbit/inch^2, for instance, has been demonstrated to be possible in magnetooptical recording [92] as well as in magnetic recording [103]. The position on the market, however, will be determined not only by the area density, the total capacity, and the data access time, but also by other factors such as handling or, most important, costs per MB. These latter aspects are probably the most important advantages of the magnetooptical disks: They are mass-storage systems at a modest price, which are very easy and convenient to handle. In the near future capacities up to 8X will be obtained by techniques, which are already available or, at least, checked with respect to their feasability [97,98]. Remarkable progress has been achieved in the last years in the field of MSR and related techniques [105,104–110] such as Magnetic Amplifying Magnetooptical Systems (MAMMOS) [107,108], which are based on thermally induced switching of magnetic anisotropy [109] or on magnetostatic coupling [110]. Visions of 100 Gbit/inch^2 density are already taken into consideration [104] These, however, will require nonconventional approaches such as the use of nanocrystalline media, magnetic force microscopy, or optical near-field detection.

Acknowledgements. The autor is indebted to T. M. Atmono, S. Becker, H. Rohrmann and F. Stobiecki for valuable contributions. The technical assistance by E. Brühl and M. Kratzer is greatly appreciated.

References

1. M.H. Kryder (Ed.) *Magnetic Information Storage Technology*, Proc. IEEE **74**, 1494–1590 (1986)
2. C.D. Mee, E.D. Daniel (Eds.): *Magnetic Recording*, Vol. I–III; (McGraw Hill, New York 1987 und 1988)
3. J.C. Mallinson: *The Foundations of Magnetic Recording* (Academic Press, San Diego 1994)
4. E. Köster: J. Magn. Magn. Mater. **120**, 1–10 (1993)
5. K. Röll: Dünne Schichten zur Informationsspeicherung, in Vakuumbeschichtung, Vol. 5, ed. by G. Kienel, (VDI, Düsseldorf 1993) pp. 114–133
6. Special Issue on Optical Memories, Jpn. J. Appl. Phys. **32**, 5185–5464 (1993)
7. A. Berkowitz (Ed.) *Magnetic Recording Materials* MRS Bulletin **XV**, 23–72 (1990)
8. S. Yamamoto, Y. Nakamura, S. Iwasaki: IEEE Trans. MAG **23**, 2070–2072 (1987)
9. M. Okuda, H. Naito, T. Matsushita: Jpn. J. Appl. Phys. **31**, 466–470 (1992)

10. M.H. Kryder: J. Appl. Phys. **57**, 3913–3918 (1985)
11. W.H. Meiklejohn: Proc. IEEE **74**, 1570–1581 (1986)
12. D.S. Bloomberg, G.A.N. Conell: Magnetooptical Recording, in Magnetic Recording Technology, ed. by C.D. Mee, E.D. Daniels (Academic Press, San Diego 1988) Vol. 3, pp. 310–404
13. M. Mansuripur: *The Physical Principles of Magneto-optical Recording* (Cambridge Univ. Press, Cambridge 1995)
14. Special Issue on Magnetooptical Recording, J. Magn. Soc. Jpn. **19**, 1–448 (1995)
15. N. Imamura, S. Tanaka, F. Tanaka, Y. Nagao: IEEE Trans. MAG **21**, 1607–1612 (1985)
16. P. Hansen, H. Heitmann: IEEE Trans. MAG **25**, 4390–4404 (1989)
17. P. Hansen: J. Magn. Magn. Mater. **83**, 6–12 (1990)
18. P. Hansen, C. Clausen, G. Much, M. Rosenkranz, K. Witter: J. Appl. Phys. **66**, 756–767 (1989)
19. P. Hansen, S. Klahn, C. Clausen, G. Much, K. Witter: J. Appl. Phys. **69**, 3194–3207 (1991)
20. P. Hansen: Magnetic Amorphous Alloys, in *Handbook of Magnetic Materials*, ed. by K.H.J. Buschow (Elsevier, Amsterdam 1991) Vol. 6, p. 289–452
21. B.S. Krusor, G.A.N. Connell: Thin-film rare earth-transition metal alloys for magnetooptic recording, in *Physics of Thin Films*, ed. by M.H. Francomb, J.L. Vossen (Academic Press, San Diego 1991) 143–217
22. W. Reim, J. Schoenes: Magneto-optical spectroscopy of f-electron systems, in *Handbook of Ferromagnetic Materials*, ed. by E.P. Wohlfahrt, K.H.J. Buschow, (North-Holland, Amsterdam 1990) Vol. 5
23. D. Weller, W. Reim: Appl. Phys. A **49**, 599–618 (1989)
24. W.H. Meiklejohn, F.E. Luborsky, P.G. Frischmann, IEEE Trans. MAG **23**, 2272–2274 (1987)
25. P. Hansen: J. Appl. Phys. **62**, 216–230 (1987)
26. W. Reim, D. Weller: IEEE Trans. MAG **24**, 2308–2310 (1988)
27. H.P.D. Shieh, M.H. Kryder: J. Appl. Phys. **61**, 1108–1122 (1987)
28. D. Raasch: Magnetische Wandenergien und Austauschkonstanten in amorphen Seltenerd-Übergangsmetall-Dünnfilmschichten, Dissertation, RWTH Aachen (1995)
29. M.H. Kryder, H.-P. Shieh, D.K. Hairston: IEEE Trans. MAG **23**, 165–167 (1987)
30. R. Sato, N. Saito, T. Morishita: IEEE Trans. MAG **24**, 2458–2460 (1988)
31. A. Hubert, G. Traeger: J. Magn. Magn. Mat. **124**, 185 (1993)
32. H. Sughara, K. Tsutsumi, K. Hashima, T. Tokunaga, Y. Fuji: IEEE Trans. MAG **24**, 2455–2457 (1988)
33. Y. Itoh, T. Takase: IEEE Trans. MAG **27**, 3756–3760 (1991)
34. D. Raasch, S. Klahn: J. Magn. Mat. **93**, 365–368 (1991)
35. D. Raasch: IEEE Trans. MAG **29**, 34–40 (1993)
36. F.J.A.M. Greidanus, W.B. Zeper: MRS Bulletin, **15**, 31–38 (1990)
37. P. Hansen, J.-P. Krumme. D. Mergel: J. Magn. Soc. Jpn., **15**, 219–226 (1991)
38. W. Reim, D. Weller: Appl. Phys. Lett. **53**, 2453–2454 (1988)
39. D. Weller, W. Reim: Fresenius Z Anal Chem **333** 481–484 (1989)
40. F.J.A.M. Greidanus, W.B. Zeper, F.J.A. den Broeder, W.F. Godlieb: Appl. Phys. Lett. **54**, 2481–2483 (1989)
41. W.B. Zeper, F.A.M. Greidanus: J. Appl. Phys. **65**, 4971–4975 (1989)

42. W.B. Zeper, F.J.A.M. Greidanus, H.W. van Kesteren, B.A.J. Jacobs, J.H.M. Spruit, P.F. Carcia: Co/Pt and Co/Pd multilayers as a new class of magneto-optical recording materials SPIE Vol. 1274 *Electro-Optic and Magneto-Optic Materials II*, 282–292 (1990)
43. W.B. Zeper, A.P.J. Jongenelis, B.A.J. Jacobs, H.W. van Kesteren: IEEE Trans. MAG **28**, 2503–2505 (1992)
44. M. Kaneko, Y. Sabi, I. Ichimura, S. Hashimoto: IEEE Trans. MAG **29**, 3766–3771 (1993)
45. H. Miyazawa, T. Ide, S. Hoshina, M. Ichikawa, S. Nebashi, T. Shimoda: IEEE Trans. MAG **29**, 3781–8783 (1993)
46. G. Bräuer, W. Dicken, P. Wirz, S. Jönsson, M. Schlott, S. Schittny: J. Magn. Soc. Jpn. **15**, 381–384 (1991)
47. G. Bräuer, W. Dicken, S. Reineck, H. Teves: J. Magn. Soc. Jpn. **15**, 377–380 (1991)
48. K. Röll, G. Bräuer: Beschichtungsverfahren mit selbstregelnder Prozeßkontrolle. BMFT-Forschungsbericht 13N5522 (1992)
49. G. Bräuer, B. Cord, K.H. Schuller, S. Schulz, J. Szczyrbowski, G. Teschner: J. Magn. Soc. Jpn. **17**, 327–332 (1993)
50. G. Bräuer, B. Cord, W. Dicken and S. Schulz: J. Magn. Soc. Jpn. **19**, 65–70 (1995)
51. E. Hellstern, S.Y. Suh, W. Wiedemann, G. Schmidt and B. Fischer: In-line production of magneto-optical quadrilayers. Ed. by *Optical Data Storage*, SPIE Proc. **1316** 230–236 (1990)
52. G. Strasser, P. Tinner, S. Voser, W. Haag, S. Candrian, H. Selhofer, F. Gassmann: J. Magn. Soc. Jpn. **17**, 321–326 (1993)
53. K. Tsutsumi, T. Fukami: J. Magn. Magn. Mat. **118** 231–247 (1993)
54. K. Tsutsumi: J. Magn. Magn. Mat. **120**, 247–253 (1993)
55. Y. Teragaki, Y. Kusumoto, S. Sumi, K. Torazawa: IEEE Trans. MAG **29**, 3796–3798 (1993)
56. M.H. Kryder, H.-P. D. Shieh: IEEE Trans. MAG **25**, 3545–3549 (1989)
57. H.-P.D. Shieh, M.H. Kryder: IEEE Trans. MAG **23**, 171–173 (1987)
58. J. Saito, M. Sato, H. Matsumoto, H. Akasaka: Jpn. J. Appl. Phys. **26**, 155–159 (1987)
59. T. Kobayashi, H. Tsuji, S. Tsunashima, S. Uchiyama: Jpn. J. Appl. Phys. **20**, 2089–2095 (1981)
60. R.J. Gambino, T.S. Plaskett R.R. Ruf: IEEE Trans. MAG **24**, 2557–2559 (1988)
61. D. Mergel: J. Appl. Phys. **69**, 4520–4522 (1991)
62. D. Mergel: J. Appl. Phys. **74**, 4072–4080 (1993)
63. F. Stobiecki, T.M. Atmono, S. Becker, H. Rohrmann, K. Röll: J. Magn. Magn. Mat. **148**, 497–502 (1995)
64. S. Shimokawato, H. Miyazawa, T. Mikoshiba T. Shimoda: IEEE Trans. MAG **27**, 5106–5108 (1991)
65. J.M. Liang: J. Magn. Soc. Jpn. **19**, 103–108 (1994)
66. S. Becker: Magnetisierungsverhalten austauschgekoppelter Schichtsysteme für die magneto-optische Datenspeicherung, Dissertation, University of Kassel (1995)
67. S. Becker, H. Rohrmann, K. Röll: J. Magn. Magn. Mat. **171**, 225–236 (1997)
68. H. Rohrmann: Kerr-mikroskopische Methoden zur Untersuchung von Ummagnetisierung und Bereichsstruktur an magnetischen Doppelschichten. Report, University of Kassel (1995)

69. D. Raasch: J. Magn. Magn. Mat. **101**, 202–204 (1991)
70. T.M. Atmono: Anisotropie und magnetooptische Eigenschaften vom amorphen FeTb und FeTbCo Einzel- und Doppelschichten. Dissertation, University Kassel (1992)
71. S. Becker, H. Rohrmann, K. Röll: J. Magn. Soc. Jpn **20**, 229–232 (1996)
72. S. Becker, P.W.M. Filbrandt, C.D. Wright, H. Rohrmann, W.W. Clegg, K. Röll: J. Magn. Magn. Mat. **148**, 241–243 (1995)
73. A. Okamura: Over-write capable magnetooptical recording medium having plastic substrate. Europ. Patent Application, No. EP 0 452 155 A2 (1991)
74. K. Aratani, M. Kaneko and Y. Mutoh, K. Watanabe, H. Makino: Overwriting on a magneto-optical disk with magnetic triple *Optical Data Storage*, ed. by SPIE Proc. **1078**, 258–264 (1989)
75. T. Kobayashi, A. Okayama, H. Morioka, S. Shiomi, M. Masuda: J. Magn. Soc. Jpn. **19**, 53–57 (1994)
76. Y. Nakaki, Y. Fujii, T. Fukami, T. Tokunaga, K. Tsutsumi, H. Shibata: IEEE Trans. MAG **26**, 1900–1903 (1990)
77. T. Tokunaga, Y. Nakaki, T. Fukami, K.T. Tsutsumi: IEEE Trans. MAG **27**, 5112–5114 (1991)
78. S. Becker, T. Lucinski, H. Rohrmann, F. Stobiecki, K. Röll: J. Magn. Magn. Mat. **140–144**, 521–522 (1995)
79. T. Otto: Wechselwirkung von mechanischen Spannungen mit strukturellen und magnetischen Eigenschaften in TbFe-Viellagenschichten. Dissertation, University of Kassel (1995)
80. K. Spörl, D. Weller, G. Rupp: J. Magn. Magn. Mat. **83**, 91–93 (1990)
81. Y. Fujiwara, X.Y. Yu, H. Watabe, S. Iwata, S. Tsunashima, S. Uchiyama: J. Magn. Soc. Jpn **19**, 161–164 (1995)
82. R.A. Hajjar, F.L. Zhou, M. Mansuripur: J. Appl. Phys. **67**, 5328–5330 (1990)
83. H. Rohrmann, M. Kratzer, K. Röll: J. Magn. Magn. Mat. **148**, 136–138 (1995)
84. S. Becker, M. Kratzer, H. Rohrmann, K. Röll: J. Magn. Magn. Mat. **155**, 376–378 (1996)
85. K. Ichihara, S. Ashida, A. Kikitsu, Y. Mizusawa: Proc. Int. Symp. on Optical Memory (1991), pp. 12–17
86. K. Aratani, A. Fukumoto, M. Ohta, M. Kaneko, K. Watanabe: SPIE **1499**, 209 (1991)
87. A. Fukumoto, S. Yoshimura, T. Udagawa, K. Aratani, M. Ohta, M. Kaneko: SPIE **1499**, 216 (1991)
88. M. Ohta, A. Fukumoto, K. Aratani, M. Kaneko, K. Watanabe: J. Magn. Soc. Jpn. **15**, 319 (1991).
89. M. Kaneko, K. Aratani, M. Ohta: Jpn. J. Appl. Phys. **31**, 568–575 (1992)
90. M. Kaneko: J. Magn. Magn. Mat. **148**, 351–356 (1995)
91. K. Takahashi, T. Inase, A. Kondo: Jpn. J. Appl. Phys. **33**, 3918–3926 (1994)
92. A. Takahashi: J. Magn. Soc. Jpn. **19**, 273–278 (1995)
93. N. Ohta: Jpn. J. Appl. Phys. **32**, 5185–5188 (1993)
94. T. Suzuki: Read/Write **95–22**, 15–22 (1995)
95. T. Iwanaga, H. Inada: Jpn. J. Appl. Phys. **31**, 580–583 (1992)
96. D.C. Cheng, J.E. Hurst, R.T. Lynch: J. Magn. Soc. Jpn. **19**, 285–288 (1995)
97. T. Iwanaga: J. Magn. Soc. Jpn. **19**, 289–294 (1995)
98. C.H. Chung, S. Hsu: J. Magn. Soc. Jpn. **19**, 279–285 (1995)
99. M. Kaneko: IEEE Trans. MAG **28**, 2495–2499 (1992)

100. Y. Nakaki, T. Fukami, T. Tokunaga, Y. Kawano, K. Tsutsumi: IEEE Trans. MAG **28**, 2509–2511 (1992)
101. J. Saito, H. Akasaka, H. Birecki, C. Perlov: IEEE Trans. MAG **28**, 2513–2515 (1992)
102. K. Tsutsumi, Y. Nakaki, T. Tokunaga, T. Fukami, Y. Fujii: IEEE Trans. MAG **29**, 3760–3765 (1993)
103. M. Futamoto, F. Kugiya, M. Suzuki, H. Takano, Y. Matsuda, N. Inaba, Y. Miyamura, K. Akagi, T. Nakao, H. Sawaguchi: IEEE Trans. MAG **27**, 5280–5285 (1991)
104. M.H. Kryder: J. Magn. Soc. Jpn. **19**, 271–272 (1995)
105. Special Issue on Magnetooptical Recording, J. Magn. Soc. Jpn. **20**, 1–456 (1996)
106. Special Issue on Magnetooptical Recording and Optical Memory, J. Magn. Soc. Jpn. **22**, (1998)
107. H. Awano, S. Ohnuki, H. Shirai, N. Ohta, A. Yamaguchi, S. Sumi, K. Torazawa: Appl. Phys. Lett. **69**, 4257–59 (1996)
108. H. Awano, S. Ohnuki, H. Shirai, N. Ohta, A. Yamaguchi, S. Sumi, K. Torzawa: SPIE **3109**, 83–86 (1997)
109. E. Stavrou, H. Rohrmann, K. Röll: IEEE Trans MAG **34**, 1988–1990 (1998)
110. N. Nishimura, T. Hiroki, T. Okada and S. Tsunashima: Jpn. J. Appl. Phys. **35**, 403–409 (1996)

3 Layered Magnetic Structures: Interlayer Exchange Coupling and Giant Magnetoresistance

P. Grünberg

The possibility of an exchange interaction between ferromagnetic films across nonmagnetic metallic interlayers has been considered for a long time. In 1986 it was clearly identified and characterized in Fe/Cr structures [1] and rare earth yttrium multilayers [2,3]. The experiments on the Fe/Cr system were confirmed shortly thereafter by means of spin-polarized electrons [4]. Furthermore, in 1990, unexpected, long-range, damped oscillations in the variation of the coupling strength with the thickness of the spacer layers were discovered [5]. Meanwhile, the coupling and its oscillations have been established on a large number of layered magnetic structures.

The interest in exchange-coupled multilayers was also enhanced in 1988 by the discovery of "Giant Magnetoresistance" (GMR), in antiferromagnetically (AF) coupled Fe/Cr systems [5–7]. Soon after this, however, it became clear that AF coupling is not a necessary prerequisite for the GMR effect. For proper materials combinations it can also occur if antialignment is obtained by other means [8] for example, by hysteresis effects.

Although from the experimental point of view it seemed that AF coupling and GMR are independent phenomena, in the theories which had been developed in the meantime there occured a connection between them. Both effects are thought to be due to the propagation of spin-polarized electrons between the ferromagnetic films across the interlayer. The similarity is best expressed by the idea that the coupling is established by a "*spin current*" between the ferromagnetic films due to the fact that along with the transport of charges there is also a transport of spins. The latter results in a torque acting on the magnetizations which is due to the fact that electrons with spins up and down have different reflection coefficients at the interfaces [9]. The torque alignes the magnetizations according to the associated ratio of reflection coefficients. For the coupling, therefore, spin-dependent reflection coefficients at the interfaces are important, whereas for GMR it is the spin-dependent diffuse scattering at the interfaces or in the bulk of the ferromagnetic films.

While these mechanisms are widely accepted, quantitative predictions for their strength are often different by orders of magnitude from the experimental results, which obscurs the relation between coupling and GMR. Furthermore, many experimental results differ from sample to sample, which has obviously to do with varying sample quality.

Despite all these difficulties, the importance of the GMR effect for applications (magnetic-field sensors) has triggered a large number of experimen-

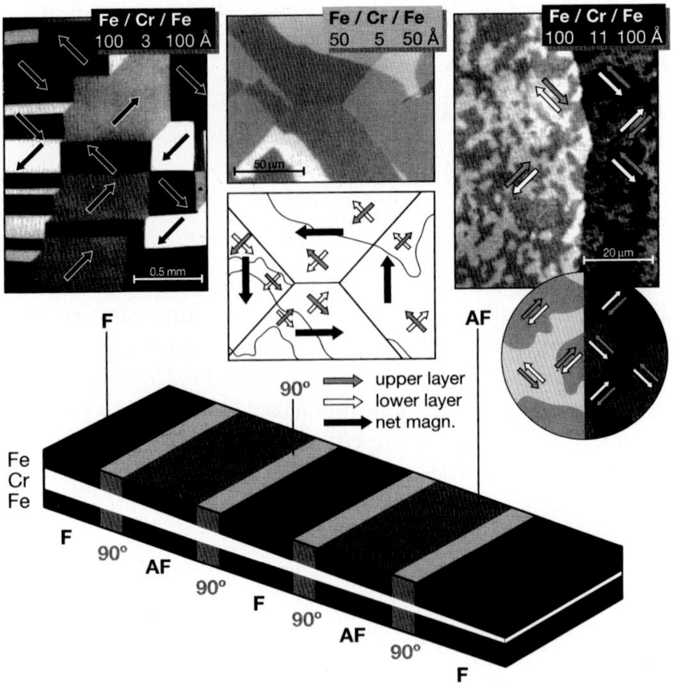

Fig. 3.1. Domain patterns due to different types of coupling in a layered magnetic structure. The pictures were taken by means of the Magneto-Optic Kerr Effect (MOKE) from different areas of a Fe double layer with a wedged Cr interlayer, as indicated in the figure

tal and theoretical research activities on both the coupling phenomena and GMR. The following is aimed to give an overview over the present status of the field and to illustrate the development by some representative results.

3.1 Qualitative Determination of the Different Types of Coupling

The exchange coupling between ferromagnetic films accross interlayers can be classified according to the different associated orientations of the magnetizations. For the 3d transition metals Fe, Co, Ni and their alloys as magnetic materials, in this sense three types of coupling have so far been found, namely ferromagnetic, antiferromagnetic, and 90°-type coupling.

A quick and efficient way to detect these coupling types qualitatively is by the observation of magnetic domains [10]. Examples are depicted in Fig. 3.1. The figure shows domain paterns in regimes of different types of coupling due to different interlayer thicknesses (note the sketch of a sample with variable interlaxer thickness in the lower left corner). The regimes of different types of coupling are marked with F (ferromagnetic), AF (antiferromagnetic) and 90° (90°-type coupling). In the left upper part a domain pattern is seen which is typical for F-type interlayer coupling. Contrary to the other pictures almost all domain walls are straight. The reason for this becomes clear from the following consideration:

In the samples under investigation here the Fe films are epitaxial and single-crystalline with an orientation of the sample plane parallel to a (100)-type crystallographic plane. Hence, they display the bulk anisotropy of single-crystalline Fe. Including the shape anisotropy, the total anisotropy has its easy axis in the sample plane along (100)-type directions. These include angles of 45° with the orientations given by the picture frame.

For the formation and the shape of domains we have to consider the local net magnetic moments. In the case of ferromagnetic coupling (left) the magnetizations of the single films and the resulting moments are along easy axis for every domain, indicated by the arrows. From one domain to the next the net moment changes its direction, and in most cases turns by 90°. This is the smallest angle different from zero, and therefore the smallest possible wall energy compatible with the crystalline anisotropy. In order to have no magnetic poles along these 90° walls their direction (here horizontal and vertical) bisects the angle between the magnetization directions in the adjacent domains. Therefore, in the case of ferromagnetic coupling the walls have well-defined orientations, which is the reason for being straight.

As expected, this changes if instead of F-type coupling one has the AF-type. The result is seen on the right-hand side of Fig. 3.1. The domain walls are now irregular. This is due to the fact that the net moment vanishes for AF-type coupling. Hence, poles do not apear for any arbitrary orientation of the wall. Therefore, there is no reason for straight domain walls and, according to their larger probability, irregular walls are formed.

For 90°-type coupling as in the middle section of Fig. 3.1 one observes straight as well as irregulary shaped domain walls. The reason for this is easy to comprehend. Irregular walls are formed between domains with the same direction of the net moment because in this case poles again do not appear for any direction of the wall. Domains do here differ by the magnetization directions in the single films, which are interchanged. Straight domain walls, on the other hand, are formed, like in the ferromagnetically coupled range, when the direction of the net moment changes.

From these examples it becomes clear that one can qualitatively recognize the different types of couplings from the shape of the domain walls.

3.2 Quantitative Description: Definition and Determination of Coupling Parameters

For a quantitative evaluation of experiments the coupling is treated by means of an expression for the interlayer exchange energy of the following kind:

$$\begin{aligned} E_i &= -J_1 \frac{\boldsymbol{M}_1 \cdot \boldsymbol{M}_2}{|\boldsymbol{M}_1| \cdot |\boldsymbol{M}_2|} - J_2 \left(\frac{\boldsymbol{M}_1 \cdot \boldsymbol{M}_2}{|\boldsymbol{M}_1| \cdot |\boldsymbol{M}_2|} \right)^2 \\ &= -J_1 \cos(\Delta\varphi) - J_2 [\cos(\Delta\varphi)]^2 \,. \end{aligned} \tag{3.1}$$

Here E_i its the interlayer coupling areal energy density, and $\Delta\varphi$ is the angle between the magnetizations of the films on both sides of the interlayer. J_1 and J_2 are parameters, describing the kind and the strength of the coupling. If J_1 dominates, then from the minima of (3.1) the coupling is F and AF for positive and negative J_1, respectively. In the same way, if J_2 dominates and is negative the minima of (3.1) correspond to 90° coupling. The first term on the left-hand side of (3.1) is often called bilinear and the second one biquadratic coupling.

If we neglect anisotropy and in (3.1) the term with J_2 we obtain for J_1 of a double layer with magnetic films of equal thickness d and magnetization M the following explicit form:

$$J_1 = \mu_0 M H_s d/2 \,, \tag{3.2}$$

where H_s is the field when the $M(H)$ curve saturates. For a multilayer system with n magnetic films in the limit $n \to \infty$, the factor of 1/2 on the right-hand side of (3.2) (which is valid for $n = 2$) has to be replaced by 1/4. In the commonly used cgs units one has to set $\mu_0 = 1$ in (3.2).

The parameters in (3.1) can be determined from measured $M(H)$ curves and from the frequencies of coupled spin-wave modes. For the $M(H)$ method one considers the total magnetic areal energy density E_{tot} of such layered structures in an external field H, including first-order cubic crystalline anisotropy of strength K_1. We assume here that the sample plane is parallel to a (100) crystallographic plane and that the external field H is along one of the inplane [100] axes which are the easy axes of the Fe-films. For a layered structure with two magnetic films of the same thickness d and magnetization M one gets

$$E_{tot} = E_j - \mu_0 H M d(\cos\varphi_1 + \cos\varphi_2) + K_1 d[\sin 2\varphi_1)^2 + (\sin 2\varphi_2)^2]/4 \,, \tag{3.3}$$

where E_j is given by (3.1), and the second and third parts represent the Zeeman energy and cubic crystal anisotropy, respectively. Here φ_1 and φ_2 are the angles between the film magnetization and the field direction. Hence, for $\Delta\varphi$ of (3.1) one has $\Delta\varphi = \varphi_1 - \varphi_2$. For a given external field the minima of (3.3) as a function of φ_1 and φ_2 yield the orientations of the magnetizations

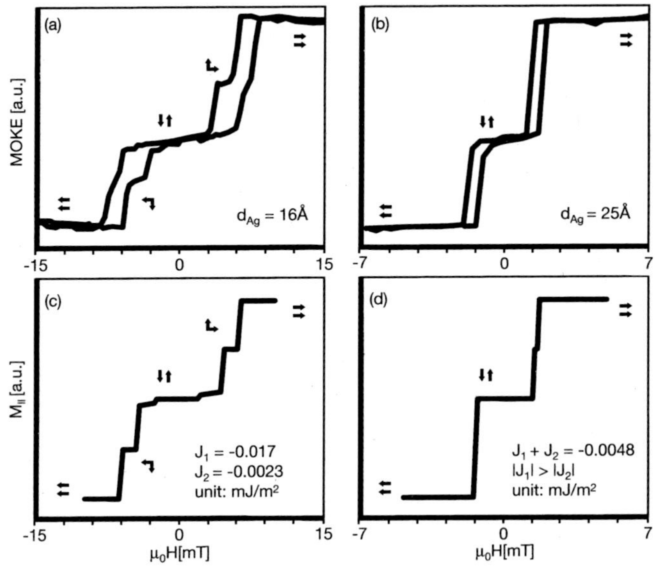

Fig. 3.2. Re-magnetization curves of layered magnetic structures which indicate various types of coupling from which the coupling paramters J_1 and J_2 can be evaluated. Together with the effect of the external field H the coupling generates the alignments of the layer magnetizations, as indicated by arrows. $M_{\|}$ is the component of the magnetization parallel to H, and MOKE is the signal of the magnetooptic effect which is here in good approximation proportional to $M_{\|}$. The curves are normalized in such a way that the maximum values of $M_{\|}$ and MOKE are approximately equal

and thus their components in any chosen direction. In this way, for a given set of parameters J_1 and J_2, any component of M can be calculated point by point as a function of the external field H. This procedure is repeated with new sets of parameters until there is optimum agreement betwen theory and experiment. This is the fit procedure for the determination of the parameters J_1 and J_2. However, it is not always possible to determine J_1 and J_2 separately, sometimes only their sum can be determined.

As an example the lower part of Fig. 3.2 shows two calculated $M_{\|}(H)$ curves together with their experimental counterparts in the upper parts. The parameters employed in the calculation are quoted. We note that apart from the hysteresis effect, which can safely be ignored, the experimental curves are nicely reproduced. However, in the case displayed on the right-hand side only a value for $J_1 + J_2$ can be extracted. In addition, we know that here $|J_2| < |J_1|$ because in the other case the $M_{\|}(H)$ curve would show an additional shoulder at $M_{\|} \approx M_s/2$, as displayed on the left-hand side. Hence without such a shoulder of 90° coupling cannot be evaluated.

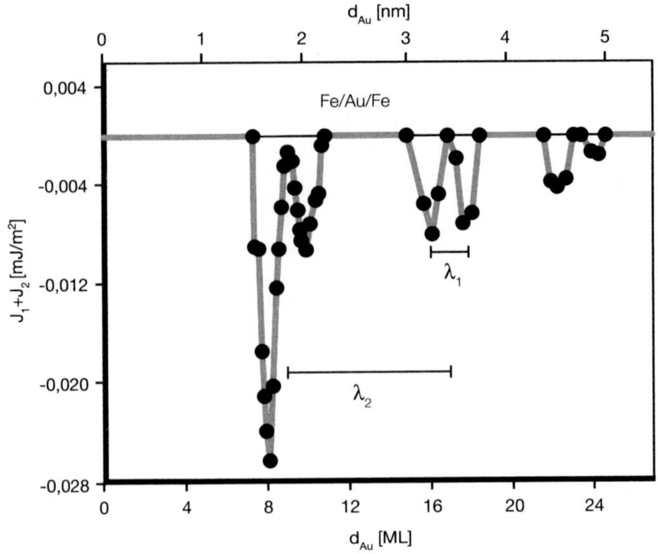

Fig. 3.3. Interlayer coupling of Fe layers across Au interlayers as a function of the Au thickness. Only the part with negative parameter $J_1 + J_2$ is shown due to the fact, that for the evaluation of positive values methods other than those used here have to be employed

Plateaus at $M_{\|} \approx 0$, as seen in Fig. 3.2, are mainly the result of AF-type coupling but are observed only if $J_1 \leq K_1 * d$. For sufficiently large J_1 or small K_1 the $M_{\|}(H)$ curves show a continuous, strongly sheared, dependence, as seen in many examples in the literature.

3.3 Multiperiodic Oscillations, Microscopic Models

A further milestone in the exploration of the interlayer-coupling phenomena was the discovery that the damped oscillations can be multiperiodic, i.e., different oscillation periods can be superimposed. This was first found for the Fe/Cr [11,12] and for the Fe/Au system [13], and later on for many others. The result for the Fe/Au system is displayed in Fig. 3.3. There are clearly two superimposed oscillation periods λ_1 and λ_2.

At the same time theories were developed which explain interlayer exchange coupling in terms of an indirect coupling of the magnetic films via the metal electrons in the interlayer. There is a strong similarity to the famous Ruderman-Kittel-Kasuya-Yosida (RKKY) interaction, which describes the interaction of isolated magnetic moments in metallic host lattices and in rare earth metals.

One important result of these theories was the discovery that the different oscillation periods are associated with certain extremal distances –

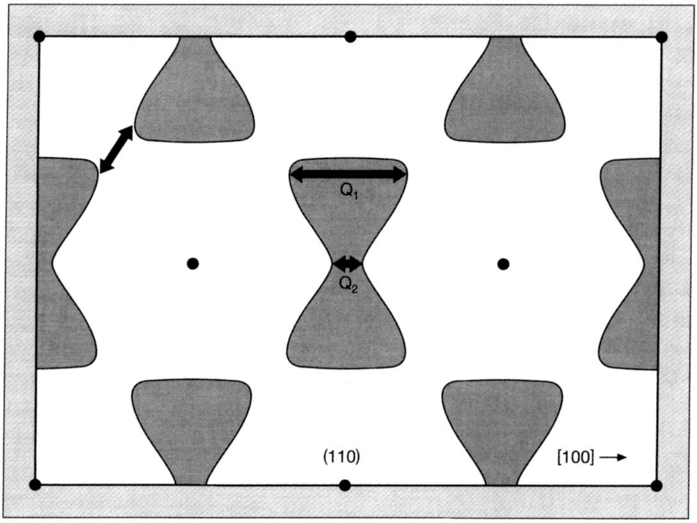

Fig. 3.4. Cross section of the Fermi surface of Au parallel to a (110) plane [15]. The two distances Q_1 and Q_2 in the "dog's bone orbit" in the [100] direction are connected with the two oscillation periods λ_1 and λ_2 visible in Fig. 3.3

also called calipers – in the Fermi surface of the interlayer material [14]. This is illustrated by a specific example, namely the coupling of magnetic films (here Fe) across interlayers of Au (Figs. 3.3 and 3.4). We first consider the case where the layer normal is parallel to a [100] direction of the Au lattice. It turns out that here the relevant calipers are the two extremal diameters Q_1 and Q_2 in the "dog's bone orbit" in the Fermi surface of the fcc noble metals (Fig. 3.4). Theory then predicts that in this case oscillatory coupling should display two periods λ_1 and λ_2, which are obtained from Q_1 and Q_2 via $\lambda_j = 2\pi/Q_j (i = 1, 2)$. If the layer normal is [100], these are the only two relevant calipers. From this we conclude that the two periods observed in (Fig. 3.3) are connected with the two calipers Q_1 and Q_2 seen in (Fig. 3.4). From recent calculations [15] of the Fermi surface of Au one gets $\lambda_1(\text{theory}) = 9.8\,\text{ML}$, $\lambda_2(\text{theory}) = 2.6\,\text{ML}$ while the experimental valuves from (Fig. 3.3) are $\lambda_1(\text{exp.}) = 8.00\,\text{ML}$, $\lambda_2(\text{exp.}) = 2.00\,\text{ML}$. Hence, there is also good quantitative agreement.

In the same way it turns out that in the [111] direction there is only one relevant caliper, in the upper part on the left-hand side of (Fig. 3.4). In experiments on samples with [111] as the layer normal, only one period should, therefore, be observed. This was confirmed experimentally [16].

For the prediction of oscillation periods λ_j from the calipers Q_j one usually considers the known Fermi surface of a quasi-infinite sample of the interlayer material. In reality, however, the interlayer has the shape of a thin film. This problem can only be solved by means of ab initio calculations which

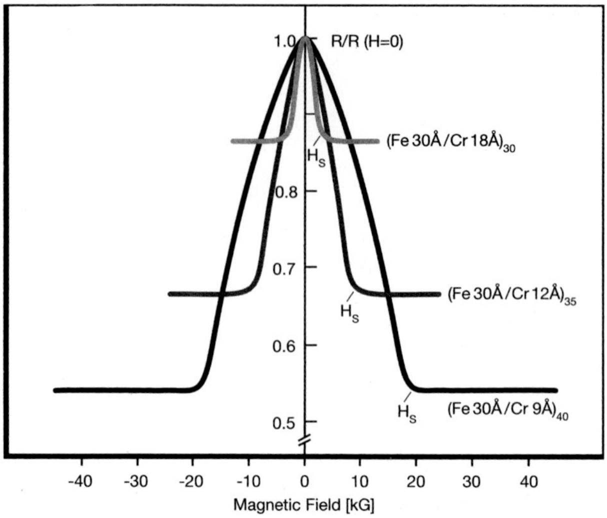

Fig. 3.5. Resistivity versus applied field for Fe/Cr (001) multilayers. H_s is the field neccessary to align the magnetic moments of the Fe layers parallel [6]

include the modifications of the band structure due to this special sample geometry. These investigations have shown [17] that for Co/Cu/Co structures the "asymptotic limit", where the calipers Q_j can be obtained from the infinitely extended crystal, is already reached at around eight monolayers of Cu.

The difference between the band structure of the interlayer material in the shape of an infinitely extended crystal and a thin film, is due to the fact that part of the propagating electron waves in the extended sample become standing waves in the film. Experimentally these "quantum well" states and their dependence on the electron spin could be directly observed by means of spin-polarized electrons [18].

Modifications of the band structure due to the sample geometry are also expected for magnetic films. As a result oscillations of the coupling strength should occur as a function of the thickness of the magnetic layers [19]. Such ascillations have indeed been detected experimentally [20].

Although it has by now been established that the reason for the oscillatory coupling is an RKKY-type mechanism. The origin of 90°-type coupling is not so clear. Like in the first observation [10] and also in (Fig. 3.1) it seems to appear predominantely in the transition zones between F- and AF-type couplings. A more precise analysis, however, reveals that this is mostly due to the fact that in these regimes the J_1 parameter changes sign, and therefore is small. Hence J_2, responsible for 90°-type coupling, dominates. Closer examination of J_2 reveals that opposite to J_1 it is more or less constant with a slight decrease when the interlayer thickness increases, but it shows a strong attenuation for increasing sample temperature.

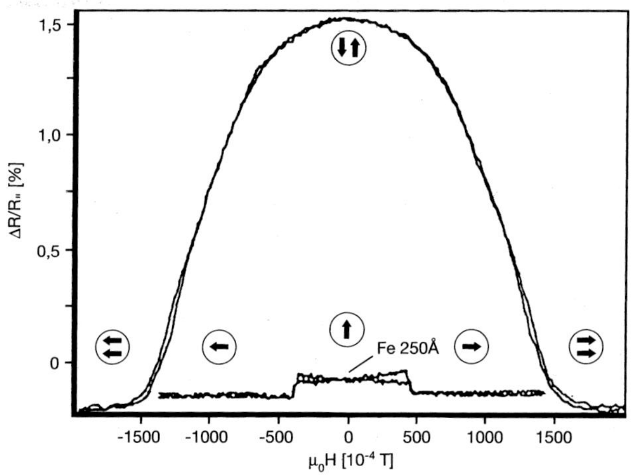

Fig. 3.6. Relative resistance change $\Delta R/R_{\|} = (R - R_{\|})/R_{\|}$ as a function of the external magnetic field H for two (011)-type Fe films separated by a (011)-type Cr interlayer (*upper curves*). Shown is also the Anisotropic Magnetoresistance effect (AMR) of a 250 Å thick Fe film (*lower curves*). The arrows indicate the associated magnetization directions. Relative to these arrows the external field H is applied horizontally, the current flows vertically. $R_{\|}$ was determined separately in the magnetically saturated state, with H applied vertically. Due to the AMR effect for both samples $\Delta R/R_{\|}$ becomes negative for sufficiently large H [7]

As far as the microscopic mechanism of 90° coupling is concerned, there are various good suggestions. They range from origins such as rough interfaces [21,22] and magnetic impurities, dissolved in the interlayers [23,24] to effects in higher order of the RKKY-type interaction [25]. In essence, however, the question if, or how much, the different mechanisms contribute, has not heen resolved. For this reason the subject is not treated in more detail here.

3.4 Giant Magnetoresistance

Figures 3.5 and 3.6 display the first observations of GMR [6,7]. The results are from antiferromagnetically coupled Fe/Cr superlattices. The resistivity drops strongly when an applied field aligns the magnetic moments of successive layers parallel. In multilayers (Fig. 3.5) the effect is much stronger than in double layers (Fig. 3.6) which is a first indication that the number of the available interfaces plays an important role.

GMR may be observed when the relative orientation of the magnetizations in adjacent layers changes. The biggest effect occurs when an antiferromagnetic arrangement is changed into a ferromagnetic one, for example, by an applied field. The antiferromagnetic arrangement can be provided by AF interlayer exchange, as in Figs. 3.5 and 3.6. In this case, oscillations of the

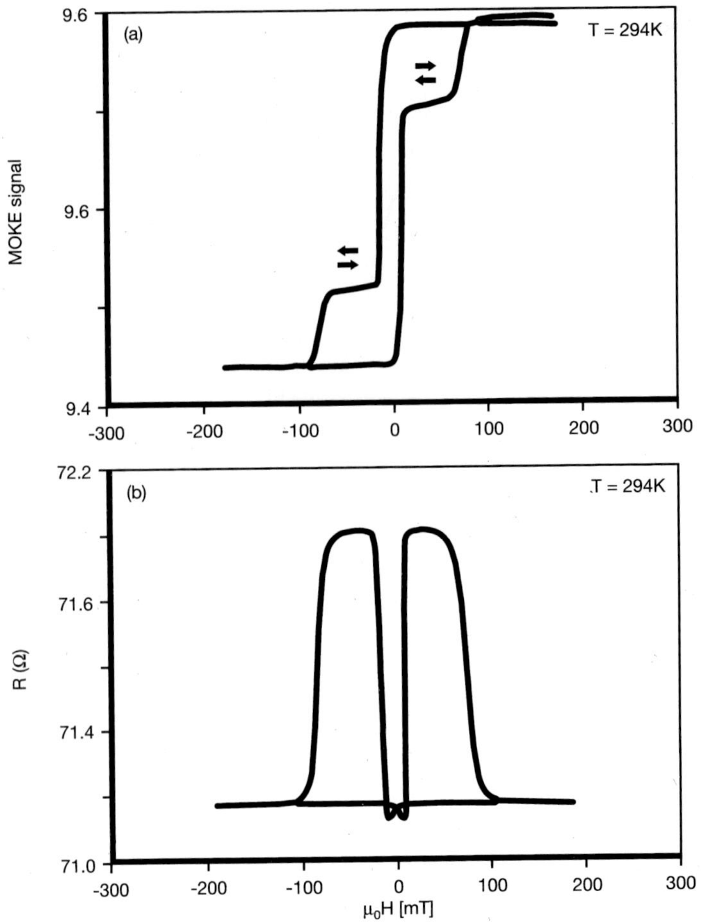

Fig. 3.7. a, b GMR in a Co/Au/Co-layered structure due to different coercivities H_c of the two Co films. (**a**) $M(H)$ curve obtained by MOKE where antialignment is indicated by arrows, and (**b**) relative resistance as a function of external field H

interlayer exchange as a function of the thickness of the nonmagnetic layers gives rise to oscillations of the GMR effect.

An antiparallel arrangement can also be obtained by other means, for example, by introducing different coercivities to successive magnetic layers, or by pinning the magnetization of some of the ferromagnetic films, for instance, by the presence of an antiferromagnetic material in direct contact, known as "exchange biasing".

If GMR is obtained via one of these methods and not via AF interlayer coupling usually the term "spin-valve system" is adopted in the literature, although there is no difference to the mechanism of GMR as produced by AF coupling.

Such a case is displayed in Fig. 3.7. The sample is a Co/Au/Co layered structure with one of the Co films deposited directly onto the GaAs substrate. Since the Co film is more strained than the other, prepared on the Au interlayer, it has a higher coercivity. During magnetization reversal of the whole structure the magnetization of the Co film on the interlayer reverses sooner as the other one so there is a small field regime where the alignment is antiparallel. In Fig. 3.7 (a) this regime is marked by antiparallel arrows. The curve in Fig. 3.7 (b) shows the associated electrical resistance R(H) and its increase due to the antiparallel alignment.

Finally, a nonsaturated GMR can be observed if the two films are magnetically decoupled so that there is an uncorrelated arrangement of the domains in successive layers at low field. For increasing fields at some point this changes to a parallel alignment, and an associated decrease of the resistance.

The GMR effect has been investigated in two different geometries, namely the "CIP" (Current In Plane) and the "CPP" (Current Perpendicular Plane) geometry. The relative effect is stronger in the CPP geometry, but due to the extremety unfavorable situation (lateral dimensions some orders of magnitude larger than the film thickness) the voltage drop across the thickness of the layers is very difficult to detect in the CPP geometry.

This problem has been solved in different ways. The first solution was to perform the experiment at low temperatures and to detect the current by means of a SQUID [26]. In an other case the sample was structured lithographically in such a way that from the extended film a "column" was cut such that the voltage drop across its length (= film thickness) became measurable with standard sensitive voltmeters [27]. Similarly, samples could be prepared in the shape of fine wires, with their axis parallel to the normal of the layered structure. For this purpose channels were first made in polycarbonate membranes by exposure to Ar^{9+} ions accelerated at 120 MeV, and followed by chemical etching. Inside the channels the layered structure in the form of a wire was prepared by electrodeposition [28]. A further mothod to measure the CPP GMR is the use of samples fabricated by evaporation onto grooved substrates [29].

As will be seen in the following, for the CIP and the CPP geometries there is an important difference concerning the related scattering lengths.

3.5 Phenomenological Description and Microscopic Mechanism of the GMR Effect

As is well known, in metals only the electrons with energies close to the Fermi edge contribute to the transport. In oder to understand GMR it is important to consider that the electrons propagate with high speed but arbitrary direction through the layered structure, as is indicated in Fig. 3.8. An applied voltage gives rise to an acceleration of the electrons in the direction of

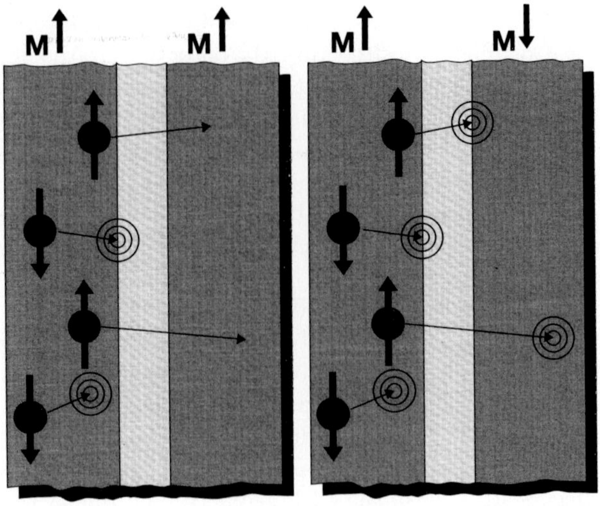

Fig. 3.8. Spin-dependent electron scattering (concentric circles) for parallel (*left*) and antiparallel (*right*) alignment of magnetic films. Shown are cross sections of layered structures. We assume here that only electrons with spin-down (antiparallel to the local magnetization) are scattered. This yields an increase of the total scattering rate at antiparallel alignment

the electric field, for example, in the layer plane (CIP) or perpendicular to it (CPP). This is repeatedly interrupted by scattering processes, but finally results in an electron drift observable as current. The scattering processes are the cause of electrical resistivity.

Let us assume that in the magnetic films and at the interfaces the scattering rate depends on the orientation of the spins with respect to the local magnetization [6,30]. Reasons for this will be given later. In the ideal case of Fig. 3.8 it is assumed that only electrons with spin down are scattered, within the ferromagnetic material as well as at the interfaces. Then it is clear that one can increase the total number of scattering events, and the resistance by turning the magnetization of one film around. Thus one changes parallel alignment (Fig. 3.8, left) into an antiparallel one (Fig. 3.8, right).

A detailed analysis shows furthermore that in the CIP and CPP geometries the dependence of the GMR effect on the interlayer thickness d_0 is rather different. In the CIP geometry the relevant length scale is given by the mean free path (approx. 10 nm) in the interlayer material. If d_0 increases beyond this value, the effect decreases rapidly. Opposite to this, in the CPP geometry the relevant thickness is given by the much larger spin-diffusion length which is on the order of 100–1000 nm. This is the mean free path of an electron between scattering events with spin reversal.

As was indicated above, the appearance of the GMR effect could be explained phenomenologically by the assumption of a spin-dependent electron

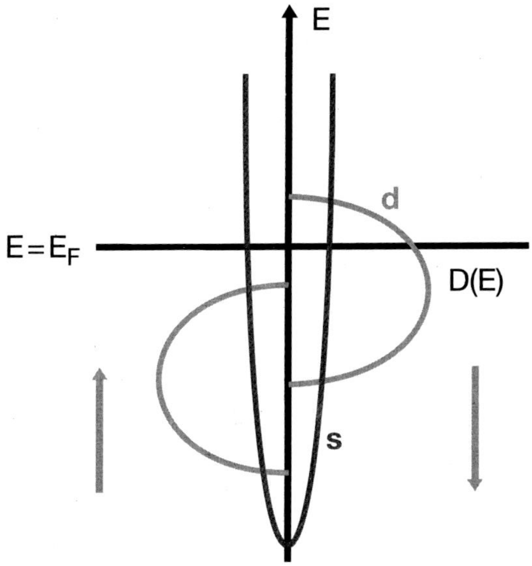

Fig. 3.9. Schematic density of states, D, as a function of energy E for 3d metals and their alloys. E_F is the Fermi energy. The splitting of the s bands is neglected. Note the high density of d states with spin-down at $E = E_F$

scattering. Hence, one has to explain what could be the reason for such a difference in scattering.

For the description of transport phenomena it is sufficient to limit the discussion to electrons with energies close to the Fermi energy E_F. Hence, we consider the band structure of magnetic 3d metals (alloys included) and focus on the range close to E_F. Figure 3.9 shows schematically the dependence of the density of states on the energy. As usual the magnetic splitting of the s bands is neglected because it has no significant effect on the density of states close to E_F.

The transport properties of 3d metals, in particular the question if predominantely s or d electrons carry the current, is still a topic of controversial discussions. A higher mobility, as documented in Fig. 3.9 by a larger vertical extent of the s band, is an argument in favor of the s electrons. A higher density of states at E_F is an argument in favor of d electrons. Current models for the GMR effect assume that s electrons mainly carry the current, but the d states are those into which the electrons are scattered. From Fig. 3.9, it becomes obvious that electrons with spin-down are scattered more than those with spin-up.

As is known, an energy diagram, as displayed in Fig. 3.9, discribes rather well the situation in the "strong" ferromagnets Co and Ni, whereas for the "weak" ferromagnet Fe the densities of states at E_F are rather equal. Hence, from this point of view Co and Ni appear to be better candidates for a

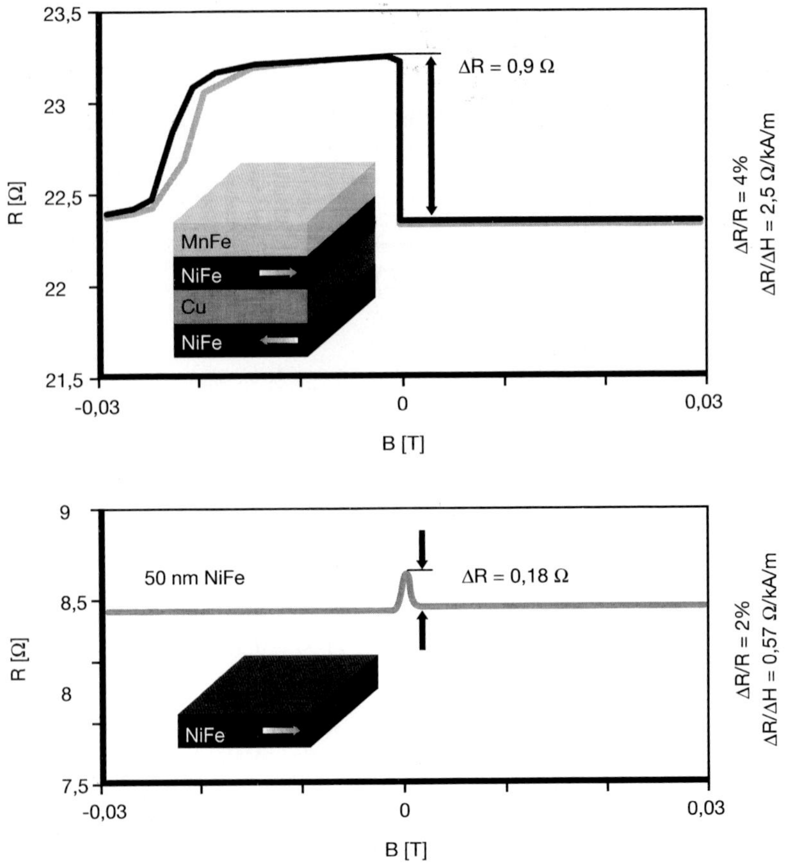

Fig. 3.10. Comparison of the performance of magnetic-field sensors based on the GMR and AMR effects (Courtesy of Philips, Eindhoven)

strong GMR effect than Fe. This is in agreement with the observation that there are a lot of layered structrues with Co, where a strong GMR effect is observed, whereas for structures with Fe so far only the Fe/Cr layered structure has yielded a big effect. The usual explanation for this its that Cr is an exception because it modifies the unfavorable band structure of Fe at the Fe/Cr interface by means of a "virtual bound state" in such a way that a high density of states with spin-up is generated. Centainly this problem requires further careful investigation especially because of the importance of GMR for applications.

3.6 Applications, Final Remarks

Finally, we shall briefly touch on the applications of GMR, but more can be found in the preceding chapters. It is obvious that GMR can be used for the detection of magnetic fields. However, the range of applications of such field sensors is surprisingly large. They can be utilized to read information from tapes, hard disks, floppies, magnetic strips in consumer electronics and in computers. Furthermore, they can be employed for sensing the position and/or the speed of moving parts. The moving part for this purpose has to be equipped with a small permanent magnet or a magnetic strip. Using a strip yields the possibility to store further information if required. Control of moving parts, for example, is of interest at many places in automobiles (e.g., ABS), in robotics, assembly lines, etc. Sensors based on GMR are not only very sensitive but can also be made very small. For sensors based on GMR in the CPP geometry, the performance even improves upon miniaturization. This definitely is of advantage for the accuracy and the resolution in position and speed control.

Figure 3.10 displays the improvement introduced by GMR in comparison with the "Anisotropic Magnetoresistance" (AMR) effect which has been used so far. AMR is based on the fact that generally the resistivity of a magnetic material depends on the direction of the current with respect to the magnetization (Fig. 3.6). Although weaker than the GMR the AMR effect has already been interesting for the application.

From Fig. 3.10 one can infer that even for the double layer GMR is already stronger than AMR. Furthermore, GMR can be realized with a higher sensitivity and a better signal-to-noise ratio.

Investigations of interlayer exchange coupling and giant magnetoresistance, therefore, establish not only a new and exciting area of research, but lead also to important applications in present and future devices.

Acknowledgements. I would like to express my gratitude and appreciation to the nomerous graduate and post-graduate students, guests and colleagues who have contributed to the work presented here.

References

1. P. Grünberg, R. Schreiber, Y. Pang, M.B. Brodsky, H. Sowers: Phys. Rev. Lett. **57** 2442 (1986)
2. M.B. Salamon, S. Sinha, J.J. Rhyne, J.E. Cunningham, R.W. Erwin, J. Borchers, C.P. Flynn: Phys. Rev. Lett. **56** 259 (1986)
3. C.F. Maykrzak, J.W. Cable, J. Kwo, M. Hong, D.B. Mc Whan, Y. Yafet, J.V. Waszczak, C. Vettier: Phys. Rev. Lett. **56** 2700 (1986)
4. C. Carbone, S.F. Alvarado, Phys. Rev. B **36** 2433 (1987)
5. S.S.P. Parkin, N. More, K.P. Roche: Phys. Rev. Lett. **64** 2304 (1990)

6. M.N. Baibich, J.M. Broto, A. Fert, F. Nguyen-Van-Dau, F. Petroff, P. Etienne, G. Creuzet, A. Friedrich. J. Chazelas: Phys. Rev. Lett. **61** 2472 (1988)
7. G. Binasch, P. Grünberg, F. Saurenbach, W. Zinn: Phys. Rev. B **39** 4828 (1989)
8. J. Barnas, A. Fuss, R.E. Camley, U. Walz, P. Grünberg, W. Zinn: Vacuum 41 1241 (1990)
   J. Barnas, A. Fuss, R.E. Camley, P. Grünberg, W. Zinn: Phys. Rev. B 42 8110 (1990)
9. J.C. Slonczewski, J. Magn. Magn. Mat. **126** 374 (1993)
10. M. Rührig, R. Schäfer, A. Hubert, R. Mosler, J.A. Wolf, S. Demokritov, P. Grünberg: Phys. Stat. Sol. (a) **125** (1991) 635
11. J. Unguris, R.J. Celotta, D.T. Pierce: Phys. Rev. Lett. **67** (1991) 140
12. S.T. Purcell, W. Folkerts, M.T. Johnson, N.W.E. McGee, K. Jager, J. aan de Steege, W.B. Zeper, W. Hoving, P. Grünberg: Phys. Rev. Lett. **67** (1991) 903
13. A. Fuss, S. Demokritov, P. Grünberg, W. Zinn: J. Magn. Magn. Mat. **103** (1992) L221
14. P. Bruno, C. Chappert: Phys. Rev. Lett **67** (1991) 1602, **67** (E), (1991) 2592
15. Mertig, T.U. Dresden: private communication
16. V. Grolier, D. Renard, B. Bartenlian, P. Beauvillain, C. Chappert, C. Dupas, J. Ferre, M. Caltier, E. Kolb, M. Mulloy, J.P. Renard, P. Veillet: Phys. Rev. Lett. **71** 3023 (1993)
    J.J. de Vries, P.J.H. Bloemen, M.T. Johnson, J. aan de Stegge, A. Reinders, W.J.M. de Jonge: J. Magn. Magn. Mat. **129** L 129 (1994)
17. P. Lang, L. Nordström, R. Zeller, P.H. Dederichs: Phys. Rev. Lett. **71** (1993) 1972
18. C. Carbone, E. Vescovo, O. Rader, W. Gudat, W. Eberhardt: Phys. Rev. Lett. **71** (1995) 2805
19. L. Nordström, F. Lang, R. Zeller, P.H. Dederichs: Phys. Rev. B **50** (1994) 13058
20. P.J.H. Bloemen, M.T. Johnson, M.T.H. van de Vrost, R. Coehoorn, J.J. de Vries, R. Jungbluth, J. aan de Steege, M. Reinders, W.J.M. de Jonge: Phys. Rev. Lett. **72** (1994) 764
21. J.C. Slonczewski: Phys. Rev. Lett **67** (1991) 3172
22. S. Demokritov, E. Tsymbal, P. Grünberg, W. Zinn, I.K. Schuller: Phys. Rev. B **49** (1994) 720
23. J.C. Slonczewski: J. Appl. Phys. **73** (1993) 5957
24. M. Schäfer, S. Demokritov, S. Müller-Pfeiffer, R. Schäfer, M. Schneider, P. Grünberg, W. Zinn: J. Appl. Phys. **77** (1995) 1
25. D.M. Edwards, J. Mathon, R.B. Muniz, M. Villeret, J.M. Ward: in *Magnetism and Structure in Systems of Reduced Dimension*, ed. by R.F.C. Farrow et al. NATO ASI Series B Vol. 309 (Plenum, New York 1993) pp. 401
26. W.P. Pratt, S.F. Lee, J.M. Slaughter, P.A. Schroeder, J. Bass: Phys. Rev. Lett. **66** (1991) 3060
27. M.A.M. Gijs, S.K.J. Lenczowski, J.B. Giesberg: Phys. Rev. Lett. **16** (1993) 740
28. L. Piraux, J.M. George, J.F. Despres, C. Leroy, E. Ferain, R. Legras, K. Ounadjela, A. Fert: Appl. Phys. Lett **65** (1994) 2484
29. T. Ono, T. Shinjo: J. Phys. Soc. **64** (1995) 363
30. P.E. Camley J. Barnas: Phys. Rev. Lett. **63** (1989) 664

4 Giant Magnetoresistance in Exchange-Biased Spin-Valve Layered Structures and its Application in Read Heads

R. Coehoorn

Exchange-biased spin-valve layered structures consist (in their simplest form) of a magnetically soft ferromagnetic layer, separated by a nonferromagnetic layer from a second ferromagnetic layer, which has its magnetization pinned by the exchange biasing interaction with a third magnetic (usually antiferromagnetic or ferrimagnetic) layer. For suitable layer compositions and nanometer-scale layer thicknesses these materials combine a fair Giant Magnetoresistance (GMR) effect with a very small field interval close to zero field in which the resistance change takes place. The resulting high sensitivity makes these materials attractive for sensor applications and other magnetoelectronic devices. In this overview, I first review experimental results on GMR of these materials, their theoretical interpretation, and the magnetic interactions which determine the magnetization reversal process. Subsequently, trends in magnetic recording and the potential benefit of exchange-biased spin-valves in yoke-type read heads for high-density digital magnetic recording are discussed. The chapter closes with an overview of unresolved issues, and with a note added in proof on developments that have taken place after the completion of the manuscript.

4.1 Introductory Comments

After the discovery in 1988 of the Giant Magnetoresistance (GMR) effect in antiferromagnetically coupled magnetic multilayers [1,2] subsequent research has led to a number of classes of materials which are suitable for low-field sensor applications. Several types of layered structures have been found which show a much higher sensitivity of the electrical resistance to applied magnetic fields than single magnetic alloy layers, which are presently used in sensor applications. Recently, prototype GMR–based read heads for high-density hard disk or tape magnetic recording have been presented [3–9]. GMR materials are also of interest for other magnetic-field sensors such as electronic compasses, and for position, velocity and acceleration sensors. An example has been presented in [10]. Another potential field of application is in Magnetic Random Access Memory (MRAM) devices [11–14].

In the following a review of the physics and materials-science aspects of the Magnetoresistance (MR), the magnetic interactions, and the magnetization-reversal processes for one specific class of low-field GMR materials, viz. the so-called exchange-biased spin-valve layered structures [15] is presented. The

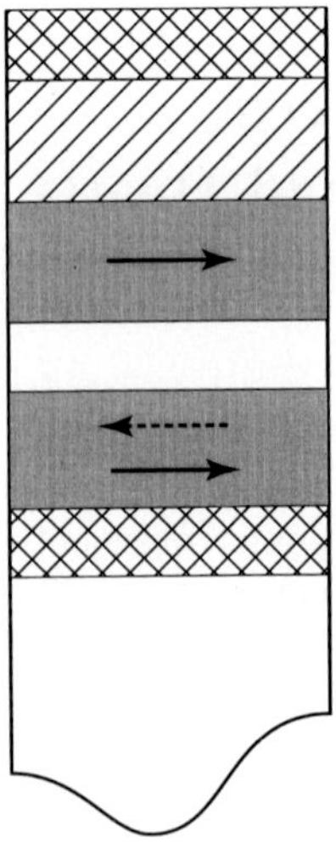

Fig. 4.1. Schematic cross-section of a "simple" exchange-biased spin-valve layered structure

approach is application-oriented. The introduction will be closed with a discussion on the application of these GMR materials in magnetic read heads. The review should make clear that the successful application of a GMR material in sensors such as heads does not only require a high GMR ratio or a high intrinsic sensitivity, but that the actual suitability is determined by a number of other properties of the GMR materials as well. Excellent reviews of the GMR effect in magnetic layered structures, including discussions of other classes of materials, have been given by Levy [16], Dieny [17], Fert and Bruno [18], Parkin [19], and by Gijs and Bauer [20].

Exchange-biased spin-valve layered structures were first described in 1990 by Dieny et al. [21–23]. In the original and simplest structure (Fig. 4.1), two Ferromagnetic (F) layers are separated by a Non-Magnetic (NM) spacer layer. A third magnetic layer of a properly chosen composition and structure, which is usually Anti-Ferromagnetic (AF), is in contact with one of the F layers. This is called a "simple" exchange-biased spin-valve layered structure, in order to distinguish it from the more complicated structure of a "dual" or "symmetric" exchange-biased spin-valve layered structure (Sect. 4.3.2). Suitable preparation conditions, for example growth in a magnetic field, lead to pinning of the magnetization of the F layer that is in contact with the AF layer, as a result of the exchange interaction across the interface. This is called the exchange biasing (or sometimes: exchange anisotropy) effect. The unpinned ("free") F layer is usually magnetically very soft. The NM spacer layer is sufficiently thick as to almost eliminate the magnetic coupling between the F layers. The functioning of this system is illustrated schematically by the magnetization and resistance curves given in Fig. 4.2. Defining the positive

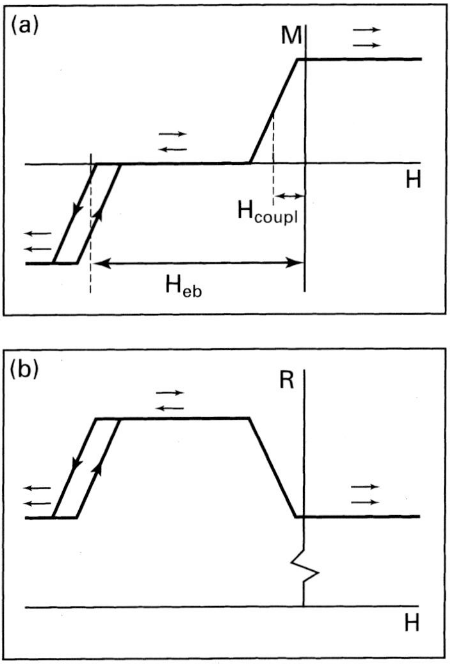

Fig. 4.2. Schematic curves of the magnetic moment (**a**) and resistance (**b**) of a simple exchange-biased spin-valve layered structure. Ferromagnetic layers with equal magnetic moments per unit of area have been assumed

field direction as the direction of the magnetization of the pinned layer, the layers have parallel magnetizations for $H > 0$. In a small field interval close to $H = 0$ the magnetization of the free F layer reverses, whereas the magnetization of the pinned F layer remains fixed. Only upon the application of a large negative field (equal to the exchange biasing field H_{eb}), the exchange biasing interaction is overcome, and the pinned layer switches, too.

The GMR effect is the increase of the resistance upon the change of the angle between neighbouring magnetic layers from a parallel to an antiparallel alignment. For combinations of F- and NM-layer compositions for which the GMR effect is high, the resistance curve of exchange-biased spin-valve structures shows a steep slope close to zero field, where the configuration of the magnetization directions switches, and stays high until, at the exchange biasing field, also the pinned layer switches. It is remarked that there is usually a small magnetic coupling between the free layer and pinned layers, leading to an offset, $\mathrm{H}_{\mathrm{coupl}}$, of the field around which the free layer switches (Fig. 4.2). Using exchange-biased spin-valve structures, it has been possible to reach very high sensitivities, combined with a very small coercivity. The sensitivity of an MR material is defined as the slope $s \equiv (\partial R/R)/\partial H$ of the relative resistance versus field curve, obtained at a suitable working point. Sensitivities up to 18%/(kA/m) (1.4%/Oe) at room temperature have been reported for spin valves using permalloy ($Ni_{80}Fe_{20}$) for the free and pinned layers, and Cu for the spacer layer [24]. This value is much larger than that for single permalloy films, which are presently employed in sensor applications such as read heads. The resistance change of these films, which originates from the Anisotropic Magnetoresistance (AMR) effect [25], is about 1.5%. The resistance change takes place in a field range of typically 0.32 kA/m (4 Oe), leading to a sensitivity of only about 5%/(kA/m).

In addition to their suitability for sensor applications, exchange-biased spin valves are very suitable for studying the origin of the GMR effect. Their design makes it possible to obtain a well-defined antiparallel state, irrespective of the ferromagnetic as well as antiferromagnetic interlayer magnetic coupling. In contrast, in the case of ferromagnetic/nonmagnetic multilayers a well-defined antiparallel alignment is only obtained for certain spacer-thickness intervals, for which the interlayer exchange coupling is antiferromagnetic, or in the case of F layers with different coercivities. Examples of the first class are Fe/Cr [26] or Co/Cu [27] multilayers. An example of the second class is the system Co/Au/Co, with different thicknesses of the two Co layers, leading to different coercive fields. From studies of the latter system Barnaś et al. [28], and Dupas et al. [29] provided the first experimental proof that antiferromagnetic interlayer exchange coupling is not required for obtaining the GMR effect. Subsequently, Speriosu et al. showed for the case of exchange-biased spin-valve structures that even the size of the GMR ratio is not affected by the sign and strength of the interlayer exchange coupling [30]. Thus, only the relative alignment of the magnetization directions of the ferromagnetic layers, parallel or antiparallel, is important.

Based on the resulting insight in the origin of the GMR effect, and in analogy to the earlier discovered spin-tunneling effect in ferromagnetic/insulator/ferromagnetic structures [31], Dieny et al. [21] have coined the name "spin-valve effect". In Sect. 4.2, the origin of the magnetoresistance in layered structures will be discussed, and it will be shown that the denotation spin-valve effect is indeed very appropriate, because it precisely describes the microscopic mechanism. The term GMR is less adequate, because the effect is not in all cases "giant" when compared to the AMR effect. Nevertheless, this has become the convention for the phenomenon, and it will be used it throughout this chapter.

Section 4.3 contains an overview of experimental results on the magnetoresistance of exchange-biased spin-valve layered structures in relation to their composition, layer thicknesses, layer stacking, and the preparational method, and of the temperature dependence of the magnetoresistance.

Modeling of magnetotransport, starting with assumptions on the spin-dependent scattering probabilities in the bulk of the layers and at the interfaces, would make it possible to "design" GMR materials in order to optimize them for a chosen application. In Sect. 4.4 an overview is given on developments in this field.

In Sect. 4.5, the physics and materials science aspects of the magnetic interactions and properties, which determine the magnetization reversal process in a spin valve will be outlined. Details of the magnetization reversal process determine, to a large degree, the device performance of a magnetic sensor. The discussion will be focussed on (i) the magnetic coupling between the free and the pinned layers, (ii) the exchange biasing interaction, and (iii) the magnetic anisotropy of the free layer. In practical systems these interactions vary along the lateral direction of the layers, causing lateral variations in the magnetization direction in the free layer, and sometimes even causing complicated domain structures. The resulting Barkhausen noise may negatively affect the signal-to-noise ratio of GMR sensors. Experimental and theoretical micromagnetic studies of these phenomena are presently being carried out by a number of groups. Results will be discussed in this section.

A major industrial driving force behind the study of the GMR effect is its application in read heads for digital magnetic recording. In Sect. 4.6 the trends in bit densities of hard disk and tape recording are sketched, and two head designs (the shielded head and the yoke-type head) are discussed. The relationship between the materials properties of the GMR material, the geometry and dimensions of a head, and the output voltage of a head, is treated for the case of a yoke-type head.

In Sect. 4.7 a summary is given of issues and challenging subjects for future research. In Sect. 4.8 a brief overview is given on recent publications.

4.2 Origin of the GMR Effect

4.2.1 Phenomenological Description

The GMR effect of magnetic layered structures is the dependence of the resistance on the angles between the magnetization directions of successive magnetic layers. The observed angular variation of the resistance of exchange-biased spin-valve structures is well described by the phenomenological expression

$$R(\theta) = R(\theta = 0) + \Delta R_{\mathrm{GMR}}(1 - \cos\theta)/2\,, \tag{4.1}$$

where θ is the angle between the magnetization directions of the free and pinned layers [21]. The ratio $\Delta R_{\mathrm{GMR}}/R(\theta = 0)$ is called the GMR ratio. Early studies for other types of layered structures have led to the same expression for the angular variation [32,33]. However, in a recent study for a multilayer structure with a relatively high GMR ratio the *conductance*, instead of the resistance, has been found to vary linearly with $(1-\cos\theta)$ [34]. The difference between the two descriptions is of second order in the GMR ratio, and is therefore most significant in the case of a high GMR ratio. This issue will

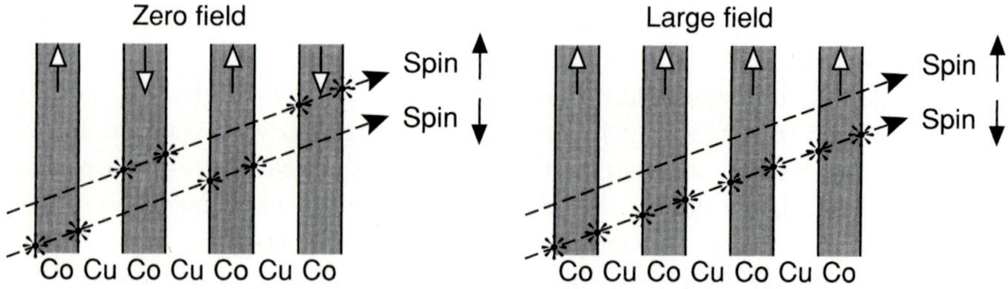

Fig. 4.3. a, b The two-current model: schematic illustration of electron transport in a Co/Cu multilayer, for antiparallel (**a**) and parallel (**b**) alignments of the magnetization directions in the successive layers. Dashed lines are electron trajectories and fanned arrows indicated diffuse electron scattering. For simplicity, only scattering at the interfaces has been indicated

again be treated in Sect. 4.4, in which results of model calculations of the angular dependence of the resistance will be discussed.

4.2.2 Simple Model

The GMR effect can be understood within the Mott two-current model for the electrical conductivity in metals [35]. As the spin quantum number of the conduction electrons is conserved in the vast majority of scattering processes, at least at low temperatures, the total conductivity can be expressed as the sum of separate contributions from spin-up and spin-down electrons.

Let us now first consider the Current In the Plane of the layers, which is the most common configuration for measuring GMR of layered structures. For this geometry the GMR effect arises when, in the case of parallel magnetizations, the layer-averaged electron mean-free path for (at least) one spin direction is larger than the multilayer period, and different from that for the other spin direction. The model is illustrated schematically in Fig. 4.3, for the Co/Cu system. For simplicity, only scattering at the interfaces is indicated. Scattering is strong for electrons with spin antiparallel to the magnetization direction of the cobalt layers, and weak for electrons with spin parallel. Each layer acts as a spin-selective valve: its magnetization direction determines whether it more easily transmits spin-up or spin-down electrons. In the case of parallel-aligned magnetic layers the resistivity for one spin channel is low, leading to a low total resistance. However, antiparallel alignment of alternate magnetic layers results in appreciable scattering for electrons in both spin directions, and hence in a larger resistance.

It is possible to obtain a simple quantitative expression for the magnetoresistance ratio of a (F/NM) multilayer (periodic repetition of a ferromagnetic layer F and a nonmagnetic layer NM), if the mean free path for *both* spin directions is much larger than the multilayer period [36]. If $R^{\uparrow}$ and $R^{\downarrow}$ are the resistances for the spin-up and spin-down channels, respectively, in the parallel case, then the resistances for the Parallel (P) and Anti-Parallel (AP) cases are

$$R_{\mathrm{P}} = \left(1/R^{\uparrow} + 1/R^{\downarrow}\right)^{-1} \tag{4.2}$$

and

$$R_{\mathrm{AP}} = (R^{\uparrow} + R^{\downarrow})/4\,. \tag{4.3}$$

Defining the scattering asymmetry parameter as $\alpha \equiv R^{\downarrow}/R^{\uparrow}$, the Magnetoresistance (MR) ratio can be expressed as

$$\frac{(R_{\mathrm{AP}} - R_{P})}{R_{\mathrm{P}}} = \frac{(1-\alpha)^2}{4\alpha}\,. \tag{4.4}$$

It follows that $R_{\mathrm{AP}} \geq R_{\mathrm{P}}$, and that the MR ratio is positive. It should be noted that the MR ratio can either be positive or negative for systems containing *different* ferromagnetic layers [37]. More generally applicable calculational methods for obtaining the CIP–GMR ratio are described in Sect. 4.4.

For the Current being Perpendicular to the Plane of the layers (CPP geometry) the proper length scale is not determined by the mean free paths for diffuse scattering (as in the CIP case), but is given by the spin diffusion length Λ_{sd} [16,38]. This is the average distance between subsequent spin-flip scattering processes. If Λ_{sd} is larger than the multilayer repetition period the MR effect may be analyzed using essentially the simple approach presented above, for arbitrary spin-conserving mean free paths. This approach has been applied successfully to a large number of systems [39]. In the CPP geometry the GMR ratio of multilayers is generally larger than in the CIP geometry [39,40]. Methods for measuring the CPP–GMR effect, and an overview of experimental results, are given in Chap. 5 by M. A. M. Gijs.

In conclusion, the GMR effect occurs if in a material under the influence of an applied magnetic field the spatial variation of the magnetization direction is changed on a scale smaller than a certain critical length. The electron scattering processes which determine this critical length depend on the measurement geometry. In the CIP and CPP geometries, the critical lengths are given by the (spin-dependent) electron mean free paths and by the the spin diffusion length, respectively.

4.2.3 Spin-Dependent Scattering

Which combinations of elemental or alloy layers are expected to yield high GMR effects? In the following the spin-dependence of scattering, first in the bulk of layers, then at interfaces will be discussed. At low temperatures, electron-phonon scattering and electron-magnon scattering may be neglected. The temperature dependence of the GMR will be treated in Sect. 4.3.1.

In transition metals with open d shells, the conductivity may be thought of arising mainly from free-electron-like s/p type states, whereas the most relevant scattering processes are from s/p to d states. The 3d states do not effectively contribute to the conductivity due to their high effective mass. However, a high 3d density of states at the Fermi level leads to a high probability of scattering to d states. Although this picture, first proposed by N. Mott, may be too simple in view of the strong hybridization between s, p and d states, it still forms a very useful basis for the understanding of spin-dependent scattering in transition metal systems. Let us first concentrate on scattering in the bulk of the layers. In magnetic transition metals the 3d density of states at the Fermi level is spin-polarized. The scattering probability (e.g., at defects) is expected to be largest for the spin direction with the largest density of 3d states at the Fermi level. Calculated densities of states for bcc Fe, fcc Co and fcc Ni are plotted in Fig. 4.4 a–c. For Co and Ni, with only minority-spin 3d states at the Fermi level, the scattering probability is therefore expected to be largest for minority spin electrons. For Fe, the spin asymmetry is in this model expected to be much smaller, because the majority-spin 3d-band is not completely occupied. In Sects. 4.3 and 4.4 experimental data will be given which support this picture.

A material of particular importance to applications is $Ni_{80}Fe_{20}$ (permalloy). In this alloy, Ni and Fe atoms are randomly or almost randomly distributed on the sites of an fcc lattice. The Curie temperature is 560 °C. The site-disorder results in a finite resistance at zero temperature, even without structural defects. The electronic structure of permalloy is very similar to that of Co and Ni: it is a strong ferromagnet, with an entirely filled majority-spin 3d band. In agreement with the discussion given above, it would therefore be expected that scattering in permalloy is strongly spin-dependent. As will be discussed in Sect. 4.4, this point of view is supported by experimental evidence from investigations for exchange-biased spin valves, whereas additional evidence is available from the analysis of the CPP and CIP GMR effect in multilayer structures (as given in [39] and [42], respectively). In contrast, Parkin has argued from the analysis of specially designed exchange-biased spin-valve structures that scattering in Ni–Fe and Ni–Co alloys is essentially *spin-independent* [43]. Based on recent very promising advances in the theory of the residual resistivity of alloys, including permalloy [44], it is expected that first principles computational studies will soon help resolve this controversy.

Diffuse interface scattering occurs if the interface is non ideal. Most quantitative theoretical discussions on interface scattering have so far focussed on the relatively simple case of coherently grown superlattices with an interface roughness due to mixing within one monolayer only. In that case scattering is the result of a lateral randomness of atomic potentials. A large spin dependence of the scattering arises if for one spin direction the atomic potentials of the two types of atoms in the mixed layer are almost identical (well matched), and strongly dissimilar for the other spin direction. Let us now first consider an A/B interface (A and B are elemental metals) and let us assume that the atomic potentials of the A and B atoms in the mixed interface layer are

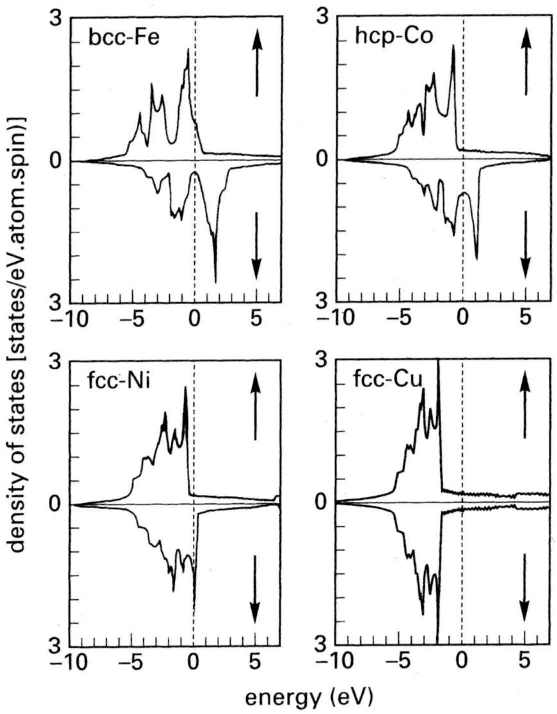

Fig. 4.4. a–d Spin-polarized densities of states for the elemental metals Fe (**a**), hcp–Co (**b**), Ni (**c**) and Cu (**d**), as obtained from selfconsistent ab initio band structure calculations using the Augmented Spherical Wave (ASW) method [41]. The energy is given with respect to the Fermi energy (*dashed line*)

the same as in the bulk of the A and B layers. It is assumed that the atoms in the mixed layer are charge-neutral, and that their magnetic moments are identical to those in the bulk of the A and B layers. This situation has been discussed by Inoue et al. for Fe/TM, Co/TM and TM/Cu systems [45], in which TM is a 3d or 4d transition metal element with an unfilled d shell. For Fe/TM and Co/TM systems the researchers argued that the scattering potential, which enters the scattering matrix element, is proportional to the absolute value of the difference between the center-of-mass energies of the d bands of the two constituent metals. A very small scattering potential (good band matching) is expected for minority spin electrons in Fe/Cr systems, whereas a much larger scattering potential (relatively large band mismatch) is expected for majority-spin electrons. This leads to high predicted values for the GMR ratio of Fe/Cr systems. In the case of TM/Cu interfaces matching is expected to be very good for majority-spin electrons if the majority-spin d band of the TM metal is fully occupied. This is the case for Ni and Co, but not for Fe, as may be seen from the calculated densities of states (Fig. 4.4 a–

c. Figure 4.4 d shows for comparison the density of states for Cu. The spin dependence of the scattering then increases with increasing mismatch for the minority spin direction, i.e., with decreasing average atomic number of the TM element [45].

Apart from electronic matching of the subsequent layers, also structural (lattice) matching is important. Lattice mismatch results in the formation of dislocations, at which scattering may take place, thereby reducing the mean free paths. Scattering at dislocations in the non magnetic spacer metal is spin-independent, and leads to a reduction of the GMR ratio. Scattering at dislocations in the ferromagnetic layers could be spin dependent (depending on structural and compositional details). Whether it would lead to an increase or a decrease of the GMR ratio depends on the presence and the strength of other spin dependent scattering processes (such as scattering at the interfaces). More generally, it should be noted that the spin dependence of scattering is not a strictly intrinsic, composition-related property, but depends on many microstructural aspects as well. Results on the spin dependence of scattering thus depend on those preparational conditions which influence the structure in the bulk of the layers and near the interfaces.

In agreement with the predictions from this simple picture, Fe/Cr and Co/Cu multilayers show very high GMR ratios, the highest values observed at 4.2 K being 220% [46] and 120% [47], respectively. Both systems reveal very good lattice as well as electronic matching. To both systems the simplifying assumptions used in the discussion given above on the spin dependent interface scattering (interface moments for interfaces with a roughness on the scale of one monolayer only essentially identical to bulk moments) apply very well [48]. However, it should be noted that in general magnetic moments at interfaces may be quite different from those in the bulk of the layers. A number of computational studies, using band-structure calculations, have addressed this issue, and the implications for the GMR ratio [48–51]. A complication is that for the case of *imperfect* interfaces magnetic moments may be drastically different from moments at the *perfect* interface, or in the bulk of the layer. An example of such a situation has been discussed by Nicholsen et al. [51], who investigated the case of mixed $Ni_{80}Fe_{20}$/Cu interfaces, with a gradual decrease of the Cu concentration from 100% to 0% within three atomic layers. From first-principles band-structure calculations they obtained a variation of the Ni moments in the mixed layer from $0.40\,\mu_B$ (closest to the permalloy layer) to $0.24\,\mu_B$ (closest to the Cu layers), versus a bulk (average) Ni moment of $0.51\,\mu_B$. Fe moments were observed to remain almost unaffected, with values close to $2.9\,\mu_B$. The presence of such a weakly magnetic intermixed layer decreases the spin-dependence of scattering, as will be discussed in Sect. 4.3.2.

4.2.4 Other Magnetoresistance Effects

The GMR effect was named "giant" in view of the fact that the relative resistance changes obtained for systems such as Fe/Cr multilayers are much larger than observed for the ordinary magnetoresistance effect in metals, and for the Anisotropic Magnetoresistance (AMR) effect in magnetic alloys. The ordinary magnetoresistance [52,53] is the increase of the resistance in an applied magnetic field (or better: in a magnetic induction B), as the result of the Lorentz force acting on the charge carriers. The AMR effect [25] is the dependence of the resistivity on the angle θ between the magnetization direction and the current density, and is (for polycrystalline materials) expressed as

$$\rho = \rho_0 + \Delta\rho_{\mathrm{AMR}} \cos^2\theta \,. \tag{4.5}$$

It is well known that the effect results from spin-orbit interaction [54], which gives rise to a dependence of the electron scattering on the angle between the electron wave vector and the magnetization direction. At room temperature the largest effect is found for $Ni_{1-x}Co_x$ alloys close to $x = 0.2$, for which the magnetoresistance ratio $\Delta\rho/\rho_0$ amounts to 6%. For $Ni_{80}Fe_{20}$, the bulk AMR ratio is about 4%. However, in thin permalloy films, which are presently used as magnetoresistive material in read heads and other sensors, the AMR effect is even smaller. It decreases with decreasing film thickness and depends on the deposition conditions. For a 30 nm thick film, the AMR ratio is typically about 2.5% [25]. The AMR effect is not always small as compared to the GMR effect. The effects can be disentangled by performing experiments for the directions of the current parallel and perpendicular to the applied magnetic field.

4.3 GMR of Exchange-Biased Spin Valves

4.3.1 Fe–Mn-Based Simple Spin Valves

The prototype of a magnetically soft exchange-biased spin-valve layered structure is the structure $Ni_{80}Fe_{20}/Cu/Ni_{80}Fe_{20}/Fe_{50}Mn_{50}$ [55]. The AF $Fe_{50}Mn_{50}$ layer leads to the required exchange biasing of the permalloy layer that is in contact with it. Unless stated otherwise, the structures discussed below have been prepared using sputter deposition. We refer to the original publications for more details on the preparation.

Dependence on the Nonmagnetic Layer Thickness and Composition. In Fig. 4.5 results by Rijks et al. are presented on the dependence of the MR ratio on the Cu-layer thickness [56]. The system studied had the composition (8 nm $Ni_{80}Fe_{20}/t_{Cu}$ nm Cu/6 nm $Ni_{80}Fe_{20}$/8 nm $Fe_{50}Mn_{50}$), grown on a 3 nm Ta underlayer on Si(100) and covered with a 3 nm Ta cap layer. Up to a critical Cu-layer thickness of about 2 nm, the Cu spacer is not sufficiently

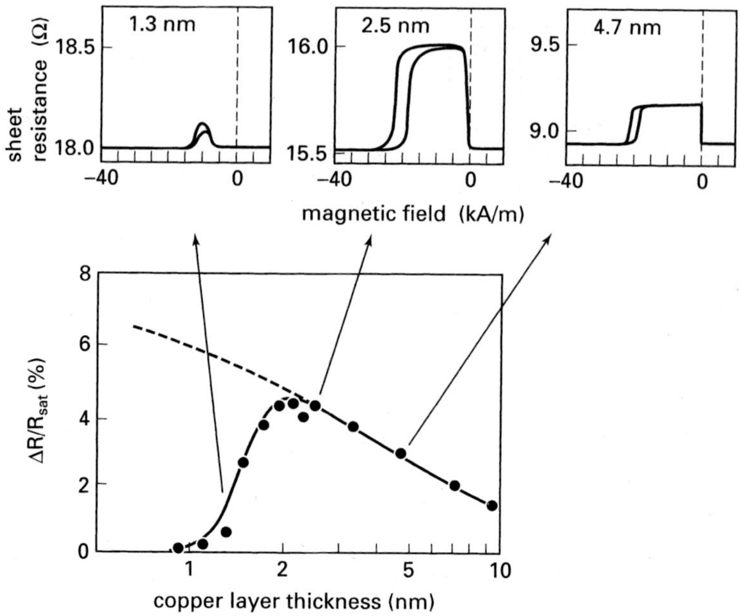

Fig. 4.5. Dependence of the MR curves and the MR ratio at room temperature on the Cu-layer thickness, for ($Ni_{80}Fe_{20}/Cu/Ni_{80}Fe_{20}/Fe_{50}Mn_{50}$) spin valves [56]. Layer thicknesses are given in the text. The *dashed line* gives a fit to the data, using (4.6), for the Cu thicknesses for which full antiparallel alignment is obtained (*see text*)

thick as to realize a perfectly antiparallel alignment of the magnetization directions in a certain field interval, leading to a decreased GMR ratio. As shown in Fig. 4.5 the MR ratio is maximum close to this critical Cu thickness, viz. almost 5%. A detailed analysis of the $R(H)$ and magnetization curves, in terms of the competing interlayer coupling and exchange biasing, has been given by Rijks et al. [56] (see also Sect. 4.5.2).

Speriosu et al. [57] have proposed the following phenomenological expression for the dependence of the MR ratio of spin valves on the spacer-layer thickness:

$$\frac{\Delta R}{R} = \left(\frac{\Delta R}{R}\right)_0 \frac{\exp(-t_{\mathrm{NM}}/\ell_{\mathrm{NM}})}{1 + t_{\mathrm{NM}}/t_0} . \tag{4.6}$$

The factor $\exp(-t_{\mathrm{NM}}/\ell_{\mathrm{NM}})$ represents the probability that an electron which potentially contributes to the GMR effect by crossing the NM layer, is *not* diffusely scattered in the NM layer. The factor in the denominator represents the shunting effect due to the NM layer. The parameter ℓ_{NM} is related to the conduction electron mean free path λ_{NM} in the NM layer. The parameter t_0 depends on the conductance of the remainder of the system (the conductance in the absence of the NM layer). Both factors contribute to the decrease

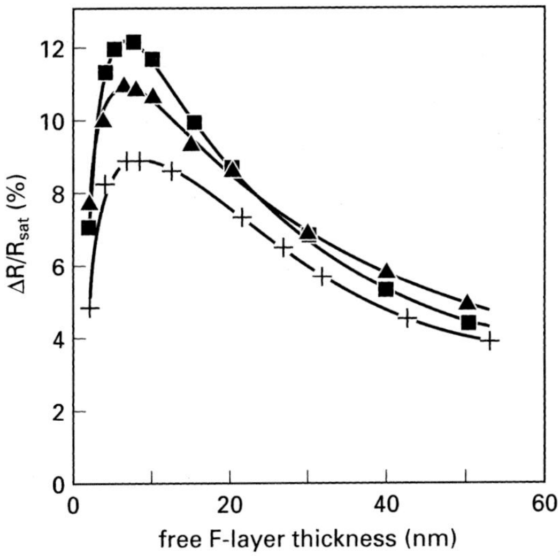

Fig. 4.6. Dependence on the noble-metal spacer-layer thickness of the MR ratio at room temperature for spin valves with the structure Si/7 nm Co/t_{NM} nm NM/4.7 nm $Ni_{80}Fe_{20}$/7.8 nm $Fe_{50}Mn_{50}$/1.5 nm NM [23]

of the GMR ratio with the NM spacer thickness. The experimental data given in Fig. 4.5 can be fitted well using the parameters $(\Delta R/R)_0 = 7.6\%$, $\ell_{Cu} = 16$ nm, and $t_0 = 4.6$ nm. One expects that ℓ_{NM} is somewhat smaller than λ_{NM}, in view of the fact that electrons which most effectively contribute to the CIP–MR effect have off-normal velocity directions. The optimal direction depends on details of the system, and is a compromise between almost parallel to the film (a large contribution to the conductivity, small sensitivity to the relative orientation of magnetization directions) and almost normal to the film (small contribution to conductivity, large sensitivity to relative magnetization directions). Dieny has proposed that for systems of practical interest, ℓ_{NM} is approximately equal to $1/2\,\lambda_{NM}$ [17]. The results discussed above then lead to $\lambda_{Cu} \approx 32$ nm.

Dieny et al. [23] have investigated the dependence of the MR ratio on the composition of the NM spacer material. For structures of the type Co/NM/ $Ni_{80}Fe_{20}$/$Fe_{50}Mn_{50}$, (NM = Cu, Ag and Au) they measured the variation of the MR ratio with the NM layer thickness (Fig. 4.6). The solid curves represent exponential fits through the data [as suggested by (4.6), but disregarding the shunting effect]. Whereas the critical thickness above which a range of full antiparallel alignment is observed is about 2 nm for Cu and Au, it is about 4–5 nm for Ag. Dieny et al. suggested that coupling through pinholes might be of importance up to much higher spacer thicknesses in the case of Ag. Other work on systems containing Ag supports this point of view [58,59]. The extrapolated MR ratio for $t_{NM} = 0$ is about 50% larger for Cu than for

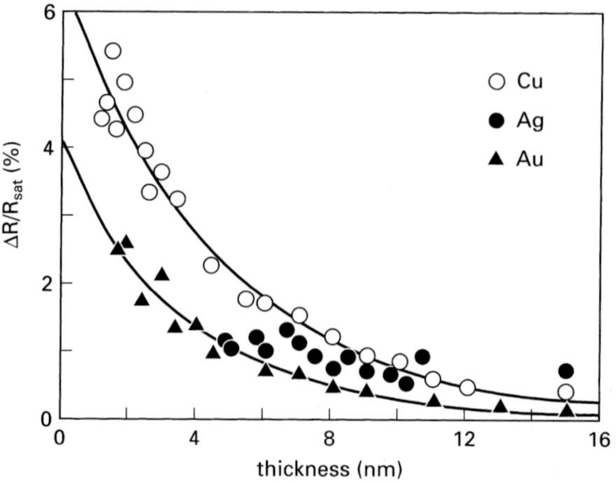

Fig. 4.7. Dependence of the MR ratio on the free magnetic-layer thickness for $(X/Cu/X/Fe_{50}Mn_{50})$ spin valves, grown on 3 nm Ta buffer layers above Si (100) substrates, with X = Co (*squares*), X = $Ni_{66}Fe_{16}Co_{18}$ (*triangles*) and (X = $Ni_{80}Fe_{20}$) (*plus signs*), measured at 5 K. Layer thicknesses: t_{Cu} = 3 nm (but 2.5 nm for F = (X = $Ni_{80}Fe_{20}$); t_F(pinned layer) = 5 nm (but 6 nm for F = $Ni_{66}Fe_{16}Co_{18}$)

Au. They ascribed this to a lower transmission of electrons for F/Au than for F/Cu interfaces. In addition, the decay length is smaller for Au than for Cu, viz. 3.6 and 4.5 nm, respectively. This is ascribed by Dieny et al. to the higher resistance of Au, although also the higher spin-orbit scattering (which may lead to spin flip), for Au (which is the heavier element) is mentioned as a possible explanation. Here it is proposed that the effect on the microstructure due to dislocations, which result from the large lattice mismatch between the F layers and Au might be an additional cause of the lower MR ratio. In [22] Dieny et al. have reported that for $F/NM/F/Fe_{50}Mn_{50}$ spin-valve structures, with F = $Ni_{80}Fe_{20}$, the spin-valve effect is also observed for NM = Pt, but is absent for NM = Ta, Al, Cr and Pd.

Dependence on the Ferromagnetic Layer Thickness and Composition. In Fig. 4.7 results obtained by Rijks et al. [60] on the dependence of the GMR ratio (at 4.2 K) on the free magnetic layer thickness are shown. The systems studied had the composition (t_F nm F/Cu/F/8 nm $Fe_{50}Mn_{50}$/3 nm Ta), grown on a 3 nm Ta underlayer above Si(100). The layer thicknesses of the Cu and pinned magnetic layers are given in the caption of the figure. The ferromagnetic layer compositions studied were F = Co, $Ni_{66}Fe_{16}Co_{18}$ and $Ni_{80}Fe_{20}$. For all F layers a rather broad maximum was found around t_F = 6–8 nm. Similar results were found by Dieny et al. [61], for $F/Cu/Ni_{80}Fe_{20}/Fe_{50}$

Mn_{50} structures. They proposed the following phenomenological expression for the variation of the MR ratio with the F layer thickness:

$$\frac{\Delta R}{R} = \left(\frac{\Delta R}{R}\right)_0 \frac{1-\exp(-t_F/\ell_F)}{1+t_F/t_0} \,. \tag{4.7}$$

The factor $(\Delta R/R)_0$ is a function of the NM–layer thickness. The factor $[1-\exp(-t_F/\ell_F)]$ is the angle-averaged probability for an electron of the type with the largest mean free path to be scattered within the ferromagnetic layer before being scattered diffusely at the outer boundary. This factor expresses the fact that if the F layers are too thin, the contrast between the mean free paths for both spin directions will be reduced, because the longer of both mean free paths is lowered by the diffuse scattering at the outer boundaries, so that the MR ratio is lowered. The denominator expresses the shunting effect due to the F layer. The parameter t_0 depends on the conductance of the remainder of the system (the system in the absence of the F layer). As argued by Dieny [17], it may be expected that $\ell_F \approx \frac{1}{2}\lambda_{F,>}$, where $\lambda_{F,>}$ is the largest of the mean free paths of both types of electrons.

Spin Valves Containing "Back Layers". Let us now consider an exchange-biased spin valve in which the free F layer is much thicker than the mean free path for minority spin electrons, but much thinner than the mean free path for majority spin electrons. As will be discussed in Sect. 4.4, this situation can be realized using, for instance, permalloy. In such a case, as predicted from (4.7), the MR ratio could be enhanced by making the free-layer thicker. However, it is also possible to increase the magnetoresistance by adding other (even nonmagnetic) "back layers" (B). Gurney et al. [62] have studied this phenomenon, in order to obtain the majority and minority spin mean free paths in the back layer. For such spin valves, with the structure B/F/NM/F/AF, the variation of the magnetoconductivity ΔG as a function of the thickness t_B of the back layer is, to a good approximation, given by [62]:

$$\Delta G = \Delta G_0 + \Delta G_B[1-\exp(-t_B/\ell_B)] \,. \tag{4.8}$$

Gurney et al. argued that the parameter ℓ_B is about equal to the majority spin mean free path in the back layer. Room-temperature data for various back layers were found to be in excellent agreement with the general form of this expression. The following values were obtained for ℓ_B: 19 nm (Cu), 2 nm ($Au_{50}Cu_{50}$), 4.6 nm ($Ni_{80}Fe_{20}$), 5.5 nm (Co) and 1.5 nm (Fe). Less directly, minority spin mean free paths were estimated from an analysis of the data, yielding $\lambda^{\downarrow} \leq 0.6$ nm ($Ni_{80}Fe_{20}$), $\lambda^{\downarrow} \leq 1$ nm (Co) and $\lambda^{\downarrow} = 2.1 \pm 0.5$ nm (Fe). The data thus provided evidence for a strong spin dependence of the scattering in permalloy and Co, and a much weaker spin dependence of the scattering in bulk Fe. This is in agreement with the qualitative picture based on the bulk densities of states (Fig. 4.4).

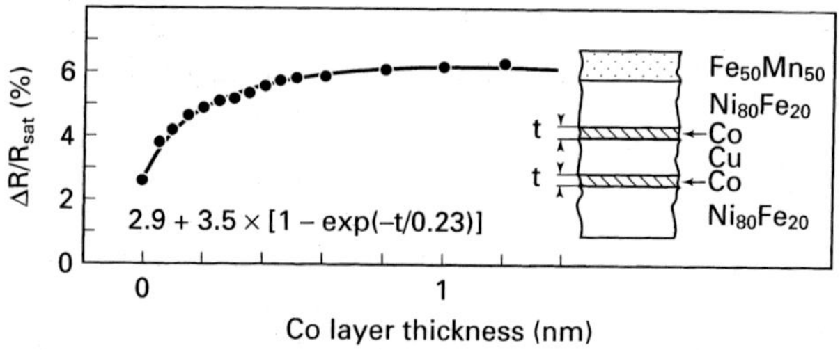

Fig. 4.8. Dependence on the Co interface layer thickness t_{Co} of the MR ratio for spin valves with the structure (Si/[5.3−t_{Co}] nm$Ni_{80}Fe_{20}$/t_{Co} nmCo/3.2 nmCu/[2.2−t_{Co}] nm$Ni_{80}Fe_{20}$/9 nm$Fe_{50}Mn_{50}$/1 nmCu) [65]. The measurements were performed at room temperature

It should be noted that the phenomenological expressions (4.6)–(4.8) for the layer thickness dependences of the GMR ratio and the magnetoconductance are, at best, only valid in restricted parts of the parameter space. A discussion of the range of validity of (4.7) and (4.8), and of the interpretation of the parameters ℓ_{F} and ℓ_{B}, was given in [63].

Temperature Dependence. Extensive studies of the temperature dependence of the MR ratio valves have been reported by Dieny et al. [61,64]. A very striking result is that for a variety of systems of the type F1/Cu/F2/$Fe_{50}Mn_{50}$ (F1 and F2 ferromagnetic metals), the MR ratio decreases *approximately linearly* with increasing temperature. For spin valves with F1 = Co, F2 = $Ni_{80}Fe_{20}$, and with NM = Cu or Au, the MR ratio was observed to extrapolate to zero at a temperature of about $T_0 \approx 560$ and 540 K, respectively, independent of the Cu or Au layer thickness. Although T_0 is much smaller than the Curie temperatures T_{C} of Co or permalloy, it is clearly correlated with T_{C}, as was shown from a study for several other combinations of F layers [61,64]: the MR ratio for Co/Cu/Co, Co/Cu/$Ni_{80}Fe_{20}$, and $Ni_{80}Fe_{20}$/Cu/$Ni_{80}Fe_{20}$ extrapolates to $T_0 \approx 630$, 510–540 and 500–530 K, respectively. As proposed by Dieny et al. [64], these relatively low values of T_0 may be understood as the result of spin-flip scattering within the ferromagnetic layers, with a predominant role of spin-flip scattering at the interfaces due to the reduced Curie temperatures of alloyed interface regions. To this valid point it is added here that even in the case of structurally perfect interfaces, magnetic interactions are weakened, due to decreased moments and a reduction of the number of magnetic nearest neighbour atoms, most likely resulting in much stronger spin-flip scattering probabilities than in the bulk of the layers. Explicit proof was given by Dieny et al. that at room temperature the mean free path in Cu or Au spacer layers is only marginally lower than at 4.2 K. For the samples studied it thus follows that in the Cu or Au layer, at room temperature, thermally induced (electron-phonon) scattering is still very weak as compared to extrinsic scattering, such as scattering at impurities or crystallographic defects.

For some applications, it is of interest that the variation with temperature of the resistance change ΔR is much smaller than the variation of $\Delta R/R$ [65].

4.3.2 Other Exchange-Biased Spin-Valve Structures

Several modifications to the simple, Fe–Mn-biased spin valve structures have been reported to yield higher GMR ratios. Parkin demonstrated that a very thin Co layer at the $Ni_{80}Fe_{20}/Cu$ interfaces in ($Ni_{80}Fe_{20}/Cu/Ni_{80}Fe_{20}/Fe_{50}Mn_{50}$) spin valves (interface "dusting") may lead to a drastic increase of the MR ratio [66]. As shown in Fig. 4.8, the MR ratio for the structure (5.3 nm $Ni_{81}Fe_{19}$/3.2 nm Cu/2.2 nm $Ni_{81}Fe_{19}$/9 nm $Fe_{50}Mn_{50}$/1 nm Cu) was found to increase (at room temperature) from 2.9% in the absence of Co, up to 6.4% for Co thicknesses above approximately 0.6 nm. Conversely, adding permalloy layers of a thickness of only 0.4 nm to the Co/Cu interfaces of Co/Cu/Co/Fe–Mn spin valves was found to decrease the MR ratio from 6.8% to 3.9% (see [66] for details of the layer structures). The beneficial effect of thin Co layers was seen to be strongly localized to the interfaces, by varying the distance d to the interface of 0.5 nm Co layers that were burried in the permalloy layers: no significant increase of the MR ratio was found for $d > 0.6$ nm. This experiment nicely demonstrated the crucial role played by the interfaces.

The effect of Co is twofold [49,51,67]. First, as discussed in Sect. 4.2, the minority spin scattering at Co/Cu interfaces is larger than at $Ni_{80}Fe_{20}/Cu$ interfaces, due to the larger mismatch of the minority spin d bands. Second, one should take into account that magnetic moments at the interface with Cu are different from those in the bulk of the layers. Experimentally, this has been studied by Speriosu et al. [57], who (at room temperature) observed a substantial decrease of the magnetization of permalloy near the interface with Cu, effectively corresponding to a "dead layer-thickness" of 0.2 ± 0.016 nm. For Co/Cu interfaces, a dead-layer thickness of only 0.1 ± 0.02 nm was observed. Theoretical and experimental evidence for a significant moment reduction at the $Ni_{80}Fe_{20}/Cu$ interface has been given by Nicholson et al. [51] (Sect. 4.2.3). It implies that the majority spin partial density of the states of magnetic atoms at the interface has 3d character at the Fermi level, resulting in enhanced scattering of majority spin electrons at the interface, thereby negatively affecting the GMR ratio. Speriosu et al. [57] have argued that at room temperature moments in the mixed $Ni_{80}Fe_{20}/Cu$ interface layers are not just reduced, but effectively zero, implying spin-independent scattering in these layers.

Spin-valves using amorphous ferrimagnetic Tb–Co layers with the approximate composition $Tb_{0.23}Co_{0.77}$ as the exchange-biasing material have been fabricated by Freitas et al. [68]. For structures of the type Si(100)/Ta/$Ni_{80}Fe_{20}$/Co/Cu/Co/$Ni_{80}Fe_{20}$/Tb–Co/Ti–W, the highest GMR ratio observed was 7.5% at room temperature [69]. The Tb–Co thickness used was 12 nm.

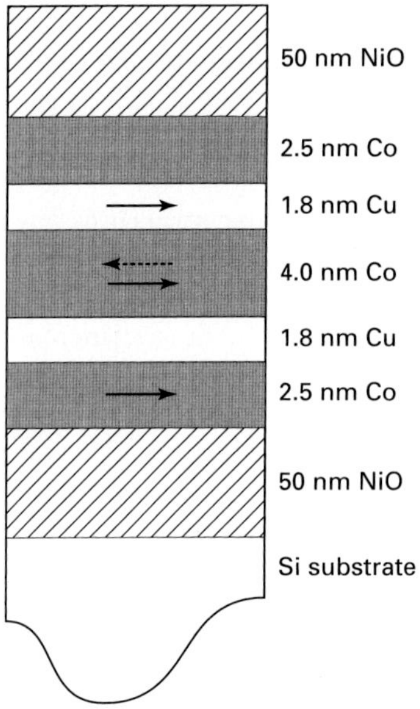

Fig. 4.9. Typical layer structure of the symmetric spin valves studied in [72]

A second alternative to $Fe_{50}Mn_{50}$ is NiO [70–72]. Anthony et al. [71] studied simple spin valves of the type $NiO/Ni_{80}Fe_{20}/Co/Cu/Co/Ni_{80}Fe_{20}$ (this could be called an "inverted" spin valve, as the pinned layer is grown first), and obtained a GMR ratio of 13% (at room temperature), for a Cu spacer thickness of 17 Å. To the best of our knowledge, the highest GMR ratio for simple spin valves has been reported by Swagten et al., [72], viz. 15% at room temperature for systems of the type (substrate/NiO/Co/Cu/Co/Cu/NiO). The second NiO layer has the function of protecting the structure from corrosion. The second Cu layer magnetically insultates the free Co layer from the second NiO layer, which would otherwise lead to exchange biasing of the free layer.

"Symmetric" or "dual" spin valves have been studied by several groups [71,73–75]. These structures contain one free magnetic layer between two pinned magnetic layers, separated by nonmagnetic spacer layers, and allow obtaining GMR ratios higher than for simple spin valves. The highest value found to date is 21% at room temperature, as reported by Egelhoff et al. [73]. The design of the structure used for this study is depicted in Fig. 4.9. The high MR ratio obtained was argued to be possibly related to a certain degree of *specular* scattering of electrons at the NiO/Co interfaces. It is noted that scattering at F/Fe–Mn or F/Tb–Co interfaces is commonly regarded as almost entirely diffuse, in view of the highly disordered atomic and magnetic structures of the biasing layers and their metallic character, which allows electrons to penetrate the AF layers. Specular scattering at F/NiO interfaces

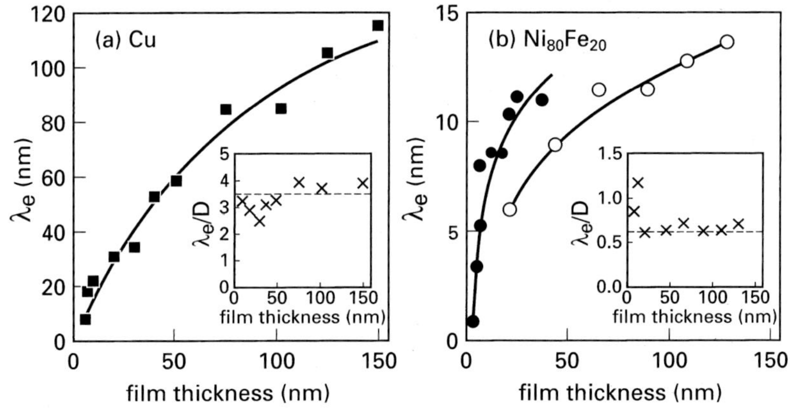

Fig. 4.10. a, b Layer-averaged extrinsic mean free path λ_e versus layer thickness at 4.2 K, for Cu (**a**) and $Ni_{80}Fe_{20}$ (**b**) films [75]. For $Ni_{80}Fe_{20}$, an intrinsic spin asymmetry ratio $\alpha = \lambda_i^{\uparrow}/\lambda_i^{\downarrow} = 4$ was assumed. The insets show the ratio of λ_e and the measured average grain size D versus film thickness. The data represented by *open* and *closed dots* in (**b**) have been obtained for separate sample series

then would be related to a high interface quality, and to the large bandgap of AF, which effectively impedes penetration of electrons. Swagten et al. considered the possibity that their record GMR values for NiO-based simple spin valves, as discussed above, are related to (partially) specular scattering at the NiO layers [72].

4.3.3 Dependence on the Microstructure

Scattering, and its spin dependence, depends on the microstructure of films. Aspects of the microstructure are the occurrence of defects (such as vacancies, dislocations, grain boundaries and interface roughness or diffusiveness), and stress. The microstructure may be influenced by the preparational method, by the preparational conditions and by the choice for particular substrates or underlayers. After deposition the microstructure may be influenced by a heat treatment.

The relation between microstructure and electrical transport was studied by Rijks et al. [76] for Cu and permalloy films from 2 to 150 nm thickness, sputter deposited on a Si(100) single crystal, using a 3 nm Ta buffer layer. The dependence of the conductivity on the film thickness was analyzed within the semiclassical Boltzmann-transport theory (Sect. 4.4), assuming two independent scattering processes: intrinsic scattering (electron-phonon scattering, and, in the case of permalloy, scattering due to the atomic site disorder) as well as extrinsic scattering (at defects such as grain boundaries). The mean free path for the latter type of scattering, λ_e, was assumed to be independent of spin and temperature. For both systems λ_e was found to increase strongly

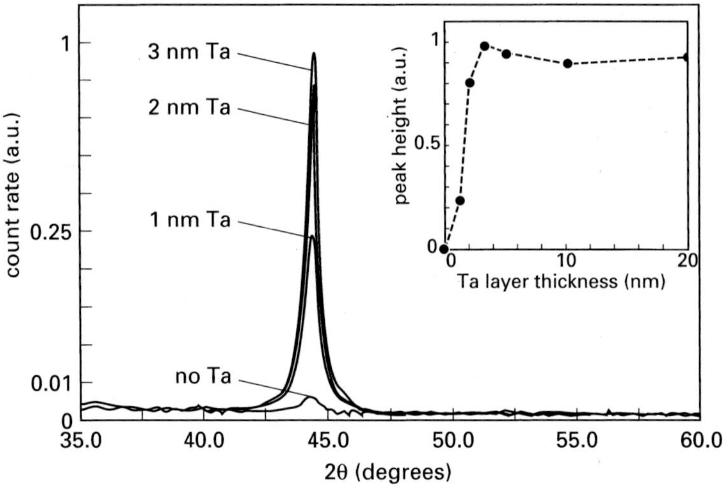

Fig. 4.11. Cu – $K_\alpha(\theta - 2\theta)$ X-ray diffraction pattern of SiO_2/t_{Ta} nm Ta/ 50 nm $Ni_{80}Fe_{20}$ layers, for various thicknesses of the Ta buffer layer [76]. Note the square-root intensity scale. The inset displays the (111) peak heights as a function of t_{Ta}

with the film thickness. The results are shown in Fig. 4.10, from which it may also be seen that, across almost the entire thickness range, λ_e was found to be proportional to the average grain diameter D. The latter quantity was obtained from plane-view Transmission Electron Microscopy (TEM) and grazing incidence X-ray diffraction. Extrinsic scattering processes are most relevant if λ_e is comparable to or smaller than the intrinsic mean free path λ_i. For Cu, this is always the case at low temperatures (no intrinsic scattering). At room temperature, extrinsic scattering is still significant for all film thicknesses studied, because λ_i was found to be approximately 33 nm. For permalloy, the experimental results yielded $\lambda_i^\uparrow + \lambda_i^\downarrow \approx 36$ nm at 4.2 K. Assuming that the spin asymmetry ratio $\alpha = \lambda_i^\uparrow/\lambda_i^\downarrow > 1$, as is expected for permalloy (Sect. 4.2.3), it follows that 18 nm $< \lambda_i^\uparrow <$ 36 nm. As λ_e was found to be smaller for all layer thicknesses studied, it may be concluded that at 4.2 K for permalloy, too, extrinsic scattering is quite significant. Although structurally more complicated, a qualitatively similar dependence of grain boundary scattering on layer thicknesses may be expected for spin-valve structures. From this study, it thus follows that quantitative analyses of spin-dependent scattering, which do not take scattering at grain boudaries or other defects into account, should be regarded with caution.

The microstructure may be influenced by the presence of thin underlayers (buffer layers) [77–80]. For example, a Ta layer of only a few nanometers thickness on top of various substrates, such as Si(100), SiO_2, and Si_3N_4 may strongly enhance the [111] texture. This is illustrated by Fig. 4.11, which

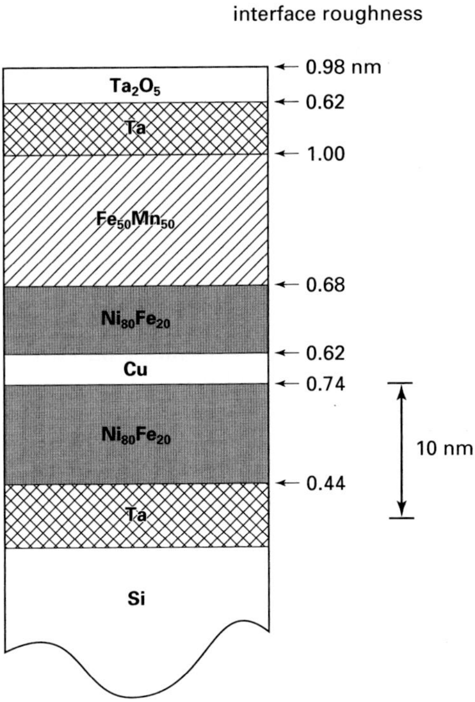

Fig. 4.12. Results of a low-angle X-ray reflectivity analysis of the interface widths in a spin-valve layered structure [80]. The numbers given are the root-mean-square values of the interface position with respect to the average plane of the interface. The Ta-oxide layer on top of the sample was formed after exposing the sample with a nominally 5 nm thick pure Ta cap layer to air

shows results of an X-ray study by Duchateau et al. [77] for a permalloy film on Si(100). A 3 nm thick Ta buffer layer leads to a full width at half maximum of the [111] rocking curve of typically 4° to 5°. The [111] texture is advantageous for a number of reasons. First, for [111]-oriented systems the strength of the exchange biasing by $Fe_{50}Mn_{50}$ is larger, and the coercivity in the exchange biasing effect is smaller than for other orientations (Sect. 4.5.2). Second, even in cases in which sufficient exchange biasing is obtained without the use of a Ta buffer layer, the MR ratio is found to be 10–15% larger for growth on a thin (2–3 nm) Ta buffer layer. The sheet resistance is then typically 20% lower [77]. The increased MR ratio may therefore be attributed to a decreased defect scattering in the spin valve. The shunting effect by the Ta buffer layer is almost negligible for the thickness used, because the resistivity of the Ta buffer layer was observed to be very high ($160 \times 10^{-8}\Omega$m). Third, as pointed out by Jérome et al. [78], the [111] texture promotes magnetic softness of the free layer. The reason is that the magnetocrystalline anisotropy energy of a cubic crystal such as permalloy is constant for magnetization di-

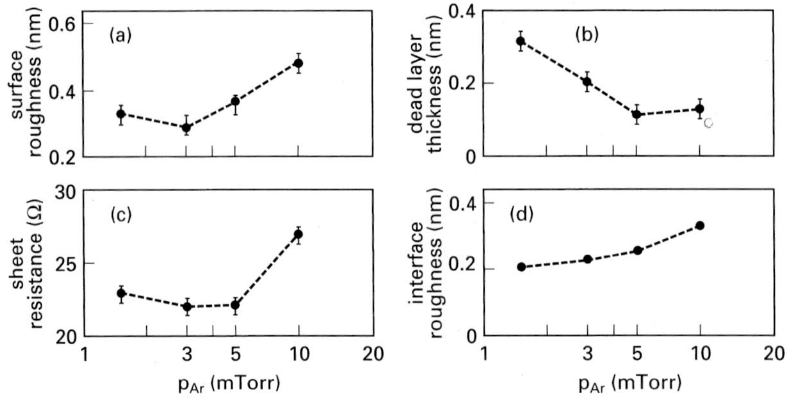

Fig. 4.13. a–d Dependence on the Ar sputter pressure of the surface roughness, as measured by AFM (**a**), the effective magnetically dead layer thickness per interface (**b**), the sheet resistance (**c**) and the interface roughness parameter h (**d**) measured for spin-valves with the structure [Si(100)/3.5 nmTa/8 nm$Ni_{80}Fe_{20}$/2.5 nmCu/6 nm$Ni_{80}Fe_{20}$/10 nm$Fe_{50}Mn_{50}$/3.5 nmTa]. The effective dead-layer thickness (**b**) has been derived from the magnetic layer-thickness dependence of the magnetization of $Ni_{80}Fe_{20}$/Cu multilayers. All measurements have been performed at room temperature. The AFM surface roughness is defined as the average of the absolute deviation from the mean surface plane. The interface roughness parameter h has been derived from the interlayer magnetic coupling. See also Sect. 4.5.1 and the inset in Fig. 4.18

rections within the (111) planes (assuming that the higher-order anisotropy constant K_2 may be neglected). As a result, there is no preferred in-plane magnetization direction for [111]-textured films (of course, uniaxial in-plane anisotropy may be induced by growth in a magnetic field, see Sect. 4.5.3).

Nakatani et al. [79] have studied the effect of other buffer layers on the the MR curves of spin valves sputter deposited on Si(100) substrates. Strong [111] textures, relatively high exchange biasing fields, and relatively high MR ratios were found using Ta, Hf, Nb, Ti, or Zr buffer layers, whereas Cu, Ag, Au and Cr did not cause this texture.

The role of Ta has been discussed in more detail by Lenssen et al. [80]. They concluded from TEM electron diffraction of MBE-grown and sputter-deposited permalloy layers on Ta buffer layers that the [111] texture is correlated with a Ta buffer layer consisting of nanometer-size randomly oriented grains in a metastable phase which is different from earlier-reported metastable β–Ta or fcc-Ta phases, or from the stable bulk bcc-type α structure. For sputtered Ta films the desired phase is observed up to tens of nanometers buffer layer thickness, for MBE-grown films it is only present up to a thickness of 1 nm. Thicker buffer layers have the stable bcc–Ta phase, which does not induce a [111] texture.

The interface quality of sputter-deposited spin valves with the structure as shown in Fig. 4.12 was studied by Huang et al. [82], using low-angle X-ray

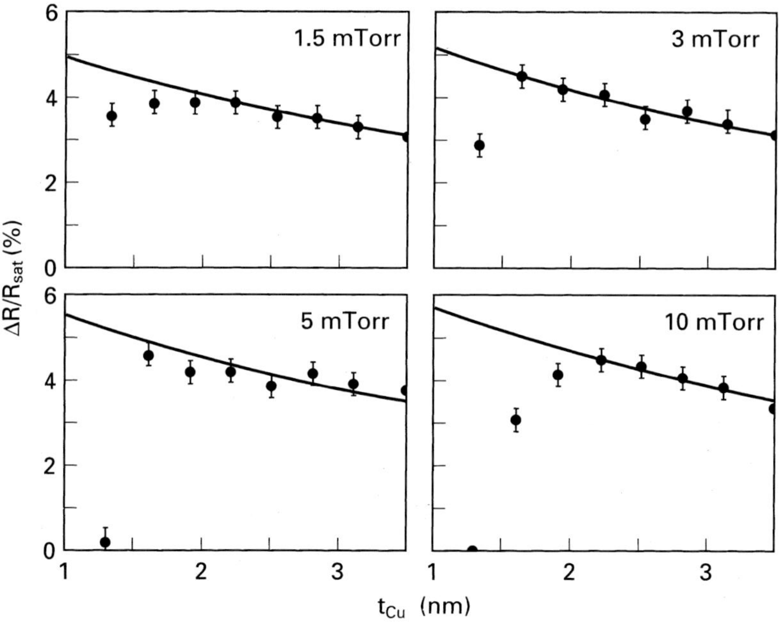

Fig. 4.14. Dependence of the MR ratio on Cu thickness and Ar sputter pressure, for spin valves of the structure given in the caption of Fig. 4.13 [83]. The full lines represent a fit to the data using (4.6)

diffraction. The measured thicknesses of the interface zones are indicated in the figure. This method is sensitive to the variation of the refraction index across the sample thickness, laterally averaged over a length scale of the order of 100 nm. As a result, it is not possible employing this method to discriminate between compositionally sharp, but wavy interfaces, and interdiffused interfaces. The interface widths obtained are therefore just upper limits to the interface width on lateral scales of atomic dimensions up to about one or a few nanometers, which are more relevant to spin-dependent scattering.

Kools [83] has investigated how interface mixing, interface magnetism, resistivity and the MR ratio of spin valves depend on the Ar pressure during dc magnetron sputter deposition. From results of Monte Carlo calculations of the kinetic-energy distribution of atoms arriving at the substrate, he showed how the Ar pressure may be used to modify this distribution. For low Ar pressures, a significant fraction of atoms arriving at the target has energies above about 15 eV, a few times the atomic binding energy, leading to atomic displacements of atoms from one layer to another over typically 2–3 atomic planes. For higher Ar pressures, such collisional mixing occurs less frequently. On the one hand, collisional mixing leads to interdiffused interfaces. On the other hand, it leads to increased atomic mobility during growth, and hence to relatively flat interfaces and a low probability for the formation of

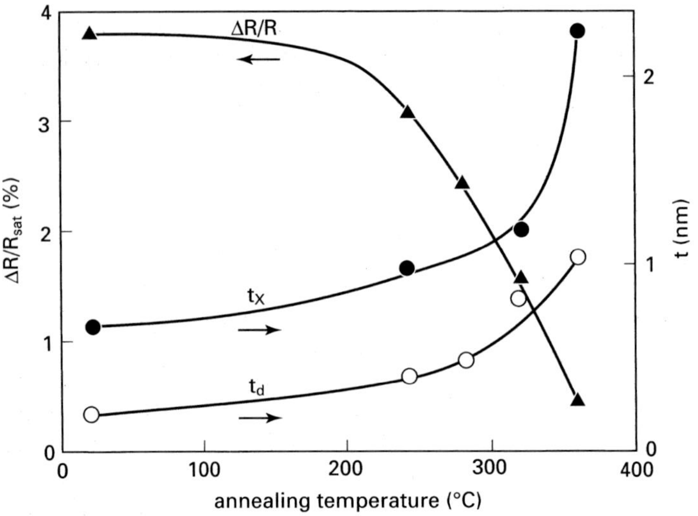

Fig. 4.15. Dependence on the annealing temperature of the MR ratio, the interface width t_X as determined from X-ray reflectivity experiments, and the effective magnetically dead-layer thickness t_d per $Ni_{80}Fe_{20}$/Cu interface, for spin valves with the structure given in Fig. 4.12 [85]. Annealing was done in a vacuum furnace for 6.5 hours. See [85] for details on the anneal sequence

voids between grains. Figure 4.13 exhibits experimental results on the surface roughness of spin valves, as measured using Atomic Force Microscopy (AFM) (a), the effective dead-layer thickness for $Ni_{80}Fe_{20}$/Cu interfaces, as measured for $Ni_{80}Fe_{20}$/Cu multilayers (b), the sheet resistance of a spin valve (c), and the roughness amplitude of the $Ni_{80}Fe_{20}$/Cu interfaces (d). Figure 4.14 shows the dependence of the MR ratio on the Ar pressure and Cu spacer thickness. The following picture was derived from these results. For Ar pressures below 5 mTorr the layers are smooth but the interfaces tend to become more intermixed with decreasing pressure. This increases the resistance and diminishes the MR ratio. For Ar pressures above 5 mTorr the interfaces are fairly sharp (low magnetic dead-layer thickness), but the waviness increases with increasing pressure. This leads to an increase of the critical Cu thickness above which a full antiparrallel alignment of the free and pinned layers is obtained, to about 2.2 nm for an Ar pressure of 10 mTorr. The solid lines in Fig. 4.14 are fits to the data, using (4.6). The parameter $(\Delta R/R)_0$ shows a gradual increase with the Ar pressure, from 5.8% at 1.5 mTorr to 6.9% at 10 mTorr, as may be interpreted as proof for the increasing transmittivity of the interfaces for majority-spin electrons [83].

Joo and Atwater [84] studied the relation between interface mixing and the MR ratio of Ni–Fe/Cu/Ni–Fe/Fe–Mn spin-valve structures that were deposited by ion-beam sputtering. This technique involves a relatively intense high-energy bombardment of the growing film surface by neutral atoms. The conditions may be compared with those for dc magnetron sputtering in the low-pressure limit (see above), and result in relatively small MR ratios (at most about 2.5%) and a relatively large interface intermixing (not quantified in [84]).

4.3.4 Effects of Annealing

For the application of spin-valve GMR systems three thermal effects are of interest. First, the variation of the GMR ratio versus temperature as a result of the thermal variation of scattering probabilities. This is a reversible effect, and has been discussed in Sect. 4.3.1. Second, the decrease with temperature of the exchange-biasing field limits its range of operation to temperatures equal to or slightly below the blocking temperature T_b above which biasing vanishes. For $Fe_{50}Mn_{50}$ and NiO, $T_b \approx 140°C$ and $\approx 180°C$, respectively. Biasing may be recovered upon cooling in a properly oriented magnetic field. Third, bringing the sample to elevated temperatures accelerates diffusion. This is an irreversible process. Studies of $Ni_{80}Fe_{20}/Cu/Ni_{80}Fe_{20}/Fe_{50}Mn_{50}$ systems have shown that interdiffusion of the $Ni_{80}Fe_{20}/Cu$ interface after 6.5 hours at 240°C [85] or 30 minutes at 250°C [86] is already significant. Results by Huang et al. [85] on the variation of the MR ratio and the width of the $Ni_{80}Fe_{20}/Cu$ interface zone, as determined from low-angle X-ray diffraction are depicted in Fig. 4.15. From a fit with (4.6) of the dependence of the MR ratio on the Cu thickness, for systems annealed at various temperatures, Speriosu et al. [57] have reported that annealing mainly decreases the prefactor $(\Delta R/R)_0$. From this observation, they argued that the main effect of annealing is to increase spin-independent scattering or spin-flip scattering at the interfaces (consistent with the observed increase of the thickness of the mixed interface zone), the changes of the conductivities of the F and NM layers being of minor importance.

Hoshiya et al. [70] observed an enhanced thermal stability of the MR ratio of Co-based spin valves, as compared to permalloy-based spin valves [70]. On the other hand, Kitade et al. [87] reported results on X/Cu/X/Fe–Mn spin valves with X = $(Ni_{80}Fe_{20})_{1-x}Co_x$ $(0 \leq x \leq 0.75)$ that reveal an opposite trend. For all alloy concentrations studied a decrease of the MR ratio was observed for annealing above 200°C. For permalloy ($x = 0$) the results reported are very similar to the results shown in Fig. 4.15. However, the relative decrease of the MR ratio was largest for the highest Co concentration studied ($x = 0.75$) (from 7.5% to about 1.5% for annealing at 300°C).

4.4 Modeling of GMR: Semiclassical Transport Model

The model for the GMR ratio presented in Sect. 4.2 is, for several reasons, too simple for most CIP systems. One of the reasons is that for most systems of interest the assumption of large mean free paths for *both* spin directions does not hold. Another complication is that there is additional electron scattering at the outer boundaries of the film (the interfaces with the substrate and the cap layer). As a result, the GMR ratio of multilayers increases with the number of repetitions, until saturation is reached when the total thickness is much larger than the largest mean free path for the case of a parallel alignment.

A more sophisticated model, which describes the electrical transport in layered structures on the basis of assumed spin-dependent scattering probabilities in the bulk of the layers, at the interfaces, and at the outer boundaries, is the semiclassical transport model proposed by Camley and Barnaś (CB) [88]. For this model, electrons are essentially regarded as pointlike particles (hence "classical") although some consequences of quantum mechanics are taken into account, such as Fermi-Dirac statistics and quantum-mechanically evaluated scattering probabilities (hence "semi").

For the CB model the electronic structure in each of the layers is treated as a spin-polarized free-electron gas. The conductivity is obtained from the electron distribution function $f^{\uparrow(\downarrow)}(\boldsymbol{v},\boldsymbol{r})$, which gives the spin-dependent probability that a state at the position $\boldsymbol{r}$ with the velocity $\boldsymbol{v}$ is occupied. A convenient notation is

$$f^{\uparrow(\downarrow)}(\boldsymbol{v},\boldsymbol{r}) = f_0^{\uparrow(\downarrow)}(\epsilon) + g^{\uparrow(\downarrow)}(\boldsymbol{v},\boldsymbol{r})\,. \tag{4.9}$$

Here $g^{\uparrow(\downarrow)}(\boldsymbol{v},\boldsymbol{r})$ is the deviation in the presence of an electric field $\boldsymbol{E}$, for spin-up ($\uparrow$) and spin-down ($\downarrow$) electrons, from the Fermi-Dirac equilibrium distribution $f_0^{\uparrow(\downarrow)}(\epsilon)$ at energy $\epsilon = \frac{1}{2}mv^2$. The electron distribution function is obtained by solving the Boltzmann transport equation in the relaxation time approximation:

$$\boldsymbol{v}\cdot\frac{\partial g^{\uparrow(\downarrow)}(\boldsymbol{v},\boldsymbol{r})}{\partial \boldsymbol{r}} - e\boldsymbol{E}\cdot\boldsymbol{v}\frac{\partial f_0^{\uparrow(\downarrow)}(\epsilon)}{\partial\epsilon} = -\frac{g^{\uparrow(\downarrow)}(\boldsymbol{v},\boldsymbol{r})}{\tau^{\uparrow(\downarrow)}}\,, \tag{4.10}$$

where e is the electron charge, m is the electron mass, and $\tau^{\uparrow(\downarrow)}$ denotes spin-dependent and layer-dependent isotropic relaxation times. The spin-dependent current density follows from an integral over velocity space of the electron distribution function:

$$\boldsymbol{j}^{\uparrow(\downarrow)}(\boldsymbol{r}) = e\left(\frac{m}{h}\right)^3 \int \mathrm{d}^3v\boldsymbol{v}g^{\uparrow(\downarrow)}(\boldsymbol{v},\boldsymbol{r})\,, \tag{4.11}$$

where h is Planck's constant.

Solutions of the Boltzmann transport equation for layered structures, with the planes perpendicular to the z axis, have, within a layer between $z = a$ and $z = b$ ($b > a$), the following form:

$$g^{+}(\boldsymbol{v},z) = -e\tau E_x v_x \frac{\partial f_0(\epsilon)}{\partial\epsilon}\left[1 - A^{+}\exp\left(\frac{a-z}{\tau v_z}\right)\right]\,, \tag{4.12}$$

for $v_z > 0$, and

$$g^{+}(\boldsymbol{v}, z) = -e\tau E_x v_x \frac{\partial f_0(\epsilon)}{\partial \epsilon} \left[1 - A^{-} \exp\left(\frac{z-b}{\tau v_z}\right)\right], \tag{4.13}$$

for $v_z < 0$. The spin (↑ or ↓) labels have been omitted from the relaxation time τ and the coefficient A. The coefficients A are determined from the boundary conditions at the interfaces or outer boudaries (transmission, reflection, diffuse scattering, see below).

As applied to bulk metals, this model has been discussed in many textbooks [89]. It leads to the well-known Drude formula $\sigma_0 = ne^2\tau/m$ for the conductivity σ_0 of an isotropic material with electron density n. Fuchs and Sondheimer (FS) [90] have first applied this model to the conductivity of homogeneous thin films. It follows from their work that the conductivity of a thin film, in which electrons are scattered diffusely at the outer boundaries, decreases with the film thickness t. As the reader may have anticipated from the form of the g functions in (4.12) and (4.13), the effect is significant for t close to and smaller than the electron mean free path λ, which is defined as $\lambda = v_F\tau$. Here v_F is the Fermi velocity. In the thick-film limit ($t \gg \lambda$), assuming the probabilities p and $1-p$ for specular reflection and diffuse scattering at the outer boundaries, respectively, the conductivity is given by the simple expression $\sigma = \sigma_0[1-(3/8)(1-p)\lambda/t]$, in which σ_0 is the bulk conductivity. In this case the coefficients A^+ and A^- are equal to $(1-p)$. The current density is close to the bulk value in the center of the film, but lowered in a zone with thickness of the order λ near the outer boundaries. A more complicated expression for the conductivity arises for thinner films, when these two boundary zones overlap [36,90].

For the more general case of a magnetic layered structure the CB model involves, apart from the spin-dependent probability p for specular reflection at the outer boundaries, two additional types of boundary conditions. First, at the interfaces three possible cases are distinguished: transmission (probability $T_{\mathrm{i,s}}$), specular reflection (probability $R_{\mathrm{i,s}}$), or diffuse scattering (probability $D_{\mathrm{i,s}}$). These probabilities depend on the interface i and the spin s, and $\mathrm{T}+\mathrm{R}+\mathrm{D}=1$.

Second, systems with different magnetization directions in the different layers are dealt with using spin-transmission coefficients $T^{\uparrow\uparrow}$, $T^{\uparrow\downarrow}$, $T^{\downarrow\uparrow}$, and $T^{\downarrow\downarrow}$ through a plane situated within the nonmagnetic spacer layer between the magnetic layers (the precise position of this plane in the nonmagnetic spacer layer does not affect the final results). These transmission coefficients determine the probability of an electron which has spin up or down with respect to the magnetization direction (quantization axis) in one magnetic layer to continue in the other magnetic layer as an electron with spin up or down with respect to the new quantization axis. The transmission coefficients are given by

$$T^{\uparrow\uparrow} = T^{\downarrow\downarrow} = \cos^2(\theta/2) \text{ and } T^{\uparrow\downarrow} = T^{\downarrow\uparrow} = \sin^2(\theta/2), \tag{4.14}$$

where θ is the angle between both magnetization directions. For example, in the case of an antiparallel arrangement of the magnetization directions $T^{\uparrow\uparrow} = T^{\downarrow\downarrow} = 0$ and $T^{\uparrow\downarrow} = T^{\downarrow\uparrow} = 1$. This means that a majority-spin electron in one magnetic layer becomes a minority-spin electron in the other magnetic layer, after traversing a nonmagnetic layer in between.

Using all boundary conditions, the coefficients A may be calculated for each layer, separately for the two spin directions. From the resulting g functions, the z-dependent current density can be calculated using (4.11), and after integrating over the film thickness, the contribution to the conductivity from each spin follows.

4.4.1 Applications of the Semiclassical Model

A first application of the model treatment presented above is the derivation of the *angular variation* of the resistance. Let us consider a structure consisting of two magnetic layers, with an angle θ between their magnetization directions, separated by a nonmagnetic spacer layer. This could be the active part of a simple exchange-biased layered structure. We assume diffuse scattering at the outer boundaries, and only transmission or diffuse scattering at the interfaces (no specular reflection). Using the expressions for the spin-transmission coefficients given by (4.14) and (4.15) it may be shown analytically that these conditions lead to a linear variation of the *conductance* with $(1 - \cos\theta)$ [91]. This result is to first order in the GMR ratio equivalent to the expression given by (4.1) (the *resistance* varies linearly with $1 - \cos\theta$). The difference is of second order in the GMR ratio, and will therefore be most significant for systems with relatively large GMR ratios. The experimental studies of the angular dependence of such structures have so far been performed for systems with small GMR ratios (an exchange-biased $Ni_{80}Fe_{20}/Cu/Ni_{80}Fe_{20}/Fe_{50}Mn_{50}$ system [21], and an antiferromagnetically exchange-coupled Fe/Cr/Fe system [32], with GMR ratios of about 3% and 0.4%, respectively), making it very difficult to detect this second order effect. A recent experimental study within which, as already mentioned in Sect. 4.2, a linear variation of the *conductance* with $(1 - \cos\theta)$ has been observed [34], has been performed for a *multilayer* system (a system of the type $Ni_{80}Fe_{20}$/Ag/Co/Ag, with discontinuous Co layers). Unfortunately, at present no analysis of the angular dependence of the resistance, based on the model presented above, is available (yet) for multilayers.

It should be pointed out that for systems containing two magnetic layers a nonlinear angular variation of the conductance with $(1 - \cos\theta)$ is expected in the case of nonzero specular reflection at the interfaces between the magnetic and nonmagnetic layers. This has been shown first by Vedyayev et al. [92], on the basis of analytical quantum-mechanical calculations.

A second application of the semiclassical transport model involves the derivation of layer- and interface-resolved spin-dependent scattering parameters from fits of the layer thickness dependence of the resistance and mag-

Table 4.1. Resistivities and scattering parameters for F/Cu/F/Fe–Mn spin valves at 1.5 K, as obtained by Dieny et al. from fits using the semiclassical transport model [93,94]. The parallel and perpendicular mean free paths are related to each other by (4.16), with $\lambda_{gr} = 40$ nm, as discussed in the text. Units: resistivities ρ in $10^{-8}\,\Omega$m, mean free paths λ in nm

	Cu	$Ni_{80}Fe_{20}$	Co	Fe
ρ	4.7	15.4	10.7	10.5
$\lambda_{\parallel}^{\uparrow}$	13	11	14	7
$\lambda_{\perp}^{\uparrow}$	20	15	21	8.5
$\lambda_{\parallel}^{\downarrow}$	13	0.6	1	7
$\lambda_{\perp}^{\downarrow}$	20	0.6	1	8.5
$T_{F/Cu}^{\uparrow}$		1	1	1
$T_{F/Cu}^{\downarrow}$		1	0.2	0.6

netoresistance. Such an analysis has been performed by Dieny et al. for F/Cu/F/Fe–Mn (F = Fe, Co, and $Ni_{80}Fe_{20}$) spin valves [93,94]. The fits were performed for data obtained at 1.5 K, where (at least in Fe and Co) spin-flip scattering may be neglected. From preliminary calculations, it was found that no satisfactory description of the resistance and the magnetoresistance was possible using the same set of parameters: parameter sets describing correctly the layer thickness dependence of the conductivities were found to underestimate the MR ratio by about 20% in relative value. This was ascribed to anisotropic scattering due to the presence of grain boundaries. In the case of a columnar grain structure, stronger scattering is expected for electrons with velocities nearly parallel to the layers (which predominantly determine the resistivity), as compared to scattering for electrons with a substantial perpendicular velocity component (whose contribution to the magnetoresistance is larger). In order to incorporate anisotropic scattering, Dieny et al. made use of an *ad hoc* extension of the semiclassical model, by using the following expression for the anisotropic mean free path for scattering in the bulk of the layers:

$$\lambda(\theta) = \lambda_{\parallel} \sin^2\theta + \lambda_{\perp} \cos^2\theta\,, \tag{4.15}$$

in which θ is now the angle of the electron-velocity vector with respect to the film normal, and in which the parallel and perpendicular mean free paths are related by

$$\frac{1}{\lambda_{\parallel}} = \frac{1}{\lambda_{\perp}} + \frac{1}{\lambda_{gr}}\,. \tag{4.16}$$

The parameter λ_{gr} is expected to be of the order of the grain diameter. Satisfactory fits were obtained with $\lambda_{gr} = 40$ nm. The same values of λ_{gr} were used for scattering in the Cu spacer, and for majority-spin electrons in the F

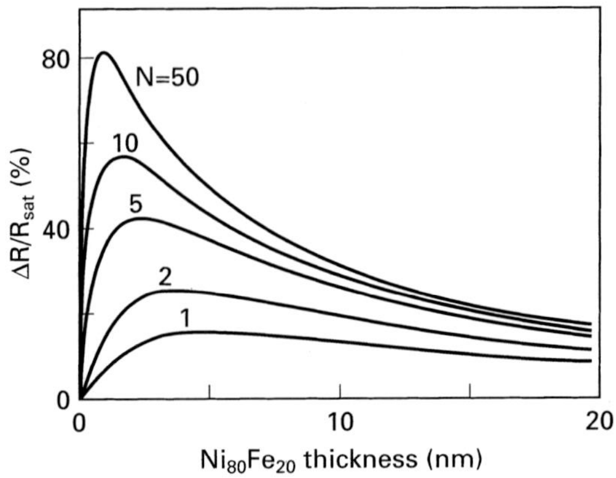

Fig. 4.16. Results of a model calculation of the MR ratio versus the thickness of the F layers for $(\mathrm{F/NM})_N/\mathrm{F}$ multilayers, with $t_{\mathrm{NM}} = 1\,\mathrm{nm}$ [93]. The mean free paths were assumed to be isotropic: $\lambda_{\mathrm{NM}} = 20\,\mathrm{nm}$, $\lambda_{\mathrm{F}}^{\uparrow} = 12\,\mathrm{nm}$ and $\lambda_{\mathrm{F}}^{\downarrow} = 0.6\,\mathrm{nm}$. At the outer boundaries, diffuse scattering was assumed, and at the interfaces $T = 1$ was assumed for both spin directions

layer. For minority-spin electrons in the F layer, very small mean free paths were found, and their (very small) anisotropy was neglected. Diffuse scattering at the outer boundaries was assumed. Fermi surfaces were assumed to be spherical.

An overview of the results on the scattering parameters for F = Co and $\mathrm{Ni_{80}Fe_{20}}$ is given in Table 4.1. The results indicate that for $\mathrm{Ni_{80}Fe_{20}/Cu}$ systems spin-dependent bulk scattering is the predominant origin of the GMR effect, whereas for Co/Cu systems spin-dependent scattering at the interfaces plays, in addition, a very important role. Within Fe layers, scattering is seen to be almost spin-independent. Upon making a comparison of the numerical values given in this table with other data, one should realize that the parameter values obtained do not only depend on details of the model used, but also on the microstructure of the samples studied. This was discussed in Sect. 4.3.3. The data given in the table pertain to spin valves grown on glass substrates, without a buffer layer. Significantly smaller resistivities (and larger GMR ratios) may be obtained by growth on Si(100) substrates and a 3 nm Ta buffer layer [60].

Assuming a more general validity of the model parameters given in Table 4.1, it would be possible to "design" GMR layered structures, in order to function optimally in a certain application. For example, the model then could be employed to predict how the F layer thickness for which the highest MR ratio is obtained varies with the number of repetitions, N, in a multilayer. Results of such a model calculation are given in Fig. 4.16, for a spacer thickness equal to 1 nm, using (for simplicity) isotropic mean free paths $\lambda_{\mathrm{F}}^{\uparrow} = 12\,\mathrm{nm}$, $\lambda_{F}^{\downarrow} = 0.6\,\mathrm{nm}$, $\lambda_{\mathrm{NM}} = 20\,\mathrm{nm}$, interface transmission coefficients

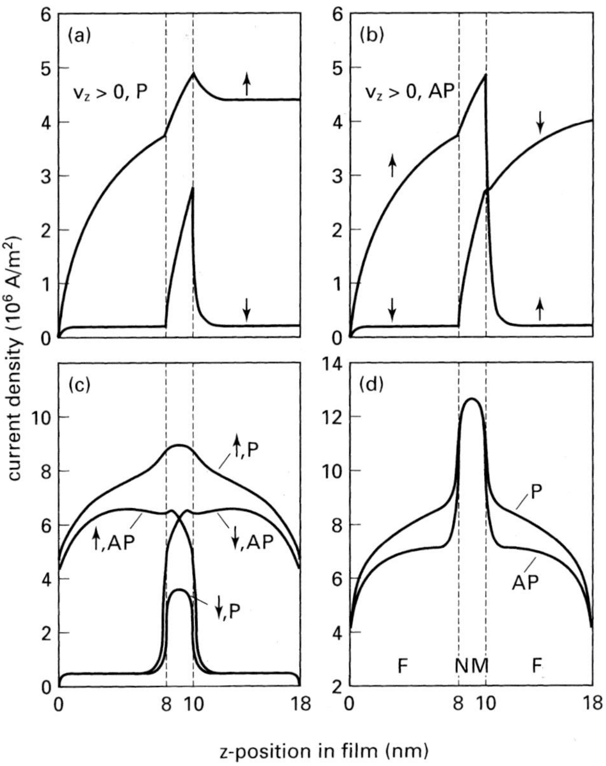

Fig. 4.17. a-d Results of a model calculation of the current density per unit electric field, in a (8 nm F/2 nm NM/8 nm F) sandwich system, using the semiclassical Boltzmann transport model [96]. The magnetization of the left F layer is fixed (up), whereas the right F layer is up/down for parallel/antiparallel alignment. The model parameters and the assumptions on scattering at the outer boundaries and at the interfaces are identical to those as given in the caption of Fig. 4.16. In (**a**) and (**b**) the contributions to the current density by electrons with $v_z > 0$ and for parallel and antiparallel configurations are shown. The contributions from electrons with $v_z < 0$ follow from a reflection in the $z = 9$ nm plane. In (**c**) contributions for both velocity directions have been added. In (**d**) contributions from both spin directions have been added. The calculated parallel and antiparallel sheet resistances are 6.82 and 7.57 Ω, and the calculated GMR ratio is 11.0%

T = 1 for both spins, and diffuse scattering at the outer boundaries. The shift with N of the maximum in the MR curves to lower F layer thicknesses reflects the strongly spin-dependent mean free paths in the F layer. For $N = 2$, the F layers should be relatively thick, in order to make it possible for the spin ↑ electrons to contribute to the conduction before scattering diffusely at the outer boundaries. On the other hand, in the case of a large number of repetitions, scattering at the outer boundaries becomes unimportant, and the F layer should simply be just thick enough to scatter spin ↓ electrons.

A third application of the semiclassical transport model is the calculation of the dependence of the current density on the position within a spin valve and on the relative alignment of the magnetization directions. Such information cannot be obtained directly by experiment, but is of practical importance to applications, as the current leads to an effective magnetic field in the magnetic layers, directed perpendicular to the current direction. For some applications, such as magnetic recording heads, this field is nonnegligible under practical operating conditions. Calculations of current distributions in spin valves have been reported by Gurney et al. [95]. In Fig. 4.17 results by Rijks and van der Kruijs are given of a semiclassical model calculation of the current density distribution in an F/NM/F sandwich system, using the same set of scattering conditions in the bulk and at the interfaces as utilized for the model calculations displayed in Fig. 4.16. The figure shows that, upon switching from the parallel to the antiparallel alignment, the largest change in the current density is found in the F layers close to the interface with the NM layers. As remarked already by Dieny [93] the current density in the NM layers is, for the model assumptions employed, the same for both alignments. This may be understood by considering that contributions to the current density for a given electron direction (positive or negative v_z) are the same in the first F layer that is traversed, and in the NM layer, and only depend on the relative alignment in the second F layer (Fig. 4.17 a–b).

4.4.2 Extensions of the Semiclassical Model

The CB semiclassical model in the form presented in Sect. 4.4 shows a number of deficiencies. A first deficiency is that it neglects anisotropic scattering resulting from grain boundaries. As shown by Dieny et al. it is possible to extend the CB model, in order to take this effect into account. In this subsection a number of other proposed extensions and modifications of the CB model are discussed.

Rijks et al. [97], and more recently Granovskiĭ et al. [98], have argued that scattering within the magnetic layers is anisotropic as a result of the anisotropic MR effect. This leads to an interference between the GMR and AMR effects: the GMR ratio may depend on the direction of the current with respect to the applied magnetic field, and the AMR ratio may depend on the relative alignment of the magnetization of successive magnetic layers. Experimental proof of such an interference has recently been obtained by Dieny et al. [99].

Hood and Falicov [100] have proposed a generalization of the CB model, by introducing layer- and spin-dependent effective masses and Fermi velocities. The resulting potential steps at the interfaces lead to reflection and refraction at the interface, with a probability that depends on the angle of incidence. No analysis of experimental results in terms of this model has yet been presented.

Several researchers have argued that scattering at interfaces is expected to depend on the angle of incidence, even in the absence of potential steps [102–104]. As pointed out by Camblong and Levy this would be the case if, for the semiclassical model as presented in Sect. 4.4, the interfaces are treated as a thin additional layer instead of a mathematically sharp plane [102], see also [101,104]). Within the semiclassical model, it would then be possible to treat interface and bulk scattering in the same way. In fact, it is quite natural to introduce thin interface layers, as for many systems of practical interest the scattering probabilities are nonbulk-like in interface zones with a thickness of more than one atomic layer. In the limiting case of an interface layer with an infinitesimal thickness ϵ, the effective specular transmission coefficient is given by $T(\theta) = \exp(-\epsilon/\lambda \cos\theta)$, in which λ is the mean free path, and θ is the angle of incidence with respect to the film normal. In this approach the ratio ϵ/λ is the only adjustable parameter. The possible influence on the GMR ratio of the dependence of interface scattering on the angle of incidence has not yet been studied. It should be noted that, similarly, specular scattering at the outer boundaries may be described more adequately using an angular-dependent specularity parameter [104].

4.4.3 Beyond the Semiclassical Model

The semiclassical model neglects certain consequences of the wave character of electrons. It treats electrons as point-like particles, instead of wave-packets, and it ignores some of the finite size effects on the electronic structure and transport properties, resulting from the presence of potential barriers and the outer boundaries. Several researchers have proposed theoretical models for the GMR effect based on a more complete quantum-mechanical description of the ground-state electronic structure of the system and/or a quantum-mechanical treatment of electrical transport. A discussion of these approaches is beyond the scope of this paper ([16,102,104–106]). Only some of the issues will be mentioned here briefly.

A first consequence of quantum-mechanics is related to the position-velocity uncertainty relation, and is sometimes called the "zero-point motion" effect. It follows from the quantization of the perpendicular wave vector due to the (partial) confinement of states between the interfaces and outer boundaries. Quantum-mechanical model calculations by Vedyayev et al. [106], and by Zhang and Butler [104] have demonstrated that it is possible to effectively take this effect into account in semiclassical calculations of the conductivity of a film with thickness t, by excluding contributions to the conductivity due to states with a wave vector component $|\, k_z \,| < \pi/t$. In other words, contributions to the conductivity due to electrons with velocities almost parallel to the plane of the layers should be excluded. For spin-valve systems with layer thicknesses close to the optimum MR ratio (a few tens of nanometers), the cut-off angle would be of the order of a few degrees [106]. In these systems, the layer thicknesses are rather large, as compared to the Fermi wavelength (which is of the order of a monolayer thickness). A more drastic effect, viz. quantum oscillations in the conductivity with the layer thickness, is only expected in the case of layer thicknesses of the order of the Fermi wavelength [104].

Vedyayev et al. [106] have demonstrated that within a quantum-mechanical model calculation, the resistance and magnetoresistance of spin valves may be described within a single set of scattering parameters. Two differences with respect to the semiclassical theory were regarded crucial: the effective cut-off angle (discussed above) and the treatment of scattering at the substrate/F layer interfaces as scattering at a very thin boundary layer (leading, as discussed above, to an angular-dependent specularity parameter). Both effects favour transport perpendicular to the film plane. Bulk mean free paths, as obtained from the fits, were only 10% higher than the $\lambda_{||}$ values obtained from the semiclassical theory.

A number of theoretical approaches focusses on effects related to the detailed band structure of the superlattice, and changes in it due to switching of the magnetization directions (e.g., the opening of superlattice gaps in the band structure of a multilayer upon switching to the antiparallel state) [107–109]. However, as yet there are no experimental reports that convincingly demonstrate any of these effects. For systems which are being considered for sensor applications, there are no indications at present that these effects are of practical interest.

4.5 Magnetic Interactions

As shown in Fig. 4.2, three magnetic interactions determine the $R(H)$ curve of a spin-valve MR material: interlayer magnetic coupling, exchange biasing, and induced magnetocrystalline anisotropy. For more simple thinfilm systems these interactions have been studied for a long time, and references will be given to this older work. Here aspects which are specific to exchange-biased spin valves will be discussed, including the interplay between these interactions.

4.5.1 Interlayer Magnetic Coupling

The magnetic coupling between the free and the pinned ferromagnetic layers is phenomenologically described by $E = -J \cos \theta$ (to first order in $\cos \theta$), E being the coupling energy (in units of J/m^2), and θ is the angle between the magnetization directions of the F layers. In the limit of a rigid magnetization of the pinned layer (infinite exchange biasing), the free layer switches at

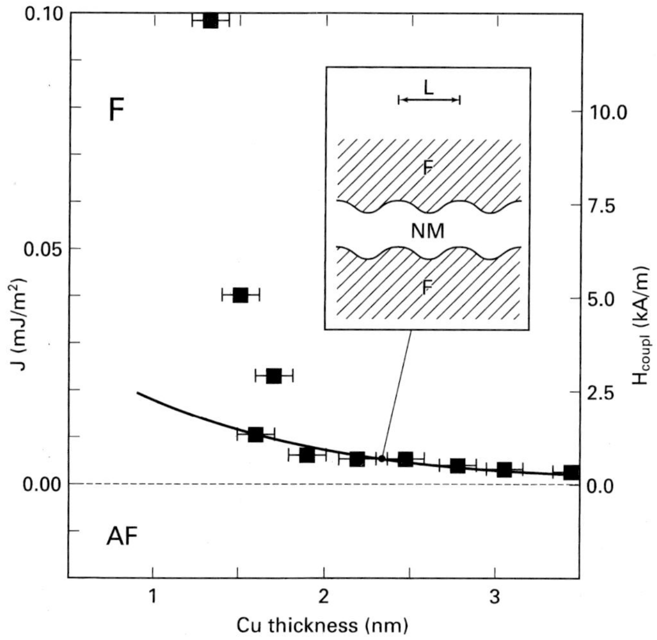

Fig. 4.18. Dependence at room temperature of the interlayer magnetic coupling parameter J (*left axis*) and the coupling field (*right axis*) on the Cu-layer thickness spin valves with the structure [Si(100)/3.5 nm Ta/8 nm $Ni_{80}Fe_{20}$/ t_{Cu} nm Cu/6 nm $Ni_{80}Fe_{20}$/8 nm $Fe_{50}Mn_{50}$/3.5 nm Ta] [110]. The full line through the data represents a fit assuming magnetostatic coupling as a result from corrugated interfaces (*see inset*) with $h = 0.4$ nm and $L = 10$ nm

the field $H_{\mathrm{coupl}} = J/\mu_0 M_{\mathrm{sat}}^{\mathrm{f0t_F^f}}$, $M_{\mathrm{sat}}^{\mathrm{f}}$ and $t_{\mathrm{F}}^{\mathrm{f}}$ being the saturation magnetization and the thickness of the free layer. For symmetric spin-valves (with the same exchange biasing direction for both pinned layers), the coupling field is twice this value. Interlayer magnetic coupling depends strongly on the composition of the F and NM layers, on the thickness of the NM layer, and on microstructural details. For Fe-Co-Ni/Cu/Fe-Co-Ni systems, coupling fields tend to increase from Ni-rich towards Co-rich systems. Three physical mechanisms have been found to contribute to the interlayer coupling [30,71,110]: direct ferromagnetic exchange coupling via pinholes (ferromagnetic bridges across the NM layer), indirect oscillatory exchange coupling, and magnetostatic coupling.

In the idealized case of flat and compositionally sharp interfaces, oscillatory exchange coupling would be the only contribution. Reviews on this subject were given in [111], and in Chap. 3 by P. Grünberg. Phenomenologically, the coupling parameter J may be written as a superposition of oscillatory terms of the form $J = J_0 \sin([2\pi t_{\mathrm{NM}}/\Lambda] + \phi)/t_{\mathrm{NM}}^2$. The period Λ depends on the spacer material and orientation. In some simple cases, such as coupling across [111]-oriented noble-metal spacer layers, only one oscillatory period is present. The parameter J_0 and the phase ϕ depend on the layer compositions, thicknesses, interface structures and on the growth orientation [112].

Speriosu et al. [30] and Anthony et al. [71] have reported oscillatory coupling for Co/Cu/Co/Fe-Mn and NiO/Ni-Fe/Co/Cu/Ni-Fe/Fe-Mn spinvalves. The observed period, about 1.0 nm, agrees well with results obtained for multilayers, and with theoretical predictions [111].

Rijks et al. [56] (Fig. 4.5), Kools et al. [83,110], and Joo and Atwater [84] studied spin valves with the composition $Ni_{80}Fe_{20}$/Cu/$Ni_{80}Fe_{20}$/$Fe_{50}Mn_{50}$, and observed ferromagnetic coupling across the entire Cu thickness range studied. A similar conclusion was reported by Hoshino et al. [113], from a study of systems containing $Ni_{66}Fe_{16}Co_{18}$, instead of permalloy. Results reported in [110] are exhibited in Fig. 4.18. Only weak indications of a superimposed oscillatory component were observed. It was concluded that pinhole coupling is likely to be responsible for the observed very strong F coupling below a Cu thickness of about 1.5 nm, whereas magnetostatic coupling, resulting from nonflat interfaces, is responsible for F coupling at larger spacer thicknesses. This so-called Néel-type coupling [114] arises as a consequence of correlated interface roughness, of the type depicted schematically by the inset in Fig. 4.18. The observed dependence of the magnetostatic ferromagnetic coupling parameter J on the Cu layer thickness t_{Cu} was modeled by Kools et al. within the Néel model, using the expression

$$J = \frac{\pi^2}{\sqrt{2}}\frac{h^2}{L}\mu_0 M_{\mathrm{sat}}^2 \exp(-2\pi\sqrt{2}t_{\mathrm{Cu}}/L)\,, \tag{4.17}$$

where h is the amplitude of a two-dimensional sinusoidal interface waviness, L is the lateral correlation length, and M_{sat} is the saturation magnetization of the permalloy layers. From a fit to the data for a series of systems deposited at an optimum Ar pressure during sputtering (5 m torr), the parameter values $h = 0.4$ nm and $L = 10$ nm were derived. The correlation length L was associated with the average grain diameter, which was about the same size in the systems investigated. Consistent with the analysis given in Sect. 4.3.3, h was observed to decrease with decreasing Ar sputter pressure, resulting in weaker F coupling [83].

4.5.2 Exchange Biasing

The interaction energy between the pinned F layer and the antiferromagnetic layer is given by $E = -E_{\mathrm{eb}} \cos\theta$, θ being the angle between the biasing direction and the magnetization of the pinned F layer. The exchange biasing field is then given by $H_{\mathrm{eb}} = E_{\mathrm{eb}}/\mu_0 M_{\mathrm{sat}}^{\mathrm{p}} t_{\mathrm{F}}^{\mathrm{p}}$, in which M_{sat}^{p} and t_{F}^{p} are the saturation magnetization and thickness of the pinned ferromagnetic layer, respectively. Such an effective unidirectional anisotropy may be obtained by growth of the AF layer on top of a magnetically saturated F layer, or by cooling the system in a field from a temperature above the blocking temperature. Several microscopic models have been put forward to explain exchange biasing. In all models an inequivalence of the spin-up and spin-down structures in AF near the interface is assumed. However, the models are different with respect to the assumed spatial variation of the spin structure in AF, the change of the magnetic structure upon changing the magnetization direction of the ferromagnet, and the perfection of the interface structure. Presumably, these aspects depend strongly on the details of the magnetic interactions and microstructure of the F/AF system involved. A brief overview of the various models has been given in [115].

The exchange biasing material that has been studied most intensively is the antiferromagnetic random substitutional fcc (γphase) alloy $Fe_{50}Mn_{50}$ [113,115–124]. Although this phase is thermodynamically unstable at room temperature, γphase films with thicknesses of a few tens of nanometers may be obtained by epitaxial growth on fcc-conforming substrates such as Cu or permalloy, to which its lattice parameter is matched within 0.4% and 2.5%, respectively. Films thicker than about 40–50 nm have been reported to show a significant fraction of the αphase, which is nonmagnetic at room temperature [118]. Fe–Mn film thicknesses of about 10 nm are sufficient for obtaining the full biasing effect, whereas the biasing field rapidly decreases to zero below film thicknesses of about 7 nm. As shown already in Fig. 4.5, the exchange biasing field for a spin valve (with 6 nm thick permalloy as the pinned layer) may be as high as 20 kA/m. In the absence of interlayer magnetic coupling, this is much larger than the field range in which the free layer switches.

Exchange biasing decreases approximately linearly with increasing temperature, down to zero at the blocking temperature. The blocking temperature depends on the film structure, thickness and growth conditions, and may become as high as about 180°C [116,123]. It increases with the Fe–Mn layer thickness, and saturates around a thickness of about 10–12 nm. For sputter-deposited Ni–Co–Fe/Cu/Ni–Co–Fe/Fe–Mn spin-valve structures with Fe–Mn layer thicknesses of the order of 10 nm, the blocking temperature has been observed to be 140°C [113]. The blocking temperature is significantly lower than the Néel temperature of bulk single crystals, which is about 230°C [122]. The increase of the blocking temperature with the Fe–Mn layer thickness has been shown to be consistent with a simple finite-size scaling relation [124]. It was already mentioned in Sect. 4.3.3 that exchange biasing by $Fe_{50}Mn_{50}$ is strongest for [111]-oriented systems, and that for sputter-deposited systems a strong [111] texture may be obtained by growth on a thin Ta buffer layer. Results on MBE-grown $Ni_{80}Fe_{20}/Fe_{50}Mn_{50}$ bilayer systems, grown on Cu single crystal substrates have firmly established the relationship between the crystallographic orientation and exchange biasing field [115]. An overview of experimental data on exchange biasing which is regarded as representative is assembled in Table 4.2.

Table 4.2. Experimental data on exchange-biased $Ni_{80}Fe_{20}$/X bilayer systems, with X = $Fe_{50}Mn_{50}$ or amorphous Tb–Co, and for NiO/$Ni_{80}Fe_{20}$ (NiO deposited first). The ordering temperatures refer to bulk materials. T_B is the blocking temperature (in the case of $Fe_{50}Mn_{50}$ and NiO), or the critical temperature above which biasing disappears irreversibly (in the case of amorphous Tb–Co). The other entries in the table refer to room temperature measurements. All given data refer to thicknesses of the biasing layers close to or larger than the layer thickness t_0 above which biasing is fully developed. The exchange biasing field given refers to 6 nm permalloy; in cases where this permalloy thickness was outside the thickness range studied, an extrapolation was made by assuming inverse proportionality of the biasing field with the permalloy thickness

	T_b (°C)	t_0 (nm)	K_{eb} (mJ/m^2)	H_{eb} (kA/m)	ρ (10^{-8} Ωm)	Ref.
$Fe_{50}Mn_{50}$	140		0.12	20		b
$T_N = 230$°C [a]	150	7	0.09	15	130	c
		7	0.10	16		d
					74	e
a–$Tb_{1-x}Co_x$	> 250	100	0.33	55		g
$T_C = 330$°C	150	40	0.28	46	236	h
for $0.67 < x < 0.83$) [f]	> 160					i
NiO	200		0.06	11	insulator	j
$T_N = 252$°C	180		0.05	8		k
	200	35	0.05	9		c

[a] [122]
[b] Sputter-deposited spin-valves, on Ta buffer layer [56,126]
[c] Sputter-deposited on glass and annealed at 240°C [125]
[d] MBE grown, [111]-oriented bilayer on Cu single crystal [115]
[e] [161]
[f] [130], p. 52
[g] Results for $x = 0.725$ [128]
[h] Results for $x = 0.77$ [68]
[i] Results for $x = 0.68$ [6]
[j] [132]
[k] [131]

For detailed reports on the dependence of the exchange biasing field of sputter-deposited systems on the F and AF layer thicknesses, temperature, deposition method, and on other interesting features we refer to [113,116–120,123,124].

For the application of exchange biasing in spin-valve sensors four aspects are of particular importance: a large exchange biasing field, a high blocking temperature, a high sheet resistance of the AF layer, and a good corrosion resistance. The minimum required blocking temperatures depend on the maximum temperature during operation (typically 80°C for a MR tape head, or 180°C for a field sensor for some automotive applications), as well as on the maximum temperature during processing of the sensor after deposition of the GMR material. For some applications the blocking temperature for $Fe_{50}Mn_{50}$ is too low. More recent studies of exchange-biased spin valves have therefore focussed on two materials with higher blocking temperatures, viz. amorphous $Tb_{1-x}Co_x$ and NiO. In Sect. 4.3.2 I have already discussed the magnetotransport properties of spin valves based on these materials. Experimental data on exchange biasing by these materials have been included in Table 4.2. Blocking temperatures ≥ 450°C have been obtained using antiferromagnetic NiMn layers [127]. However, the formation of the proper antiferromagnetic phase requires an annealing treatment of at least several hours at about 255°C, which in spin-valve layered structures could lead to substantial interface interdiffusion. Exchange biasing by NiMn will not be discussed below.

Exchange biasing by amorphous $Tb_{1-x}Co_x$ alloy thin films, with $x \approx 0.725$, has been studied intensively by Cain et al. [128]. For Tb–Co thicknesses above 100 nm, a large room-temperature biasing field was found. Smith and Cain have proposed a micromagnetic model for this phenomenon [129]. A disadvantage of the use of Tb–Co for the purpose of exchange biasing is the sensitivity with respect to the alloy composition. It is noted that this may be related, in part, to the large variation of the compensation temperature with the composition, from 20°C to 200°C from $x = 0.77$ to $x = 0.70$ [130], combined with the difficulties of controlling the stoichiometry of the film deposited. At the compensation temperature, the antiparallel Tb and Co sublattice magnetizations are equal, leading for a single film to infinite coercivity. Freitas et al. [68] found high biasing fields for a slightly different alloy concentration, around $x = 0.77$, using Tb–Co layers of only 40 nm thickness. These results make the application of Tb–Co in spin valves more attractive, in view of the smaller shunting effect. However, the temperature above which the biasing effect becomes zero was reported to be only 150°C. It should be noted that, unlike in the case of Fe–Mn, biasing by Tb–Co films which have been heated to a temperature above this critical temperature cannot be restored by a cool-down in a magnetic field [8]. It is not known whether only magnetic or (also) structural changes take place above the critical temperature. Therefore it is not clear whether the critical temperature may be called a "blocking temperature".

In Sect. 4.3.2 the transport properties of spin valves containing NiO have been discussed. NiO is a cubic insulating antiferromagnetic compound with a Néel temperature of 525 K. Exchange biasing in coupled films of permalloy on NiO was first demonstrated by Carey and Berkowitz [131], who deposited NiO by reactive sputtering, and subsequently by Soeya et al. [132] and Lin et al. [125]. Advantages as compared to Fe–Mn are the larger blocking temperature, smaller electrical shunting (being an insulator) and no corrosivity. In

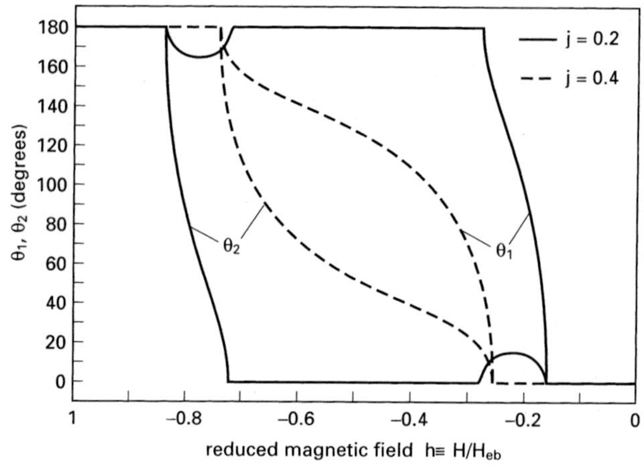

Fig. 4.19. Results of a model calculation of the angles of the magnetization directions in the free and biased F layers of a F1/NM/F2/AF spin valve with respect to the exchange biasing direction, as a function of the magnetic field [56]. The parameter j is defined as $j \equiv H_{\rm coupl}/H_{\rm eb}$. The F1 and F2 layer thicknesses are equal

addition, as discussed in Sect. 4.3.2, part of the high GMR ratios observed for NiO-based systems may be due to specular scattering at the F/AF interface. A disadvantage is the smaller exchange biasing interaction (at room temperature) and the much larger coercivity in the $M(H)$ curve. An increase of the exchange biasing field at room temperature may be obtained by replacing NiO by NiO/CoO multilayers, however, at the expense of a decrease of the blocking temperature [133,134].

The exchange biasing field should be much larger than all other magnetic interactions acting on the pinned layer. In the case of a patterned stripe, the shape anisotropy of the pinned layer and the magnetostatic interaction between the free and pinned layers complicate the analysis, and may mostly require a higher exchange biasing field than anticipated from studies of unpatterned samples. In the case of a large sample (not patterned), the only competing interactions are the interlayer magnetic coupling and the in-plane induced anisotropy of the pinned layer. Neglecting the induced anisotropy of the free and pinned layers, the interplay between coupling and biasing has been investigated theoretically and experimentally by Rijks et al. [56]. A result of their calculations, for a spin valve with equal free and pinned magnetic layer thicknesses, is depicted in Fig. 4.19. The calculations assumed single-domain rotational magnetization reversal processes for both layer. The figure shows the variation with the applied field of the angles of the magnetization in the free and pinned layers with respect to the biasing direction, for two different cases: strong biasing (dimensionless parameter $j \equiv H_{\rm coupl}/H_{\rm eb} = 0.2$) and weak biasing ($j = 0.4$). In the first case, there is still a field range in which

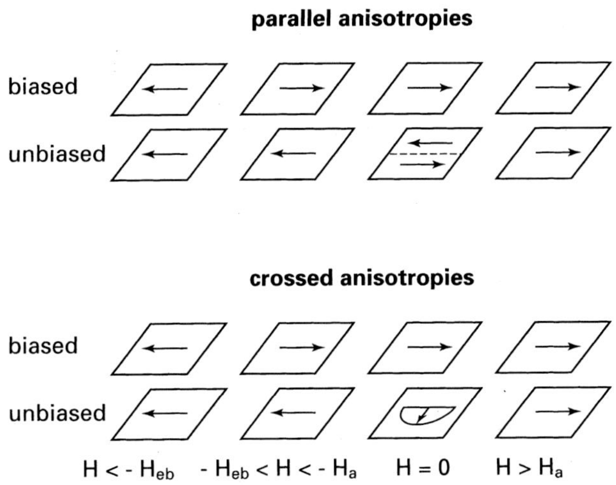

Fig. 4.20. Schematic representation of the magnetization reversal in the F layers of a spin valve structure with parallel and crossed anisotropies. H_a and H_{eb} are the induced anisotropy and exchange biasing fields, respectively. Interlayer magnetic coupling is neglected

the F layers are antiparallel, so the MR ratio is unaffected. However, there is already an appreciable broadening of the field range in which the free layer switches. In the latter case, the high resistance plateau in the $R(H)$ curve has disappeared completely, as the free and pinned layers rotate simultaneously, so the MR ratio is diminished. This occurs for $j > 0.25$. The analysis of experimental data essentially confirmed this picture ([56], see also [113]).

4.5.3 Magnetic Anisotropy

For most sensor applications, switching of the free layer should preferably be free of hysteresis. As demonstrated by Rijks et al. [24], this can be accomplished for spin valves within which the free layers have an in-plane, uniaxial anisotropy, with the easy axis perpendicular to the unidirectional anisotropy axis of the pinned layer. In such systems with crossed anisotropies, with the external field parallel to the biasing direction, the magnetization of the free layer switches by a coherent rotation. A schematic representation of the magnetization reversal processes for the cases of crossed and parallel anisotropies is presented in Fig. 4.20. In the latter case, magnetization reversal is the result of domain wall movement, leading to hysteresis [24]. As a result, strong Barkhausen noise will be superimposed on the sensor output signal due to pinning and depinning of domain walls.

Nishioka et al. [81] reported that the hysteresis (coercivity) observed for an exchange-biased spin valve with parallel anisotropies, with the applied field (almost) parallel to the exchange-biasing direction, decreases with decreasing Cu-spacer layer thickness. They attributed this to the increasing ferromagnetic interlayer coupling, and analysed their results in terms of model calculations of the magnetization-reversal mechanism which (in contrast to the calculations by Rijks et al. discussed above [56]) included the induced magnetic anisotropy of the magnetic layers. Even more general model calculations for the magnetization reversal, including the case of crossed anisotropy directions, have been presented by Parker et al. [135].

Let us now turn to the materials science-aspects of magnetic anisotropy. Magnetic anisotropy in Ni–Co–Fe alloys can be induced by growth of the film in an external magnetic field, or by annealing afterwards in a magnetic field [136,137]. The latter method is less suitable for most layered structures, as it involves diffusion in the bulk of the layers, which would, at the same time, lead to interface interdiffusion. The anisotropy is thought to be the result of directional ordering of certain atomic pairs [138–141]. In order to realize this ordering a magnetic field which saturates the film is required. For permalloy, with an anisotropy field of about 0.4 kA/m, applied fields during deposition at room temperature of typically 5–15 kA/m are sufficient. For Ni–Co–Fe thin films, the anisotropy field which results from the deposition in a field depends strongly on the composition [137]. For films with the composition $(Ni_{80}Fe_{20})_{1-x}Co_x$, the anisotropy field increases rapidly with increasing x, up to values between 1.5 and 2.5 kA/m for $x > 0.4$.

The following three aspects are of particular importance for the application of spin-valve structures. First, for very thin permalloy films, sputter-deposited at room temperature, the anisotropy field is smaller than the bulk value, and extrapolates to zero in the limit of zero film thickness. Saturation is reached at about 20 nm film thickness [142]. The effect is more generally observed for ternary Ni–Co–Fe alloys [143]. No completely satisfactory microscopic models have yet been presented for this phenomenon. Second, the anisotropy field of the free layer may be affected by annealing the spin valve after deposition. Significant effects are already observed at temperatures of about 150°C, which may occur during further processing of the material, in order to make a device. An example of such an effect has been discussed by Rijks et al. [24]. Third, the magnetization direction in the free layer is in actual spin valves not quite uniform [144]. Apart from large structural defects which are clearly visible in, for example, TEM cross sections, such as grain boundaries, there are defects or inhomogeneities on a nanometer or subnanometer scale, which are very difficult to detect. As a result, the easy-axis direction in magnetic alloy films is generally not uniform, causing lateral variations ("ripple") of the magnetization direction [136]. Gillies et al. [145] have recently found indications from Lorentz-microscopy studies of the ripple pattern of permalloy films below 20 nm film thickness that the average ripple angle significantly increases with decreasing film thickness. The results suggest that ripple angles are much larger in films with nanometer-scale thicknesses (as used in spin valves) than for films which are one order

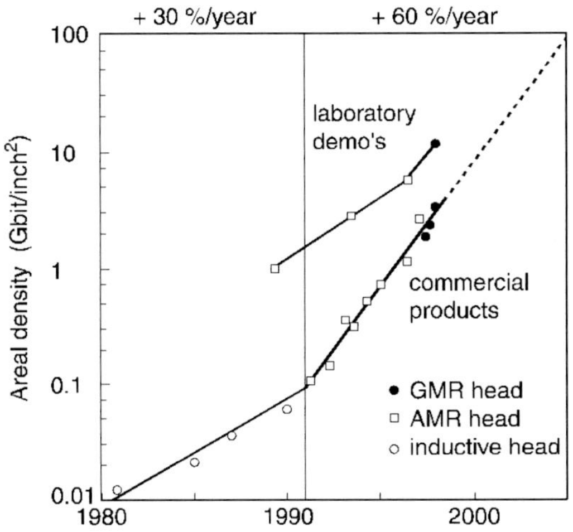

Fig. 4.21. Increase with time of the areal bit density in hard-disk recording, for commercial IBM systems [148], updated with additional data up to 1998

of magnitude thicker as used, for example, in AMR sensors. The effect of ripple on the magnetization and magnetoresistance curves becomes particularly significant close to the anisotropy field, where the ripple angle diverges [146]. The presence of ripples leads to less ideal hard axis $M(H)$ and $R(H)$ curves, being rounded off near the field where saturation is reached. It should be noted that, in a spin valve, additional lateral variations of the direction of the magnetization of the free layer may be found if the interlayer magnetic coupling with the biased layer is not laterally uniform. This could result from the presence of pinholes, or be due to the magnetic stray field from the pinned layer due to an imperfect interface with the spacer layer. The effect of ripple on much more prominent magnetization variations, related to the presence of domain walls, is at present a subject of intensive study [147].

4.6 Application in Read Heads

4.6.1 Trends in Magnetic Storage

The major industrial driving force behind the study of the GMR effect is its suitability for application in read heads for digital magnetic recording. As may be seen from Fig. 4.21, in hard-disk recording the introduction of read heads making use of the AMR effect, at the beginning of the 1990s, has increased the yearly growth of the areal bit density from about 30% per year to about 60% per year. The application of GMR materials is expected to help sustain this annual growth after its projected introduction in commercial systems in the second half of this decade [148,149]. In commercial tape systems, AMR read heads were first introduced in 1980 by Philips: as an

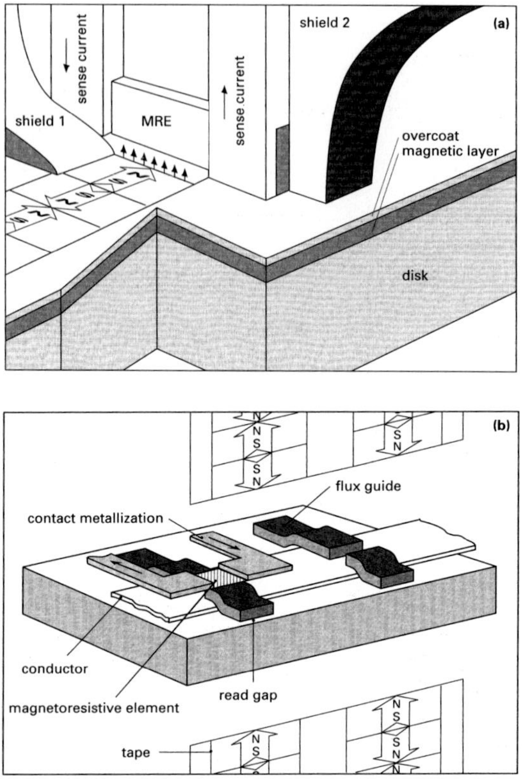

Fig. 4.22. a, b Schematic views of the read principle for a shielded head (**a**) [149] and a multiple track yoke-type head (**b**). For the purpose of clarity the bit length shown in (**b**) is not to scale, and the MRE and contact metallization have not been drawn for one of the heads in (**b**)

additional read head in compact cassette systems, they were used to retrieve auxiliary low-frequency information from the tape, written superimposed on the two audio channels (Computer Coded Search, CCS system [150,151]). An early application (1985) of multichannel AMR heads was the professional 32-track Philips Communication Logging System (CLS) [151,152]. Another application is the Digital Compact Cassette (DCC) consumer audio system, which has been introduced in the beginning of the nineties [153]. Thinfilm MR heads are produced by photolithographic processing. Whereas hard disk systems combine a very high areal density with a very short access time, data tapes have the advantage of being removable, having a very high volume bit density, and having a very low price per bit stored. For certain applications magnetic-tape systems are therefore expected to remain or become of interest. Examples of an existing and a potential future application are backup of computer data and digital video recording, respectively. As in the case of

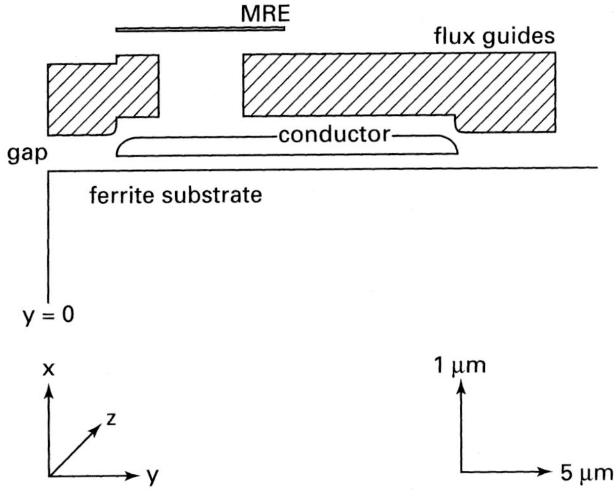

Fig. 4.23. Schematic cross section of a yoke-type MR head [5]

hard-disk recording, the introduction of GMR-based read heads could lead to a further increase of the bit density. For portable applications, the higher sensitivity of GMR-based heads is also of interest as a means of reducing the power consumption of the head.

4.6.2 Head Designs

The Magnetoresistive Element (MRE) in a MR read head generally consists of a stripe-shaped MR thin film, contacted by a pair of leads through which the sense current is applied, and over which the output voltage is measured. Depending on the application, different head designs and head dimensions are used. Perspective views of two designs, the "shielded" head and the "yoke-type" head, are shown in Fig. 4.22. A schematic cross section of a yoke-type head is presented in Fig. 4.23. Below the conventional terminology for the dimensions in the direction perpendicular to the medium ("height", y), parallel to the medium / perpendicular to the tracks ("width", z), and parallel to the medium / parallel to the tracks ("length", x) is used.

In the case of shielded heads, the minimum bit length which may be resolved is determined by the distance between two relatively thick soft-magnetic layers ("shields"), and by the distance from the MRE to the medium. This design has been used successfully in prototype hard-disk heads for gigabit-density recording using AMR thin films [154]. In commercial hard-disk systems present (1996) bit lengths and track widths are down to about 0.3 μm and 3 μm, respectively. The smallest present geometrical dimensions for the active part of the MRE stripe are $2 \times 1\,\mu m^2$, with a distance between the shields of 0.25 μm and a distance between the MRE and the medium of 100 nm.

Yoke-type AMR heads have been employed successfully in tape recording, for instance, in CLS and DCC multichannel heads. Within this design, the MRE is part of a yoke, and bridges a narrow gap between soft magnetic flux guides. In the design sketched in Fig. 4.23, the substrate is made of a soft-magnetic material, and hence it forms part of the yoke. The read gap, which is the distance between the soft magnetic flux guides close to the tape bearing surface (the surface that is in contact with the tape) determines the minimum bit length that can be read by the head. In commercial DCC heads, presently using AMR films, the read width is 70 μm, the stripe height is 10 μm, and the gap length is typically 0.4 μm.

Prototype GMR sensor-in-gap type read heads based on exchange-biased spin-valve layered structures have been offered so far by IBM [3,4], Hitachi [7], INESC [8] and Fujitsu [9], and a prototype GMR yoke-type read head has been presented by Philips [5,6]. First results on read heads of a third type, with the long direction of the MRE stripe orthogonal to the medium ("end-on"), have been presented by NVE Inc. [155]. In this design, a symmetric exchange coupled F/NM/F sandwich type GMR film is employed.

4.6.3 Yoke-Type Heads

The advantages of a GMR layered structure in a head, instead of an AMR film, depend on a rather complicated interplay between the properties of the MR material and the geometry (shape and dimensions) of the head. In this subsection the case of yoke-type heads [150,156,157] is discussed. The head design considered is sketched in Fig. 4.22 b, with the dimensions as indicated in the cross-section given in Fig. 4.23. The magnetically active parts are the ferrite substrate, the stripe-shaped MRE and the soft magnetic flux guides. The integrated conductor makes it possible to apply a magnetic field at MRE, which may be used (i) for testing heads on the wafer, and (ii) for applying a dc bias field during operation. These parts are separated by oxide layers. The surface of the head that is in contact with the tape is covered with a protective coating (not shown in the figure). In present DCC heads the MRE is a single-layer alloy film, such as permalloy, showing the AMR effect. The easy axis is parallel to the long axis of the stripe. This configuration, with the current direction parallel to the long axis of the stripe, is by itself not very favorable for sensor applications, because the variation of the resistance with the applied field is symmetric around zero field. It has the form of a parabola, see (4.5), with its top at $H = 0$, resulting in zero sensitivity at $H = 0$. In order to linearize the $R(H)$ curve at $H = 0$, the so-called "barber pole" method can be used. This method is illustrated in Fig. 4.24. It should be noted that linearization of the AMR effect is realized at the expense of a decrease of the resistance of the element, and therefore of the head output voltage.

A comparison will be made with the response from GMR heads, based on exchange-biased spin valves with crossed anisotropy axes: the exchange

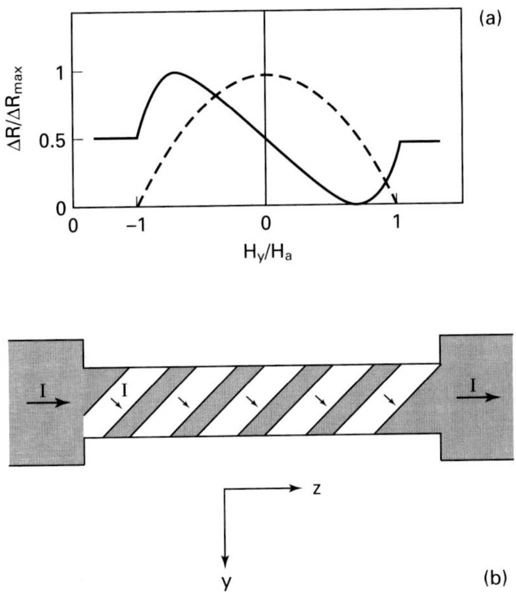

Fig. 4.24. (**a**) Relative resistance change of a AMR stripe, as a function of the ratio of the applied field to the anisotropy field, in the absence of the barber pole (*dashed curve*) and with the barber pole (*solid curve*). (**b**) Barber pole structure. The *shaded areas* represent a highly conducting overlayer, the *white areas* represent the uncovered AMR film

biasing direction of the pinned layer and the easy direction of the free layer are parallel and perpendicular, respectively, to the height-direction, in order to obtain hysteresis-free, low-noise magnetic switching of the free layer (Sect. 4.5.3). In addition to the GMR effect, spin valves also show the AMR effect. As shown by Rijks et al. [97], the total magnetoresistance curve is, to a very good approximation, given by the sum of the curves for the two separate effects. They have shown how the AMR ratio for a layered structure can be calculated within a semiclassical transport model. In these calculations the scattering of an electron in a magnetic layer is assumed to depend on the direction of its spin *and* of its velocity with respect to the magnetization direction of that layer. Schematic $\Delta R(H)$ curves are plotted in Fig. 4.25 for different ratios between the anisotropic and giant magnetoresistance ratios. The AMR effect makes the sensitivity curves asymmetric. For these schematic curves, the largest sensitivity is found at the anisotropy field $H = +H_a$. However, in realistic samples there are lateral anisotropy fluctuations (Sect. 4.5.3), resulting in a rounding-off of the $R(H)$ curves near the average anisotropy field and to a shift of the point of maximum sensitivity. Typical MR ratios for ($8\,\mathrm{nm Ni_{80}Fe_{20}}/2.5\,\mathrm{nm Cu}/6\,\mathrm{nm Ni_{80}Fe_{20}}$) spin valves are $\Delta R/R_{\mathrm{GMR}} = 4.5\%$ and $\Delta R/R_{\mathrm{AMR}} = 1.1\%$ [56]. It follows from Fig. 4.25 that as a result of the

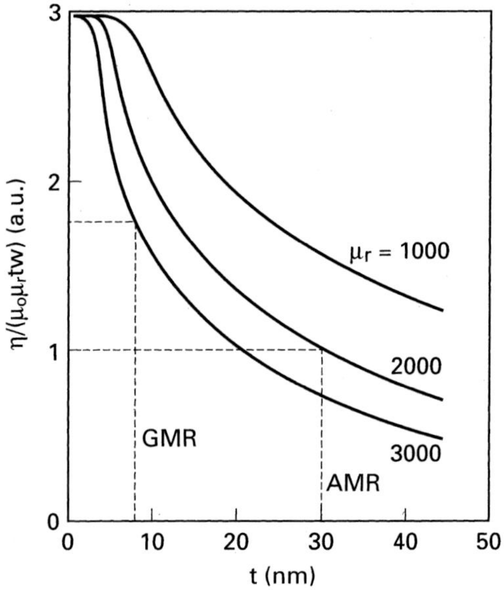

Fig. 4.25. Dependence on the ratio r $\equiv \Delta R_{\mathrm{AMR}}/\Delta R_{\mathrm{GMR}}$ of the magnetoresistance curves for an exchange-biased spin valve with crossed anisotropies

additional AMR effect the sensitivity of these spin valves, as calculated for $H = 0.5\,H_{\mathrm{a}}$, is almost 50% higher than expected from the GMR effect only.

Let us consider the situation in which a flux Φ_0 enters the yoke in the $y = 0$ plane. The y component of the flux in MRE varies across the height (y) direction, and is given by

$$\Phi(y) = \mu_0\mu_r H(y)tw \cong \mu_0 M_y(y)tw\,, \tag{4.18}$$

where M_y is the y component of the magnetization, t is the thickness of the sensitive magnetic layer (the free layer thickness in the case of a spin valve), w is the width of the active part of the stripe, h is the stripe height, μ_{r} is the relative permeability of the sensitive layer and $H(y)$ is the internal magnetic field. The angle θ between the local magnetization direction and the y axis is now determined by cos $\theta(y) = M_y/M_{\mathrm{sat}}$. In order to calculate the change of the stripe resistance as a result of the "incoming" flux Φ_0, it is necessary to calculate $M_y(y)$.

Now suppose that the local sheet resistance $R_{\Box}(y)$ varies linearly with $\Phi(y)$. It follows from (4.1) that this assumption is strictly valid for a GMR spin valve, whereas it follows from (4.5) (with θ replaced by $\theta - \pi/4$ in order to take the linearization by the barber pole into account) that this is a good approximation for AMR films for fluxes far below saturation. Using (4.18),

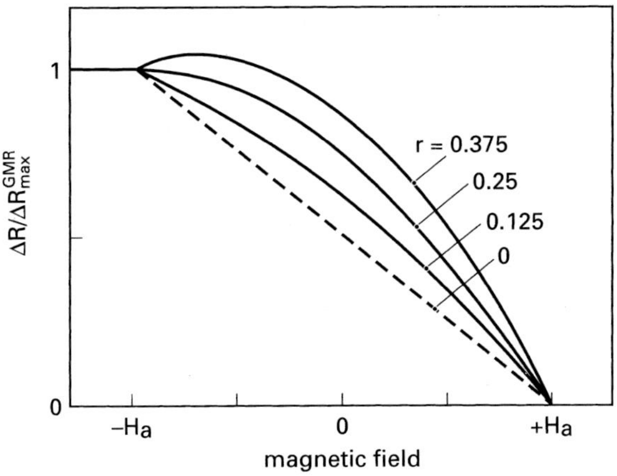

Fig. 4.26. Dependence on the F layer thickness and relative permeability of $\eta/(\mu_0\mu_r tw)$, the change of the average magnetic field in MRE per unit incoming flux, as obtained from finite-element calculations for the head geometry shown in Fig. 4.23 [158]. For the relative permeabilities of the substrate and the flux guides, $\mu_r = 2000$ was taken

the change of the local sheet resistance $\Delta R_\square(y)$ which results from a change of the local flux $\Delta\Phi(y)$ is then given by

$$\Delta R_\square(y) = \frac{1}{\mu_0\mu_r tw}\frac{\partial R_\square}{\partial H}\Delta\Phi(y)\,. \tag{4.19}$$

Subsequently use is made of the fact that for AMR as well as spin-valve GMR systems the relative resistance changes are very small (much smaller than 100%). The total resistance change of the stripe is then given by the average over the stripe height of the change of the local sheet resistance, multiplied by the geometrical factor w/h (stripe width over stripe height). With (4.19), it follows that the change of the resistance is proportional to the change of the average flux in the MRE, $\Delta\overline{\Phi}$:

$$\Delta R = \frac{1}{\mu_0\mu_r tw}\left(\frac{1}{R}\frac{\partial R}{\partial H}\right)\left(\frac{w}{h}R_\square\right)\Delta\overline{\Phi}\,. \tag{4.20}$$

Finally, assuming a *linear response*, i.e., $\Delta\overline{\Phi} = \eta\Phi_0$, with η a head efficiency factor, the voltage change ΔV which results from a flux Φ_0 entering the head is given by [5]

$$\Delta V = \frac{\eta}{\mu_0\mu_r tw}s\left(\frac{w}{h}R_\square\right)I_{\text{sense}}\Phi_0\,, \tag{4.21}$$

where s is the sensitivity, defined as $s \equiv (\partial R/R)/\partial H$. The factor $\eta/(\mu_0\mu_r tw)$ is the change of the average internal magnetic field in MRE per unit incoming flux.

Table 4.3. Comparison of the voltage output of AMR (30 nm$Ni_{80}Fe_{20}$, linearized with a barber pole) and GMR(8 nm$Ni_{80}Fe_{20}$/2.5 nmCu/6 nm$Ni_{80}Fe_{20}$/8 nm$Fe_{50}Mn_{50}$) yoke-type heads, for the design given in Fig. 4.23. As discussed in the text, the results follow from a model calculation assuming the operation of MRE under conditions of linear response. In the GMR case the contribution to the sensitivity by the AMR effect has been neglected. The relative efficiency η has been derived from the data given in Fig. 4.26. For AMR heads the field interval ΔH_R in which the resistance changes is equal to $\sqrt{2}H_a$, H_a being the anisotropy field. For exchange-biased GMR spin valves, $\Delta H_R = 2H_a$. The resistance values given refer to $70 \times 10\,\mu m^2$ MRE stripes, and have been taken from [5]

	μ_r	t (nm)	$\eta/(\mu_0\mu_r tw)$ (rel. units)	$\Delta R/R$ (%)	ΔH_R (kA/m)	s (%/(kA/m))	R (Ω)	ΔV (rel. units)
AMR	2000	30	1	1.5	0.56	2.7	90	1
GMR	3000	8	1.75	4.0	0.53	7.5	160	8.7

For AMR-based heads containing single alloy films, and linearized by a barber pole, $\Phi(y) = 0$ across the entire element height for zero incoming flux. If linearized by a barber pole, the approximation by a linear response is well satisfied provided the incoming flux Φ_0 is sufficiently small. As shown by Folkerts and Ruigrok from finite element and transmission line calculations, the efficiency factor $\eta(\mu_r, t)$ may, to a very good approximation, be viewed as a function of the product $\mu_r t$ [158]. Results of the calculations, for the head dimensions given in Fig. 4.23, are shown in Fig. 4.26. The shape of the curves may be explained as follows. For very thin films, the presence of MRE does not alter the magnetic field in the gap in the yoke which is bridged by MRE. Therefore, the read flux and hence the efficiency increases proportional to the product of the permeability and the thickness of the film. As a result, in the limit of very thin films and for a fixed stripe width w, the factor $\eta/(\mu_0\mu_r tw)$ is a constant. This is roughly the case when the effective shape anisotropy of the stripe is much smaller than the induced magnetic anisotropy. In order to obtain the proper criterion for the thickness above which the demagnetization effect becomes important, it is necessary to consider the geometry of the entire system, i.e., the stripe *and* its soft-magnetic environment. From the calculations by Folkerts and Ruigrok it follows that for the geometry considered the effect of the soft-magnetic environment is to effectively reduce the demagnetizing field of the stripe by a factor of about two (relative to that of a *bare* stripe in a homogeneous field) [158]. This leads to the criterion that the efficiency increases proportionally to the thickness of the film as long as $t \ll 2h/\mu_r$. Thick permalloy films typically have $\mu_r \approx 2000$, corresponding to an anisotropy field of 0.4 kA/m. With $h = 10\,\mu m$, $2h/\mu_r$ is equal to 10 nm. Indeed, the results show that at and above this thickness, η/t decreases with t: the sensor output is negatively affected by the increasing effective shape anisotropy of the MRE stripe. It thus follows that for a head with given shape and dimensions the voltage output is determined by the interplay between the materials properties ($s = (\partial R/R)/\partial H$, $R_\square$ and μ_r) and the MRE dimensions (t, w and h). In the case of single-layer permalloy films $\eta/(\mu_0\mu_r tw)$ and $R_\square$ decrease with increasing layer thickness t. On the other hand, $s = (\partial R/R)/\partial H$ increases with t, until a thickness much larger than the majority-spin mean free path (see Sect. 4.4 and [25]). The optimum output is found for film thicknesses around 24 nm [159]. Typical data for heads with a 30 nm permalloy film with barber pole are given in Table 4.3.

For spin-valve GMR-based MREs, the response is complicated by the ferromagnetic interlayer coupling, and by the magnetostatic interaction between the pinned and the free layer due to the field generated by the pinned layer at the edges of the stripe. The latter interaction tends to align the free and pinned layers antiparallel. As a result of these interactions, the magnetization in the film is already nonzero for zero incoming flux. It varies with y, and in extreme cases, parts of the stripe may even be fully saturated ($M_y/M_{sat} = \pm 1$), resulting in a lowered and highly nonlinear sensor output, and in a high level of Barkhausen noise. Optimization with respect to sensor stability thus requires balancing these interactions in such a way that at any point across the stripe height the magnetization is far from saturation. If necessary, an additional biasing field from the integrated conductor may be used. It should be noted that in a design with the free layer closest to the soft magnetic substrate the effective field in the free layer that is generated by the sense current is very small under usual operating conditions [6]. For analyses of the magnetization distribution in single-film and layered test structures without flux guides, it is referred to the theoretical and experimental studies by Folkerts et al. [6] and by Kools et al. [160].

Folkerts et al. [6] have formulated design rules which lead to a fair balance between all interactions affecting the magnetization of the free layer, resulting in a nonsaturated state for zero incoming flux and to a linear response. The voltage output of MRE may then be estimated using (4.21) and the numerical results for the efficiency, given in Fig. 4.26. Typical data for MRE, based on a (8 nm$Ni_{80}Fe_{20}$/2.5 nmCu/6 nm$Ni_{80}Fe_{20}$/8 nm$Fe_{50}Mn_{50}$) spin valve, are given in Table 4.3. As compared to AMR-based heads, the output voltage is expected from this approach to be one order of magnitude higher. In the table, the additional contribution from the AMR effect to the sensitivity is neglected. As has been discussed at the beginning of this section, the AMR effect is expected to increase the voltage output by about 40%. Based on slightly different materials parameters, Folkerts et al. [5] arrived at very similar results, viz. a predicted 10–14 times larger output than for a reference AMR-based head [5]. Experimentally, the benefit was typically found to be a factor of 10 [5]. It should be noted that taking benefit of the additional AMR effect, by operating the element around an average internal field that is shifted from zero (Fig. 4.25), limits the maximum driving amplitude. It follows from the table that the larger output voltage of GMR heads, as compared to AMR-based heads, is not only due to the higher MR ratio and higher permeability, but also to the higher stripe resistance and the lower shape anisotropy of MRE (lower thickness of the free layer). It should be emphasized that there are many other aspects, besides the voltage output, which determine the suitability of an MR material for an application, including linear range, stability, lifetime (related to, e.g., wear, corrosion resistance, and thermal stability) and some aspects affecting the complexity and robustness of the manufacturing process. However, a discussion of these aspects is outside the scope of this chapter.

4.7 Issues and Challenges

One decade after the discovery of the GMR effect, various types of GMR layered structures have been found which show a high MR ratio at low magnetic field. From the overview given above, it may have become clear that exchange-biased spin-valve layered structures show a combination of properties that make these materials attractive for the application in low-field sensors, such as read heads in hard-disk and tape recording. In spite of intensive research on these materials, there are several unresolved issues. An overview of those issues which are regarded here as the most interesting and challenging from a fundamental physics point of view or for applications, is given below.

The first issue concerns the quantitative description of the variation of the resistance and magnetoresistance of spin valves with the various layer thicknesses. Results from theoretical studies as well as the analyses of experimental results using transport models have not yet resulted in a widely accepted quantitative picture on spin-dependent scattering in the bulk of the layers, and at the interfaces. For example, the spin dependence of scattering in bulk permalloy is still a matter of controversy. In addition, an important issue to be resolved is how anisotropic defect scattering (e.g., at grain boundaries) may be taken into account properly. This requires detailed knowledge of the microstructure of the samples, taking into account that many microstructural aspects, such as grain sizes, vary with the thickness of the layers deposited.

Interestingly, record GMR ratios have recently been reported for spin valves containing NiO layers for exchange biasing. The high MR ratios observed have been argued to be possibly due to (partially) reflective electron scattering at the metal/NiO interfaces. More work is required to investigate this issue.

Some challenging aspects of the materials research on GMR materials, such as the relation between, on the one hand, the preparational method and microstructure, and, on the other hand, the GMR ratio and the interactions which determine the magnetization reversal processes have been discussed. Most of the studies reviewed are fairly recent. An issue which is just beginning to be explored is the role of interface and bulk defects on the micromagnetic properties, for example, the creation of domain walls upon switching and the enhanced magnetization ripple for films with nanometer-scale thicknesses. Another issue is the variation with temperature of the exchange biasing effect. Higher blocking temperatures than may be obtained for $Fe_{50}Mn_{50}$ are required for some applications. It is not yet clear to what extent other materials, such as NiO or amorphous Tb–Co, may be regarded as realistic alternatives.

The performance of an MR device is determined by the interplay between materials properties and design-related aspects, as has been discussed in Sect. 4.6 for the specific example of a yoke-type read head. Optimization of the performance may benefit from the input from micromagnetic modeling of the magnetic response in a device, studied as a function of geometrical dimensions and material properties. Micromagnetic modeling of the performance of microstructured or nanostructured devices based on layered magnetic materials introduces the new challenge of properly taking into account the interactions between the magnetic layers. The possibility to simulate the performance of microstructured and nanostructured devices is also expected to be of crucial importance in view of future developments towards radically new designs of devices, which even more fully exploit the high GMR ratios and the tunability of magnetic interactions in layered magnetic materials.

4.8 Notes Added in Proof

Since the completion of the manuscript new results have been published on almost all issues that have been discussed in this chapter. We give here a brief selection. References [162] and [163] are overview papers.

Sect. 4.2.3: Oparin et al. [164] have presented results of first principles calculations of the magnetic structure at interdiffused $Cu/Ni_{80}Fe_{20}$ interfaces, allowing noncollinear spin structures. Such structures were found to be more stable than collinear spin structure, investigated in an earlier publication by the same group [51]. The reduced magnetization at the interface would correspond with approximately one "magnetically dead" layer at the interface.

Sect. 4.3.1: A quantitative analysis of the experimentally observed thermal variation of the GMR ratio in exchange biased spin valves has been carried out by Dieny et al. [99] by extending the Camley and Barnaś model [88] in order to take thermal effects into account: (i) spin-intermixing in the bulk of the ferromagnetic layers by electron-magnon scattering, (ii) interfacial scattering by paramagnetic interfacial layers which may form as the temperature is increased, and (iii) electron-phonon scattering in the nonmagnetic spacer layer.

Sect. 4.3.2: More recent publications by Egelhoff et al. on symmetric spin valves with two NiO pinning layers, showing an MR ratio above 20%, are [165] and [166]. Sugita et al. have obtained the highest MR ratio for a symmetric spin valve reported so far, viz. 27.8%, for an α–Fe_2O_3/Co/Cu/Co/Cu/Co/α–Fe_2O_3 system. The system is grown epitaxially on α–Al_2O_3 [167]. The antiferromagnetic α–Fe_2O_3 layers, with a thickness of 50 nm, give rise to a large effective coercivity of the Co layers which are in exchange contact with them. The authors regard the high MR ratio as an indication of specular electron reflection at the metal/insulator Co/α–Fe_2O_3 interfaces.

Sect. 4.3.3: A high resolution electron microscopy study of Co/Cu/Co/Fe–Mn spin valves has been carried out by Bayle-Guillemaud et al. [168]. The relationship between the presence of a Ti seed layer, [111]-texture, and the exchange bias field was studied, as well as the interface diffusiveness. Intermixing at the interfaces over two monolayers was observed.

Sect. 4.3.4: A TEM study of the thermal stability of the structure of T/Co/Cu/Ci/Fe–Mn/Ti spin valves, as related to the magnetoresistance, has been presented by Portier et al. [169].

Sect. 4.4.2: Extensions of the semiclassical transport model in order to include spin-mixing have been presented by Dieny et al. [99] and by Chen and Hershfield [170]. At finite temperatures spin mixing can occur upon electron-magnon scattering. However, in disordered systems spin-mixing can already take place upon scattering at very low temperatures, viz. due to spin-orbit interaction. The significance of this effect for permalloy has been demonstrated by Steenwyk et al. [171] from a study of the CPP–GMR effect in permalloy-based spin valves. An overview of extensions of the Camley and Barnaś model that would be required to be able to optimize the resistance and magnetoresistance of exchange biased spin valves has been given by Dieny et al. [172].

Sect. 4.4.3: First-principles calculations of the magnetoconductance of spin valves of the type Co/Cu/Co, taking the full band structure into account, and adjusting scattering rates within each of the layers phenomenologically, have been published by Butler and coworkers in a number of publications [173–176]. This work addresses issues such as the layer thickness dependence of the conductance and magnetoconductance, spin dependent transmission and reflection of electrons at the interfaces, channeling of electrons in the Cu layer (waveguide modes), and the dependence of the conductance on the angle between the magnetization directions.

Sect. 4.5: The magnetic interactions discussed in Sect. 4.5 are all of influence on the magnetization reversal process in an applied field. Imaging of this process in exchange biased spin valves, using Lorentz transmission electron microscopy, has been reported in [144,177–182].

Sect. 4.5.2: The subject of exchange biasing has been reviewed by Nogues and Schuller [183]. As compared to the systems discussed in Sect. 4.5.2, many more antiferromagnetic alloys and compounds have been investigated for their suitability as exchange biasing materials. Of particular interest in view of their high blocking temperature are, amongst other systems, NiMn, Ir–Mn, Pt–Pd–Mn and Cr–Pt–Mn. An overview of the properties of some Mn-based antiferromagnets that are suitable for exchange biasing as been given by Lederman [184]. Studies of the thermal relaxation of the exchange biasing interaction, which may be a limiting factor in some applications, have been presented in [185–187]. As the exchange biasing field depends on the effective saturation magnetization of the pinned layer, successful attempts have been reported to increase the field range within which the pinned layer and free layer magnetizations are antiparallel by making use of an "artificial antiferromagnet" (AAF) as the pinned layer, coupled to an antiferromagnetic (AF) layer. Such "AAF + AF" structures have been proposed for applications in robust magnetic field sensors [188] and have been used in GMR read heads [189].

Sect. 4.5.3: With the purpose of obtaining a low coercivity (due to a low magnetocrystalline anisotropy) of the free magnetic layer, a thinner layer structure and a material with a higher resistance, spin valves have been fabricated on the basis of amorphous alloys such as Co–Fe–B [190].

Sect. 4.6: Results of electronic noise measurements of spin valve MR elements have been reported by van de Veerdonk et al. [191], who observed a field-dependent $1/f$ noise power spectrum on top of the white thermal (Johnson) noise. The field dependence of the $1/f$ noise spectrum has been modeled in terms of thermally induced magnetization fluctuations, which are translated into fluctuations of the resistance via the GMR effect. Electronic noise in the MR element is one of the contributions to the overall noise level in hard disk [192] and tape systems. The fabrication process of exchange biased spin valves on an industrial scale, repeatability and thickness uniformity have been discussed by Schwartz et al. [193]. The requirements for recording applications on the control over the layer thickness and the layer properties have been discussed by Gurney et al. [194].

Sect. 4.6.2: Shielded spin valve GMR hard-disk read heads have been introduced commercially in 1997. Results on laboratory tests and demonstrations of recording at densities up to 10 Gbit/inch^2 have been reported in [195–200]. A demonstration of recording at a density of 12 Gbit/inch^2 has been reported by Tsang et al. in [189]. Shielded spin valve GMR tape read heads have been fabricated by Oliveira et al. [201].

Sect. 4.6.3: The maximum output voltage change that can be obtained from an MR head is proportional to the maximum sense current density allowed (4.21). The sense current is limited by irreversible degradation processes such as electromigration, of which the rate at any given current density increases strongly with the element temperature. The current density in the active part of spin valves used in shielded read heads is (averaged over the layers) typically 2×10^{11} to 6×10^{11} A/m^2 [189,195–200]. The temperature increase is then typically a few tens of K. In a shielded head (Fig. 4.22 a), the field induced by the sense current on the free magnetic layer can be significant. This effect, and the effect of the magnetostatic coupling between the pinned and free layers, has been studied using Lorentz transmission electron microscopy of stripe-shaped spin valves ([202,203] and references therein). We remark that in a yoke-type read head, as discussed in this section, the current through the stripe-shaped spin valves does not generate a significant field in the free layer when the free layer is closest to the soft magnetic substrate. The output signal and noise of GMR spin-valves in test structures similar to yoke-type read heads has been discussed by Kools et al. [204].

Acknowledgements

I am very grateful to G. E. W. Bauer, B. Dieny, P. Grünberg and S. S. P. Parkin, and to my colleagues of the Philips Research Laboratories and of the Eindhoven University of Technology, in particular to P. Beliën, P. J. H. Bloemen, W. Folkerts, M. A. M. Gijs, M. J. M de Jong, W. J. M. de Jonge, J. C. S. Kools, K.-M. H. Lenssen, T. G. S. M. Rijks, J. J. M. Ruigrok and K. M. Schep, for useful discussions and comments on the manuscript. This work was carried out as part of the European Community ESPRIT3 Basic Research Project "Study of Magnetic Multilayers for Magnetoresistive Sensors" (SmMmS).

References

1. G. Binasch, P. Grünberg, F. Saurenbach and W. Zinn, Phys. Rev. B **39**, 4828 (1989)
2. M. N. Baibich, J. M. Brote, A. Fert, F. Nguyen Van Dau, F. Petroff, P. Etienne, G. Creuzet, A. Friederich and J. Chazelas, Phys. Rev. Lett. **61**, 2472 (1988)
3. D. E. Heim, R. E. Fontana Jr., C. Tsang, V. S. Speriosu, B. A. Gurney, and M. L. Williams, IEEE Trans. Magn. **30**, 316 (1994)
4. C. Tsang, R. E. Fontana, T. Lin, D. E. Heim, V. S. Speriosu, B. A. Gurney, and M. L. Williams, IEEE Trans. Magn. **30**, 3801 (1994)
5. W. Folkerts, J. C. S. Kools, T. G. S. M. Rijks, R. Coehoorn, M. C. de Nooijer, G. J. H. Somers, J. J. M. Ruigrok and L. Postma, IEEE Trans. Magn. **30**, 3813 (1994)
6. W. Folkerts, J. C. S. Kools, M. C. de Nooijer, J. J. M. Ruigrok, L. Postma, K.–M. H. Lenssen, G. H. J. Somers and R. Coehoorn, IEEE Trans. Magn. **31**, 2591 (1995); W. Folkerts and J. C. S. Kools, Sensors and Actuators **A64**, 237 (1998)
7. Y. Hamakawa, H. Hoshiya, T. Kawabe, Y. Susuki, R. Arai, K. Nakamoto, M. Fuyama and Y. Sugita, IEEE Trans. Magn. **32**, 149 (1996)
8. P. ten Berge et al., IEEE Trans. Magn. **31** 2603 (1995)
9. H. Kanai, J. Kane, K. Aoshima, M. Kanamine and Y. Uehara, IEEE Trans. Magn. **31**, 2612 (1995)
10. J. Daugton, J. Brown, E. Chen, R. Beech, A. Pohm and W. Kude, IEEE Trans. Magn. **30**, 4608 (1995)
11. J. L. Brown, IEEE Trans. Comp. Pack. Manuf. Techn. **A17**, 373 (1994)
12. A. V. Pohm, R. S. Beech, P. A. Bade, E. Y. Chen and J. M. Daughton, IEEE Trans. Magn. **30**, 4650 (1994)
13. D. D. Tang, P. K. Wang, V. S. Speriosu, S. Le and K. K. Kung, IEEE Trans. Magn. **31**, 3206 (1995)

14. Y. Irie, H. Sakakima, M. Satomi and Y. Kawawake, Jpn. J. Appl. Phys. **34**, L415 (1995)
15. In this chapter the term "layered structure" will be used for the general case of a thin film consisting of a stack of different layers, whereas the term "multilayer" will be used for the more specific case of a layered structure within which a sequence of layers is periodically repeated.
16. P. M. Levy, in *Solid State Physics* **47**, ed. D. Turnbull and H. Ehrenreich, (Acad. Press, San Diego, 1994) p. 367–462.
17. B. Dieny, J. Magn. Magn. Mater. **136**, 335 (1994)
18. A. Fert and P. Bruno in *Ultrathin magnetic structures*, ed. J. A. C. Bland and B. Heinrich (Springer-Verlag, Berlin Heidelberg New York 1994), Vol. 2, ch. 2
19. S. S. P. Parkin in *Ultrathin magnetic structures*, ed. J. A. C. Bland and B. Heinrich (Springer-Verlag, Berlin Heidelberg New York 1994), Vol. 2, ch. 2
20. M. A. M. Gijs and G. E. W. Bauer, Adv. Phys. **46**, 285–445 (1997)
21. B. Dieny, V. S. Speriosu, S. S. P. Parkin, B. A. Gurney, D. R. Wilhoit and D. Mauri, Phys. Rev. B **43**, 1297 (1991)
22. B. Dieny, V. S. Speriosu, B. A. Gurney, S. S. P. Parkin, D. R. Wilhoit, K. P. Roche, S. Metin, D. T. Peterson and S. Nadimi, J. Magn. Magn. Mater. **93**, 101 (1991)
23. B. Dieny, V. S. Speriosu, S. Metin, S. S. P. Parkin, B. A. Gurney, P. Baumgart and D. R. Wilhoit, J. Appl. Phys. **69**, 4774 (1991)
24. T. G. S. M. Rijks, W. J. M. de Jonge, W. Folkerts, J. C. S. Kools and R. Coehoorn, Appl. Phys. Lett. **65**, 916 (1994)
25. T. R. McGuire and R. I. Potter, IEEE Trans. Magn. **11**, 1018 (1975)
26. S. S. P. Parkin, N. More and K. P. Roche, Phys. Rev. Lett. **64**, 2304 (1990)
27. D. H. Mosca, F. Petroff, A. Fert, P. A. Schroeder, W. P. Pratt and R. Loloee, J. Magn. Magn. Mater. **94**, L1 (1991)
28. J. Barnaś, A. Fuss, R. E. Campley, U. Walz, P. Grünberg and W. Zinn, Vacuum **41**, 1241 (1990)
29. C. Dupas, P. Beauvillain, C. Chappert, J. P. Renard, F. Trigui, P. Veillet, E. Vélu and D. Renard, J. Appl. Phys. **67**. 5680 (1990)
30. V. S. Speriosu, B. Dieny, P. Humbert, B. A. Gurney and H. Lefakis, Phys. Rev. B **44**, 5358 (1991)
31. M. Julliere, Phys. Lett **54 A**, 225 (1975); S. Maekawa and U. Gafvert, IEEE Trans. Magn. **18**, 707 (1982)
32. A. Chaiken, G. A. Prinz and J. J. Krebs, J. Appl. Phys. **67**, 4892 (1990)
33. T. Shinjo, H. Yamamoto, T. Anno and T. Okuyama, Appl. Surf. Science **60,61**, 798 (1992)
34. J. L. Duvail et al., J. Magn. Magn. Mater. **151**, 324 (1995),
35. N. Mott, Proc. Roy. Soc. **156**, 368 (1936)
36. A. Fert, in *Materials Science Forum*, **59–60**, 439–480 (1990)
37. For systems of the type (F1/NM/F2/NM) an *increase* of the resistance upon switching the magnetization directions of the ferromagnetic layers F1 and F2 from an antiparallel to a parallel alignment is obtained if, effectively, scattering at the layers F1 and F2 is characterized by *opposite* spin asymmetry parameters (J. M. George et al. Phys. Rev. Lett. **72**, 408 (1994); J.–P. Renard et al., Phys. Rev. B **51**, 12821 (1995))
38. T. Valet and A. Fert, J. Magn. Magn. Mater. **121**, 378 (1993); T. Valet and A. Fert, Phys. Rev. B **48**, 7099 (1993)

39. W. P. Pratt, Jr., S. F. Lee, J. M. Slaughter, R. Loloee, P. A. Schroeder and J. Bass, Phys. Rev. Lett. **66**, 3060 (1991); W. P. Pratt, Jr., S.–F. Lee, P. Holody, Q. Yang, R. Loloee, J. Bass and P. A. Schroeder, J. Magn. Magn. Mater. **126**, 406 (1993); Q. Yang, P. Holody, R. Loloee, L. L. Henry, W. P. Pratt, Jr., P. A. Schroeder and J. Bass, Phys. Rev. B **51**, 3226 (1995)
40. M. A. M. Gijs, S. K. J. Lenczowski and J. B. Giesbers, Phys. Rev. Lett. **70**, 3343 (1993)
41. Figure partly taken from R. Coehoorn, in *Supermagnets, Hard Magnetic Materials*, G. J. Long and F. Grandjean (eds.), (Kluwer, 1991), p. 133–170.
42. S. K. J. Lenczowski, M. A. M. Gijs, R. J. M. van de Veerdonk, J. B. Giesbers and W. J. M. de Jonge, MRS Symp. Proc., to be published (1995)
43. S. S. P. Parkin, Digest of the 14th Int. Coll. on Magn. Films and Surfaces, p. 27 (Düsseldorf, 1994)
44. J. Banhart and H. Ebert, Europhys. Lett. **32**, 517 (1995)
45. J. Inoue, A. Oguri and S. Maekawa, J. Phys. Soc. Jpn **60**, 376 (1991); J. Inoue, H. Itoh and S. Maekawa, J. Phys. Soc. Jpn **61**, 1149 (1992)
46. R. Schad, C. D. Potter, P. Beliën, G. Verbanck. V. V. Moschalkov and Y. Bruynseraede, Appl. Phys. Lett. **64**, 3500 (1994)
47. S. S. P. Parkin, Z. G. Li and S. J. Smith, Appl. Phys. Lett. **58**, 2710 (1991)
48. For Fe/Cr interfaces this is far from trivial; competing antiferromagnetic Fe–Cr and Cr–Cr interactions between atoms at the interface play a crucial role (R. Coehoorn, J. Magn. Magn. Mater. **151**, 341 (1995))
49. R. Coehoorn, J. Magn. Magn. Mater. **121**, 432 (1993)
50. H. Itoh, J. Inoue and S. Maekawa, Phys. Rev. B **47**, 5809 (1993)
51. D. M. C. Nicholson, W. H. Butler, X.–G..Zhang, J. M. MacLaren, B. A. Gurney and V. S. Speriosu, J. Appl. Phys. **76**, 6905 (1994)
52. M. Kohler, Ann. Physik **6**, 18 (1949)
53. J. P. Jan, in *Solid State Physics* **5**, eds F. Seitz and D. Turnbull, (Acad. Press, New York, 1957) p. 1–96.
54. J. Smit, Physica **16**,, 612 (1951)
55. The layers are given in the order of deposition.
56. T. G. S. M. Rijks, R. Coehoorn, J. T. F. Daemen and W. J. M. de Jonge, J. Appl. Phys. **76**, 1092 (1994)
57. V. Speriosu, J. P. Nozières, B. A. Gurney, B. Dieny, T. C. Huang and H. Lefakis, Phys. Rev. B **47**, 11579 (1993)
58. B. Rodmacq, G. Palumbo and P. Gerard, J. Magn. Magn. Mater. **118**, L11 (1993)
59. T. L. Hylton, K. R. Coffee, M. A. Parker and J. K. Howard, Science **261**, 1021 (1993)
60. T. G. S. M. Rijks, unpublished.
61. B. Dieny, P. Humbert, V. S. Speriosu and B. A. Gurney, Phys. Rev. B **45**, 806 (1992)
62. B. A. Gurney, V. S. Speriosu, J.–P. Nozières, H. Lefakis, D. R. Wilhoit and O. U. Need, Phys. Rev. Lett. **71**, 4023 (1993)
63. T. G. S. M. Rijks, PhD thesis, Eindhoven University of Technology (1996)
64. B. Dieny, V. S. Speriosu and S. Metin, Europhys. Lett. **15**, 227 (1991)
65. B. Dieny, J. P. Nozières, V. S. Speriosu, B. A. Gurney and D. R. Wilhoit, Appl. Phys. Lett. **61**, 2111 (1992)
66. S. S. P. Parkin, Phys. Rev. Lett. **71** 1641 (1993); S. S. P. Parkin, Appl. Phys. Lett. **61** 1358 (1993)

67. W.H. Butler, X.–G. Zhang, D.M. Nicholson and J.M. MacLaren, J. Magn. Magn. Mater. **151**, 354 (1995)
68. P.P. Freitas, J.L. Leal, T.S. Plaskett, L.V. Melo and J.C. Soares, J. Appl. Phys. **75**, 6480 (1994)
69. J.L. Leal, N.J. Oliviera, L.M. Rodrigues, A.T. Sousa, and P.P. Freitas, IEEE Trans. Magn. **30**, 3831 (1994)
70. H. Hoshiya, M. Komuro, K. Mitsuoka and Y. Sugita, Nippon Oyo Jiki Gakkaishi **18**, 355 (1994)
71. T.C. Anthony, J.A. Brug and S. Zhang, IEEE Trans. Magn. **30**, 3819 (1994)
72. H. Swagten, G.J. Strijkers, P.J.H. Bloemen, M.M.H. Willekens and W.J.M. de Jonge, Phys. Rev. B **53**, 9108 (1996); H.J.M. Swagten, G. Strijkers, R.H.J.N. Bitter, W.J.M. de Jonge and J.C.S. Kools, IEEE Trans. Magn. **34**, 948 (1998)
73. W.F. Egelhoff, Jr. et al. J. Appl. Phys. **78**, 273 (1995)
74. P.M. Baumgart, B. Dieny, B.A. Gurney, J.–P. Nozières, V.S. Speriosu and D.R. Wilhoit, U.S. Patent 5287238 (1994)
75. S. Noguchi, R. Nakatani, K. Hoshino and Y. Sugita, Jpn. J. Appl. Phys. **33**, 5734 (1994)
76. T.G.S.M. Rijks, R.L.H. Sour, D.G. Neerinck, A.E.M. de Veirman, R. Coehoorn, J.C.S. Kools, M.F. Gillies and W.J.M. de Jonge, IEEE Trans. Magn. **31**, 3865 (1995)
77. J.P.W.B. Duchateau, unpublished.
78. R. Jérome, T. Valet and P. Galtier, IEEE Trans. Magn. **30**, 4878 (1994)
79. R. Nakatani, K. Hoshino, S. Noguchi and Y. Sugita, Jpn. J. Appl. Phys. **33**, 133 (1994)
80. K.–M.H. Lenssen, J.C.S. Kools, A.E.M. De Veirman, J.J.T.M. Donkers, M.T. Johnson, A. Reinders and R. Coehoorn, J. Magn. Magn. Mater. **156**, 63 (1996)
81. K. Nishioka, S. Gangopadhyay, H. Fujiwara and M. Parker, IEEE Trans. Magn. **31**, 3949 (1995)
82. T.C. Huang, J.–P. Nozières, V.S. Speriosu, H. Lefakis and B.A. Gurney, Appl. Phys. Lett. **60**, 1573 (1992)
83. J.C.S. Kools, J. Appl. Phys. **77**, 2993 (1995)
84. H.S. Joo and H.A. Atwater, IEEE Trans. Magn. **31**, 3946 (1995)
85. T.C. Huang, J.–P. Nozières, V.S. Speriosu, B.A. Gurney and H. Lefakis, Appl. Phys. Lett. **62**, 1478 (1993)
86. J.C.S. Kools, unpublished.
87. Y. Kitada, H. Kikuchi, H. Kishi, M. Otagiri and K. Kobayashi, IEEE Trans. Magn. **31**, 2600 (1995)
88. R.E. Camley and J. Barnaś, Phys. Rev. Lett. **63**, 664 (1989)
89. See, e.g., N.W. Ashcroft and N.D. Mermin, *Solid State Physics*, (Holt, Rinehart and Winston, New York, 1976) and J.S. Dugdale, *The electrical Properties of Metals and Alloys*, (Edward Arnold, London, 1977)
90. K. Fuchs, Proc. Cambridge Philos. Soc. **34**, 100 (1938); E.H. Sondheimer, Adv. Phys. **1**, 1 (1952)
91. M.J.M. de Jong, unpublished.
92. A. Vedyayev, B. Dieny, N. Ryzhanova, J.B. Genin and C. Cowache, Europhys. Lett. **25**, 465 (1994)
93. B. Dieny, Europhys. Lett. **17**, 261 (1992); (4.15) on the anisotropic mean free path: B. Dieny, private communication.

94. B. Dieny, V. S. Speriosu, J.–P. Nozières, B. A. Gurney, A. Vedyaev and N. Ryzhanova, in *"Magnetism and Structure in Systems of Reduced Dimensions"*, eds R. F. C. Farrow et al., Plenum, New York (1993)
95. B. A. Gurney, V. S. Speriousu, H. Lefakis and D. R. Wilhoit, J. Appl. Phys. **76**, 6616 (1994)
96. T. G. S. M. Rijks and R. van de Kruijs, unpublished.
97. T. G. S. M. Rijks, R. Coehoorn and W. J. M. de Jonge, in *Magnetic Ultrathin Films*, ed. B. T. Jonker et al., MRS Symposia Proceedings No. 313 (MRS, Pittsburg, 1993), p. 283; T. G. S. M. Rijks, R. Coehoorn, M. J. M. de Jong and W. J. M. de Jonge, Phys. Rev. B **51**, 283 (1995)
98. A. B. Granovskii, A. V. Vedyaev and A. V. Kalitsov, Phys. Sol. State **37**, 183 (1995)
99. B. Dieny, A. Granovsky, A. Vedyaev, N. Ryzhanova, C. Cowache and L. G. Pereira, J. Magn. Magn. Mater. **151**, 378 (1995)
100. R. Q. Hood and L. M. Falicov, Phys. Rev. B **46**, 8287 (1992)
101. B. L. Johnson and R. E. Camley, Phys. Rev. B **44**, 9997 (1991)
102. H. E. Camblong and P. M. Levy, Phys. Rev. Lett. **69**, 2835 (1992)
103. A. Brataas and G. E. W. Bauer, Europhys. Lett. **26**, 117 (1994)
104. X.–G. Zhang and W. H. Butler, Phys. Rev. B **51**, 10085 (1995)
105. P. Levy, S. Zhang and A. Fert, Phys. Rev. Lett. **65**, 1643 (1990); Phys. Rev. B **45**, 8689 (1992)
106. A. Vedyayev, B. Dieny and N. Ryzhanova, Europhys. Lett. **19**, 329 (1992)
107. T. Oguchi, J. Magn. Magn. Mater. **126**, 519 (1993)
108. K. M. Schep, P. J. Kelly and G. E. W. Bauer, Phys. Rev. Lett. **74**, 586 (1995)
109. P. Zahn, I. Mertig, M. Richter and H. Eshrich, Phys. Rev. Lett. **75**, 2996 (1995); I. Mertig, P. Zahn, M. Richter, H. Eshrich, R. Zeller and P. H. Dederichs, J. Magn. Magn. Mater. **151**, 363 (1995)
110. J. C. S. Kools, T. G. S. M. Rijks, A. E. M. de Veirman and R. Coehoorn, IEEE Trans. Magn. **31** 3918 (1995)
111. See the contributions by K. B. Hathaway, by A. Fert and P. Bruno, by D. T. Pierce, J. Unguris and R. J. Celotta and by S. S. P. Parkin in *Ultrathin Magnetic Structures*, ed. J. A. C. Bland and B. Heirrich (Springer, Berlin, 1994), Vol. 2, Ch. 2.
112. See, e.g., M. T. Johnson et al., Phys. Rev. Lett. **68**, 2688 (1992); M. T. Johnson et al., Phys. Rev. Lett. **69**, 969 (1992); P. J. H. Bloemen et al., Phys. Rev. Lett. **72**, 764 (1994); J. J. de Vries et al., Phys. Rev. Lett. **75**, 4306 (1995); M. T. Johnson et al., Phys. Rev. Lett. **75**, 4686 (1995)
113. K. Hoshino, S. Noguchi, R. Nakatani, H. Hoshiya and Y. Sugita, Jpn. J. Appl. Phys. **33**, 1327 (1994)
114. L. Néel, C. R. Acad. Sci. **255**, 1676 (1962)
115. R. Jungblut, R. Coehoorn, M. T. Johnson, J. aan de Stegge and R. Reinders, J. Appl. Phys. **75**, 6659 (1994)
116. C. Tsang and K. Lee, J. Appl. Phys. **53**, 2605 (1982)
117. R. D. Hempstead, S. Krongelb and D. A. Thompson, IEEE Trans. Magn. **14**, 521 (1978)
118. C. Tsang, N. Heiman and K. Lee, J. Appl. Phys. **52**, 2471 (1981)
119. J. C. Scott, J. Appl. Phys. **57**, 3681 (1995)
120. C. Schlenker, S. S. P. Parkin, J. C. Scott and K. Howard, J. Magn. Magn. Mater. **54–57**, 801 (1986)

121. C. Hwang, R.H. Geiss and J.K. Howard, J. Appl. Phys. **64**, 6115 (1988)
122. Y. Endoh and Y. Ishikawa, J. Phys. Soc. Jpn, **30**, 1614 (1971)
123. O. Allegranza and M. Chen, J. Appl. Phys. **73**, 6218 (1993)
124. S.S.P. Parkin and V.S. Speriosu, in *Magnetic Properties of Low-Dimensional Systems II*, Springer Proceedings in Physics **50**, eds. L.M. Falicov, F. Mejía-Lira and J.L. Morán-López (Springer, Berlin, 1990)
125. T. Lin, C. Tsang, R.E. Fontana and J.K. Howard, IEEE Trans. Magn. **31**, 2585 (1995)
126. T.G.S.M. Rijks, unpublished.
127. T. Lin, D. Mauri, N. Staud, C. Hwang, J.K. Howard and G.L. Gorman, Appl. Phys. Lett. **65**, 1183 (1994)
128. W.C. Cain, W.H. Meiklejohn and M.H. Kryder, J. Appl. Phys. **61**, 4170 (1987); W.C. Cain, J.W. Lee, P.V. Koeppe and M. Kryder, IEEE Trans. Magn. **24**, 2609 (1988); W.C. Cain and M. Kryder, IEEE Trans. Magn. **26**, 2412 (1990); W.C. Cain and M.H. Kryder, J. Appl. Phys. **67** 5722 (1990)
129. N. Smith and W.C. Cain, J. Appl. Phys. 69, 2471 (1991)
130. P. Hansen, in *Landolt-Börnstein*, Vol. 19g, ed. H.P.J. Wijn (Springer, Berlin, 1988), p. 57.
131. M.J. Carey and A.E. Berkowitz, Appl. Phys. Lett. **60**, 3060 (1992)
132. S. Soeya, S. Tadokoro, T. Imagawa, M. Fuyama and S. Narishige, J. Appl. Phys. **74**, 6297 (1993)
133. M.J. Carey and A.E. Berkowitz, J. Appl. Phys. **73**, 6892 (1993)
134. J. Fujikata, K. Ishihara, K. Hayashi, H. Yamamoto and K. Yamada, IEEE Trans. Magn. **31**, 3936 (1995)
135. M.R. Parker, H. Fujiwara, S. Hossain and W.E. Webb, IEEE Trans. Magn. **31**, 2618 (1995)
136. E. Pugh, in *Physics of Thin Films*, Vol. 1, ed. G. Hass, (Acad. Press, New York, 1963), p. 277–334
137. C.H. Wilts and F.B. Humphrey, J. Appl. Phys. **19**, 1191 (1968)
138. L. Néel, Compt. Rend. Acad. Sci. **237**, 1468, 1613 (1953); J. Phys. Radium **15**, 225 (1954)
139. S. Taniguchi and M. Yamamoto, Sci. Rep. Inst. Tohoku Univ. **A6**, 330 (1954); S. Taniguchi, Sci. Rep. Inst. Tohoku Univ. **A7**, 269 (1955)
140. S. Chikazumi, J. Phys. Soc. Jpn **11**, 551 (1956)
141. Reviews containing discussions on pair ordering in Ni–Fe alloys have been written by J.C. Slonczewski, in *Magnetism*, eds G.T. Rado and H. Suhl, Vol. I (Acad. Press, New York, 1963), p. 205–242, and by D.O. Smith, in *Magnetism*, eds G.T. Rado and H. Suhl, Vol. III(Acad. Press, New York, 1963), p. 465–523.
142. M. Goto, H. Tange and T. Kamimori, J. Magn. Magn. Mater. **62**. 251 (1986)
143. T.G.S.M. Rijks, R.F.O. Reneerkens, R. Coehoorn, J.C.S. Kools, M.F. Gillies and J.N. Chapman, J. Appl. Phys. **82**, 3442 (1997)
144. M.F. Gillies, J.N. Chapman and J.C.S. Kools, J. Appl. Phys. **78**, 5554 (1995)
145. M.F. Gillies, J.N. Chapman and J.C.S. Kools, J. Magn. Magn. Mater. **140–144** 721 (1995)
146. H. Hoffmann, IEEE Trans. Magn. **4**, 32 (1968)
147. M.F. Gillies, thesis, University of Glasgow (1996)
148. C.H. Bajorek and C.D. Mee, Data Storage **1**, (1) 23 (1994)
149. E. Grochowski and D.A. Thompson, IEEE Trans. Magn. **30**, 3797 (1994)
150. W.F. Druyvesteyn, J.A.C. van Ooyen, L. Postma, E.L.M. Raemaekers, J.J.M. Ruigrok and J. de Wilde, IEEE Trans. Magn. **17**, 2884 (1981)

151. W. Folkerts, Read/Write **18**, 8 (1994)
152. M. G. J. Heijman, J. H. W. Kuntzel and G. H. J. Somers, Philips Technical Review **44**, 169 (1988)
153. V. Zieren, G. Somers, J. J. M. Ruigrok, M. de Jongh, A. van Straalen, W. Folkerts, E. Draaisma, F. Pronk and T. Mitchell, IEEE Trans. Magn. **29**, 3064 (1993)
154. C. Tsang, M. Chen and T. Yogi, Proc. of the IEEE, **81**, 1344 (1993)
155. E. Y. Chen, A. V. Pohm, J. M. Daughton, J. Brown and W. C. Black, IEEE Trans. Magn. **30**, 3816 (1994)
156. J. J. M. Ruigrok, Philips J. Res. **36**, 289 (1981); J. J. M. Ruigrok, J. Appl. Phys. **53**, 2599 (1982)
157. J. J. M. Ruigrok, *Short-Wavelength Magnetic Recording: New Methods and Analyses*, (Elsevier, Oxford, New York, 1990)
158. W. Folkerts and J. J. M. Ruigrok, unpublished
159. W. Folkerts, unpublished
160. J. C. S. Kools, J. J. M. Ruigrok, L. Postma, M. C. de Nooijer and W. Folkerts, IEEE Trans. Magn. **33**, 4513 (1997)
161. T. G. S. M. Rijks, unpublished
162. J. C. S. Kools, IEEE Trans. Magn. **32**, 3165 (1996)
163. R. Coehoorn, J. C. S. Kools, T. G. S. M. Rijks and K.–M. H. Lenssen, Philips J. Research **51**, 93 (1998)
164. A. B. Oparin, D. M. C. Nicholson, X. G. Zhang, W. H. Butler, W. A. Shelton, G. M. Stocks and Yang-Wang, J. Appl. Phys. **85**, 4548 (1999)
165. W. F. Egelhoff, Jr. et al., J. Appl. Phys. **79**, 5277 (1996)
166. W. F. Egelhoff, Jr. et al., J. Appl. Phys. **82**, 6142 (1997)
167. Y. Sugita, Y. Kawawake, M. Satomi and H. Sakakima, Jap. J. Appl. Phys. **37**, 5984 (1998)
168. P. Bayle-Guillemaud, A. K. Petford-Long, T. C. Anthony and J. A. Brug, IEEE Trans. Magn. **32**, 4627 (1996)
169. X. Portier, A. K. Petford-Long, P. Bayle-Guillemau, T. C. Anthony and J. A. Brug, Appl. Phys. Lett. **72**, 118 (1998)
170. J. Chen and S. Hershfield, Phys. Rev. B **57**, 1097 (1998)
171. S. D. Steenwyk, S. Y. Hsu, R. Loloee, J. Bass and W. P. Pratt, Jr., J. Magn. Magn. Mater. **170**, L1 (1997)
172. B. Dieny, L. G. Peirera, R. H. Taylor and S. Y. Yamamoto, IEEE Trans. Magn. **34**, 942 (1998)
173. W. H. Butler, X. G. Zhang, T. C. Schulthess, D. M. C. Nicholson, J. M. MacLaren, V. S. Speriosu and B. A. Gurney, Phys. Rev. B **56**, 14574 (1997)
174. W. H. Butler, X. G. Zhang, J. M. MacLaren, IEEE Trans. Magn. **34**, 927 (1998)
175. R. H. Brown, D. M. Nicholson, W. H. Butler, X. G. Zhang, W. A. Shelton, T. C. Schulthess and J. M. MacLaren, Phys. Rev. B **58**, 11146 (1998)
176. W. H. Butler, X. G. Zhang, T. C. Schulthess, D. M. C. Nicholson, A. B. Oparin and J. M. MacLaren, J. Appl. Phys. **85**, 5834 (1999)
177. J. N. Chapman, M. F. Gillies and P. P. Freitas, J. Appl. Phys. **79**, 6452 (1996)
178. X. Portier, A. K. Petford-Long, R. C. Doole, T. C. Anthony and J. A. Brug, IEEE Trans. Magn. **33**, 3574 (1997)
179. J. P. King, J. N. Chapman and J. C. S. Kools, J. Magn. Magn. Mater. **177–181**, 896 (1998)
180. J. N. Chapman, P. R. Aichison, K. J. Kirk, S. McVitie, J. C. S. Kools and M. F. Gillies, J. Appl. Phys. **83**, 5321 (1998)

181. A. K. Petford-Long, X. Portier, E. Y. Tsymbal, T. C. Anthony and J. A. Brug, IEEE Trans. Magn. **35**, 788 (1999)
182. C. C. Yu and A. K. Petford-Long, J. Appl. Phys. **85**, 5753 (1999)
183. J. Nogues and I. K. Schuller, J. Magn. Magn. Mater. **192** 203 (1999)
184. M. Lederman, IEEE Trans. Magn. **35** 794 (1999)
185. P. A. A. van der Heijden, T. T. M. M. Maas, W. J. M. de Jonge, J. C. S. Kools, F. Roozeboom and P. J. van der Zaag, Appl. Phys. Lett. **72**, 492 (1998)
186. J. Fujikata, K. Hayashi, H. Yamamoto and M. Nakada, J. Appl. Phys. **83** 7210 (1998)
187. R. H. Taylor, R. O–Barr, S. Y. Yamamoto and B. Dieny, J. Appl. Phys. **85** 5036 (1999)
188. K.–M. H. Lenssen, A. E. T. Kuiper and F. Roozeboom, J. Appl. Phys. **85** 5531 (1999)
189. C. Tsang, M. Pinarbasi, H. Santini, E. Marinero, P. Arnett, R. Olson, R. Hsiao, M. Williams, R. Payne, R. Wange, J. Moore, B. Gurney, T. Lin and R. Fontana, IEEE Trans. Magn. **35**, 689 (1999)
190. Y. Shirota, S. Tsunashima, R. Imada, Y. Nomura, S. Iwata and M. Jimbo, Jap. J. Appl. Phys. Part 1, **38**, 714 (1999)
191. R. J. M. van de Veerdonk, P. J. L. Beliën, K. M. Schep, J. C. S. Kools, M. C. de Nooijer, M. A. M. Gijs, R. Coehoorn, and W. J. M. de Jonge, J. Appl. Phys. **82**, 6152 (1997)
192. K. G. Ashar, *Magnetic Disk Drive Technology*, IEEE Press, New York (1997)
193. P. V. Schwartz, R. Bubber, A. P. Paranjpe and J. C. S. Kools, J. Vac. Science and Technology **A16**, 3080 (1998)
194. B. A. Gurney et al., J. Appl. Phys. **81**, 3998 (1997)
195. H. Yoda, H. Iwasaki, T. Kobayashi, A. Tsutai and M. Sahashi, IEEE Trans. Magn. **32**, 3363 (1996)
196. H. Kanai et al., IEEE Trans. Magn. **32**, 3368 (1996)
197. K. Nanamoto et al., IEEE Trans. Magn. **32**, 3374 (1996)
198. H. Kanai et al., IEEE Trans. Magn. **33**, 2872 (1997)
199. H. Tong et al., IEEE Trans. Magn. **33**, 2884 (1997)
200. A. Veloso, P. P. Freitas, N. J. Oliveira, J. Fernandes and M. Ferreira, IEEE Trans. Magn. **34**, 2343 (1998)
201. N. J. Oliveira, P. P. Freitas, S. Li and V. Gehanno, IEEE Trans. Magn. **35**, 734 (1999)
202. X. Portier, A. K. Petford-Long, T. C. Anthony and J. A. Brug, J. Appl. Phys. **83**, 6840 (1998)
203. X. Portier, A. K. Petford-Long, T. C. Anthony and J. A. Brug, J. Appl. Phys. **85**, 4120 (1999)
204. J. C. S. Kools, R. Coehoorn, W. Folkerts, M. C. de Nooijer and G. H. J. Somers, Philips J. Research **51**, 93 (1998)

5 Experiments on the Perpendicular Giant Magnetoresistance in Magnetic Multilayers

M. A. M. Gijs

Since the discovery of the Giant Magnetoresistance (MR) effect in Fe/Cr magnetic multilayers [1], most of the experiments have been carried out with the measuring current in the plane of the multilayer, the so-called Current-In-Plane (CIP) geometry. The reason is simple: it is very easy to measure the fairly large resistance of a thin film or multilayer with the current in the plane of the sample because the length of the sample is typically orders of magnitude larger than the film thickness. The opposite is true for the situation where the measuring current is perpendicular to the multilayer plane (the so-called CPP geometry): the length of the sample is determined by the layer thickness which is usually much smaller than the typical lateral dimension. So one needs extremely sensitive measuring techniques to measure ultra-small resistances or nano- and microfabrication technologies to reduce the cross-section of the sample, giving rise to resistances that can be measured by conventional methods. Despite these experimental difficulties, the perpendicular MR effect appeared particularly attractive to experimentalists, because it typically is a factor two to five larger than the corresponding current-in-plane MR effect. This makes the subject of interest both for fundamental physical reasons and also for possible applications.

Although theoretically the CPP–MR effect has attracted considerable attention, it is not the intention of this chapter to present an overview of all theoretical models for the perpendicular magnetoresistance. Models will only be discussed as far as they help in clarifying the experimental data presented in this chapter. The physics of the giant MR effect has been reviewed by Fert and Bruno [2], and recently by Levy covering the field up to the first half of 1993 [3]. Dieny has outlined the giant MR effect in spin-valve sandwiches and multilayers, and has highlighted the experimental and theoretical results which are of particular interest for applications of these systems, especially in magnetic recording technology [4]. The latter subject has also been discussed by R. Coehoorn in Chap. 4. Another review specifically on the CPP–MR effect with an extensive experimental and theoretical review section and a discussion of new developments, such as increased understanding of band-structure effects on the transport has been published [5].

We intend to discuss various experimental approaches for studying the CPP–MR effect that have been developed in a recent period. The chapter is organized as follows: In Sect. 5.1, first the physics of the giant MR effect at low temperature is discussed, where the electronic transport through the multilayer can be simply described in terms of two parallel conducting spin-channels. In addition, CPP–MR experiments performed at 4.2 K will be treated here. More particularly, we present the superconducting contacting technique, where the sample is not microstructured and the very small perpendicular multilayer resistance is measured using an ultra-sensitive voltage detection system. The merit of these experiments is that they have led to the determination of all relevant spin-dependent scattering parameters at 4.2 K, and thereby have contributed a lot to the understanding of the physics of MR in the CPP geometry. In Sect. 5.2, other research is reviewed, dealing with the investigation of the CPP–MR effect up to room temperature. First of all, the modification of the two-channel model due to temperature effects will be discussed. Subsequently, three experimental techniques which have led to CPP–MR measurements up to room temperature will be introduced. In a first experimental approach, advanced microlithographic techniques were utilized to realize "pillar" microstructures, which due to their relatively large resistance are measurable as a function of temperature by conventional means. Secondly, the growth of multilayer pillars from an electrolyte into nanoporous membranes is treated. Finally, another experimental method, where the multilayer is evaporated at an angle on grooved substrates in order to realize a CPP-geometry is outlined. The advantages/disadvantages of the various techniques and their application potential will be discussed. The conclusions will be summarized in Sect. 5.3.

5.1 Perpendicular Magnetoresistance at 4.2 K: Exploring the Physics of Spin-Dependent Scattering

5.1.1 The Two-Channel Model at Low Temperatures

The intention is here to give a short theoretical introduction to the physics of the Giant MR (GMR) effect in general and of the CPP–MR effect in particular. In view of the recent reviews [2–4] and the discussion in Chap. 4, we will present here only the main qualitative ideas for describing the GMR effect in the CIP-geometry. We elaborate a little bit more on the CPP geometry and the differences between the CIP–MR and CPP–MR effect, more particular with respect to the critical physical length scales. After that, the theoretical model of Valet and Fert [6] will be discussed in somewhat more detail.

Overview of Theoretical Models. Necessary conditions for the occurrence of a GMR effect in a magnetic multilayer are (i) the transition of an antiparallel-like state of the magnetizations of the magnetic layers to a parallel state under the influence of the magnetic field and (ii) the existence of two non equivalent channels of electrical conduction. In a ferromagnetic film or a magnetic multilayer, these channels are associated with the two kinds of spin an electron can have with respect to the local magnetization [7], whereby the spin-up or spin-down electrons conventionally have the z component of their spin parallel or antiparallel to the local magnetization, when the z axis is chosen as the quantization axis.

In the original investigation on the GMR effect in the CIP geometry [1], spin-dependent scattering of the conduction electrons is proposed as being the explanation for a difference in resistance for the two spin channels. This is still believed to be the correct explanation of the observed phenomena, although it has become clear that in reality the asymmetry between the two spin channels is of more complex nature and is also determined by potential steps at interfaces [8] and by band structure effects [9,10]. Moreover the relative contribution of these various effects to the spin asymmetry is different if one is in the CIP or CPP geometry. Still most theories explain the GMR effect by spin-dependent scattering at defects either in the bulk or at interfaces and thereby allow to determine phenomenological scattering parameters which give good insight into the origin of the GMR effect. For the CIP geometry, Boltzmann-like models [11–16], so-called quantum-models [17,18] or tight-binding models [19] have been adopted to explain the experimental MR data of the various multilayer systems. The application of these theories to experimental data have led to the conclusion that interfaces in the multilayer contribute most to the spin-dependent scattering and hence to the GMR effect (see, for instance, [20,21]). Also for the GMR in the CPP geometry several models were developed, like the quantum model [22,23], which is based on the Kubo formalism, or the Landauer-Büttiker-type model of Bauer et al. [24,25]. There is also a number of models that include realistic band structures in a theory of electronic transport. Thereby, they indicate that band-structure effects indeed are important [26,9,10,27], especially for the CPP geometry. Finally, Boltzmann-type models [6,16] exist for the CPP–MR, in which spin-dependent scattering at interfaces and in the bulk of the magnetic layers is parametrized. All these models are applicable at $T = 0\,\mathrm{K}$ where the two spin channels conduct current in parallel; mixing of the two channels, which can take place at higher temperature, for instance, by scattering by spin-waves, is mostly not considered. Also a common feature is that the CPP–MR, calculated following a certain model, is larger than the corresponding CIP–MR.

The models developed are very generally based on the different conduction properties of the spin-up (majority) and spin-down (minority) electrons in ferromagnetic metals [7]. This holds both for the CIP and for the CPP case: If the resistivity mean free path λ would be much larger than the individual layer thicknesses, CIP– and CPP–MR would be comparable according to the models, that do not take into account potential steps and/or band structure. However, there are important fundamental differences between the CPP– and CIP–MR: a CPP geometry can be thought of as a collection of layer resistances in series, while for the CIP-geometry layer conductivities add in parallel. This results in an important "short-circuit" effect in the CIP geometry and therefore a reduction of the CIP–MR, once λ is smaller than the normal-metal individual-layer thickness. For the CPP geometry no such effect exists, and λ is not a critical length scale. However, it was pointed out that the spin-accumulation and spin-relaxation effects in a direction perpendicular to the layer plane, occurring in the CPP and not in the CIP geometry, give rise to a different relevant physical length scale of the conduction process [6,23]. While for the CIP geometry the perturbation in the resistance caused by an interface does not extend further than the resistivity mean free path λ, it is the spin-flip diffusion length l_{sf} which is the relevant length scale for the CPP geometry. This is an important difference, since l_{sf} for typical multilayer systems can be of the order of 100 nm at low temperature, while λ is of the order of 10 nm. As the interface is the dominant source for spin-dependent scattering, this means that CIP–MR vanishes once typical normal-metal thicknesses exceed λ, while CPP–MR vanishes only when these thicknesses exceed l_{sf}. This makes CPP–MR more "robust" than CIP–MR, i.e., it will depend less critically on individual layer-thickness constraints.

The Valet–Fert Model. For a thorough discussion of the different theoretical models in the CPP geometry and for a critical evaluation of their potential to explain the experimental data, we refer to the review by Gijs and Bauer [5]. We will discuss here in more detail the model of Valet and Fert [6,28,29], which has most widely been used for comparison with CPP–MR experiments. The model is quite general in the sense that the effect of finite spin-flip diffusion lengths both in the normal metal ($l_{\mathrm{sf}}^{\mathrm{N}}$) and in the ferromagnet ($l_{\mathrm{sf}}^{\mathrm{F}}$) is taken into account. In the limit that $l_{\mathrm{sf}}^{\mathrm{N}}$ and $l_{\mathrm{sf}}^{\mathrm{F}}$ are much larger than the multilayer period, simple expressions for the perpendicular resistance are obtained which can be used straightforwardly to separate the volume and interface contributions to the spin dependent scattering. But also the spin-flip diffusion length itself can be studied from a comparison of this theory with experimental CPP–MR data of multilayers, in which a certain concentration of paramagnetic impurities or spin-orbit scatterers is introduced in the normal metal [30–32]. By combining the Valet-Fert theory with theoretical models for the various spin-relaxation times [33], the fitting of the experimental CPP–MR data can als give important information on the fundamental scattering processes going on in the multilayer.

In the model an infinite multilayer is considered where single-domain Ferromagnetic metal (F) layers of thickness t_{F} alternate with normal-metal (N) layers of thickness t_{N}. A simplified single parabolic conduction band is assumed, with the effective mass m and the Fermi velocity v_{Fermi}. The initial theory [6,28,29] was restricted to zero temperature, where electron-magnon spin-flip scatterings are frozen out so that the only spin-flip scatterings are through spin-orbit interactions on defects or impurities. In the model a given current density J flows along the z axis which is perpendicular to the plane of the layers, and the resistance is calculated in the two extreme configurations of anti-Parallel (AP) and Parallel (P) alignments of the magnetizations. The CPP–MR ratio is defined as $[r^{(\mathrm{AP})} - r^{(\mathrm{P})}]/r^{(\mathrm{P})}$, $r^{(\mathrm{i})}$ being the CPP resistance per unit area of one bilayer in the magnetic configuration (i).

The basic transport equations in the z direction, in the limit $\lambda_\mathrm{s}/l_\mathrm{s} \ll 1$, are given by

$$\frac{e}{\sigma_\mathrm{s}}\frac{\partial J_\mathrm{s}}{\partial z} = \frac{\overline{\mu}_\mathrm{s} - \overline{\mu}_{-\mathrm{s}}}{l_\mathrm{s}^2} \tag{5.1}$$

and

$$J_\mathrm{s} = \frac{\sigma_\mathrm{s}}{e}\frac{\partial \overline{\mu}_\mathrm{s}}{\partial z}\,. \tag{5.2}$$

In previous equations the electronic mean free path $\lambda_\mathrm{s} = v_\mathrm{Fermi}\tau_\mathrm{s}\tau_\mathrm{sf}/(\tau_\mathrm{s} + \tau_\mathrm{sf})$ and the spin-flip diffusion length $l_\mathrm{s} = [\frac{1}{3}(v_\mathrm{Fermi}\lambda_\mathrm{s})\tau_\mathrm{sf}]^{1/2} = (D_\mathrm{s}\tau_\mathrm{sf})^{1/2}$ were introduced. Here, D_s is the diffusion constant, τ_s the intra-spin channel-relaxation time and τ_sf the spin-flip relaxation time. $\overline{\mu}_\mathrm{s} = \mu_\mathrm{s}(z) - e\Phi(z)$ is the electrochemical potential, $\Phi(z)$ being the electrostatic potential and $\mu_\mathrm{s}(z)$ the part of the electrochemical potential related to spin accumulation. $-e$ is the charge of the electron, σ_s is the conductivity, and J_s the current density for spin s.

Spin-flip scattering by spin-orbit interactions has extensively been studied by electron-spin resonance [34], which provides an estimate of the numerical value for the spin-flip diffusion lengths in the various metals. For 3d elements in noble metals the ratio between the spin-flip and non-spin-flip cross sections is around 10^{-2}, and therefore one expects $l_\mathrm{s} \approx 10\,\lambda_\mathrm{s}$ for structures composed of 3d elements with Cu or Ag. This shows that the condition $\lambda_\mathrm{s}/l_\mathrm{s} \ll 1$ is fulfilled in most experimental samples, even if the layer thicknesses are comparable with the mean free path.

From the coupled equations (5.1) and (5.2), $r^{(\mathrm{P})}$ and $r^{(\mathrm{AP})}$ can be obtained:

$$r^{(\mathrm{P,AP})} = r_0 + 2\,r_\mathrm{SI}^{(\mathrm{P,AP})} \tag{5.3}$$

with $r_0 = (1-\beta^2)\rho_\mathrm{F}^* t_\mathrm{F} + \rho_\mathrm{N}^* t_\mathrm{N} + 2r_\mathrm{b}^*$ and $r_\mathrm{SI}^{(\mathrm{P,AP})}$ a "spin-coupled" interface resistance given by

$$\begin{aligned} r_\mathrm{SI}^{(\mathrm{P})} = &\left[\frac{(\beta-\gamma)^2}{\rho_\mathrm{N}^* l_\mathrm{sf}^\mathrm{N}}\coth\left(\frac{t_\mathrm{N}}{2l_\mathrm{sf}^\mathrm{N}}\right) + \frac{\gamma^2}{\rho_\mathrm{F}^* l_\mathrm{sf}^\mathrm{F}}\coth\left(\frac{t_\mathrm{F}}{2l_\mathrm{sf}^\mathrm{F}}\right) + \frac{\beta^2}{r_\mathrm{b}^*}\right] / \\ &\left\{\frac{1}{\rho_\mathrm{N}^* l_\mathrm{sf}^\mathrm{N}}\coth\left(\frac{t_\mathrm{N}}{2l_\mathrm{sf}^\mathrm{N}}\right)\frac{1}{\rho_\mathrm{F}^* l_\mathrm{sf}^\mathrm{F}}\coth\left(\frac{t_\mathrm{F}}{2l_\mathrm{sf}^\mathrm{F}}\right) + \right. \\ &\left.\frac{1}{r_\mathrm{b}^*}\left[\frac{1}{\rho_\mathrm{N}^* l_\mathrm{sf}^\mathrm{N}}\coth\left(\frac{t_\mathrm{N}}{2l_\mathrm{sf}^\mathrm{N}}\right) + \frac{1}{\rho_\mathrm{F}^* l_\mathrm{sf}^\mathrm{F}}\coth\left(\frac{t_\mathrm{F}}{2l_\mathrm{sf}^\mathrm{F}}\right)\right]\right\} \end{aligned} \tag{5.4}$$

and

$$r_{\mathrm{SI}}^{(\mathrm{AP})} = \left[\frac{(\beta-\gamma)^2}{\rho_{\mathrm{N}}^* l_{\mathrm{sf}}^{\mathrm{N}}} \tanh\left(\frac{t_{\mathrm{N}}}{2l_{\mathrm{sf}}^{\mathrm{N}}}\right) + \frac{\gamma^2}{\rho_{\mathrm{F}}^* l_{\mathrm{sf}}^{\mathrm{F}}} \coth\left(\frac{t_{\mathrm{F}}}{2l_{\mathrm{sf}}^{\mathrm{F}}}\right) + \frac{\beta^2}{r_{\mathrm{b}}^*}\right] / \left\{\frac{1}{\rho_{\mathrm{N}}^* l_{\mathrm{sf}}^{\mathrm{N}}} \tanh\left(\frac{t_{\mathrm{N}}}{2l_{\mathrm{sf}}^{\mathrm{N}}}\right) \frac{1}{\rho_{\mathrm{F}}^* l_{\mathrm{sf}}^{\mathrm{F}}} \coth\left(\frac{t_{\mathrm{F}}}{2l_{\mathrm{sf}}^{\mathrm{F}}}\right) + \frac{1}{r_{\mathrm{b}}^*}\left[\frac{1}{\rho_{\mathrm{N}}^* l_{\mathrm{sf}}^{\mathrm{N}}} \tanh\left(\frac{t_{\mathrm{N}}}{2l_{\mathrm{sf}}^{\mathrm{N}}}\right) + \frac{1}{\rho_{\mathrm{F}}^* l_{\mathrm{sf}}^{\mathrm{F}}} \coth\left(\frac{t_{\mathrm{F}}}{2l_{\mathrm{sf}}^{\mathrm{F}}}\right)\right]\right\}, \tag{5.5}$$

where both bulk and interfacial spin-dependent scatterings are taken into account, introducing a bulk spin-asymmetry coefficient β (obviously $\beta_{\mathrm{F}} \neq 0$ and $\beta_{\mathrm{N}} = 0$) in the expressions of the resistivities: $\rho_{\uparrow(\downarrow)}^{\mathrm{N,F}} = 2\rho_{\mathrm{N,F}}^*[1-(+)\beta_{\mathrm{N,F}}]$, and an interfacial spin-asymmetry coefficient γ for the boundary resistance at the interfaces: $r_{\uparrow(\downarrow)} = 2r_{\mathrm{b}}^*[1-(+)\gamma_{\mathrm{N,F}}]$. Here the quantities marked by $*$ are implicitly defined. They are the following functions of the measurable resistivities $\rho^{\mathrm{N,F}}$ and the interface resistance per unit surface r_{b}: $\rho_{\mathrm{N,F}}^* = \rho^{\mathrm{N,F}}/(1-\beta^2)$ and $r_{\mathrm{b}}^* = r_{\mathrm{b}}/(1-\gamma^2)$. The “average” spin-flip diffusion length which appears, $l_{\mathrm{sf}}^{\mathrm{N,F}}$, is defined by $(l_{\mathrm{sf}}^{\mathrm{N,F}})^{-2} = (l_{\uparrow}^{\mathrm{N,F}})^{-2} + (l_{\downarrow}^{\mathrm{N,F}})^{-2}$ as already mentioned by van Son et al. [35]. Equations (5.4) and (5.5) have a clear physical meaning if one considers their asymptotic behaviors at large and small values of $t = t_{\mathrm{N}} = t_{\mathrm{F}}$. In the limit of layer thicknesses much larger than the spin-flip diffusion lengths, both $r_{\mathrm{SI}}^{(\mathrm{P})}$ and $r_{\mathrm{SI}}^{(\mathrm{AP})}$ reduce to $2r_{\mathrm{SI}}^0$, two times the spin-coupled interface resistance of an isolated F/N interface given in [36]. This is not surprising, because the perturbation at successive interfaces no longer interferes; and, as one expects, CPP–MR vanishes as $\exp(-t/2l_{\mathrm{sf}})$ in this limit. On the contrary, in the limit $t \ll l_{\mathrm{sf}}$, where the spin is conserved over a period, one recovers the prediction of a two-current model in the limit of independent channels being in parallel. In this case there is no longer any special relation with the situation of an isolated F/N interface. In fact, the whole picture of spin-coupled interface resistance becomes irrelevant in the limit $t \ll l_{\mathrm{sf}}$, and one writes more conveniently the bilayer resistance as

$$r^{(\mathrm{P,AP})} = \left(\frac{1}{r_{\uparrow}^{(\mathrm{P,AP})}} + \frac{1}{r_{\downarrow}^{(\mathrm{P,AP})}}\right)^{-1}, \tag{5.6}$$

with

$$r_{\uparrow(\downarrow)}^{(\mathrm{P})} = 2\rho_{\mathrm{F}}^*[1-(+)\beta]t_{\mathrm{F}} + 2\rho_{\mathrm{N}}^* t_{\mathrm{N}} + 4r_{\mathrm{b}}^*[1-(+)\gamma] \tag{5.7}$$

and

$$r_{\uparrow(\downarrow)}^{(\mathrm{AP})} = \frac{r_{\uparrow}^{(\mathrm{P})} + r_{\downarrow}^{(\mathrm{P})}}{2}. \tag{5.8}$$

This type of expressions, rigorously derived for the more general CPP case, was first assumed by the Michigan State University group [37,38] to account for their data. From their estimate of the electron mean free paths, $l_{sf} \approx 10\lambda$, there is no doubt that $t \ll l_{sf}$ is effectively the relevant limit to consider. A detailed discussion on (5.6)–(5.8) in relation with the existing experimental results will be given in Sect. 5.1.2.

One should here emphasize the appealing ability of CPP–MR experiments to clearly separate surface and bulk spin-dependent scatterings contributions. Indeed, from (5.6)–(5.8), one obtains

$$\sqrt{(R^{(AP)} - R^{(P)})R^{(AP)}} = \beta \frac{t_F}{t_F + t_N} \rho_F^* L + 2\gamma r_b^* N \,, \tag{5.9}$$

where $R^{(i)} = Nr^{(i)}$ is the total areal resistance of a multilayer of the total thickness $L = N(t_F + t_N)$ and the total number of bilayers N, in magnetic configuration i. Equation (5.9) has the great advantage of a right-hand side made up of additive contributions from bulk and interface spin-dependent scatterings. The plot of $\sqrt{(R^{(AP)} - R^{(P)})R^{(AP)}}$ versus N *at constant* L is thus of great interest to identify where the spin scattering comes from; this was first done by Lee et al. [37] and Pratt et al. [38]. After independent measurements of ρ_F^* and r_b^*, the intercept with the vertical axis gives β, while the slope is characteristic of γ.

5.1.2 Experiments Based on the Superconducting Contact Technique

One way of measuring CPP–MR of magnetic multilayers is to fabricate samples of a macroscopic cross section – thereby avoiding complicated microfabrication techniques – and then utilze a very sensitive voltage measuring technique. This is because the perpendicular sample resistance is so extremely small for these samples. Such technique employing a sensitive SQUID-based system, capable of measuring small resistances with nΩ resolution, was pioneered by the Michigan State University group and has become a powerful tool for investigating spin-dependent scattering processes in magnetic multilayers.

Experimental Method. The geometry of the samples is exhibited in Fig. 5.1 [39,40]. First a bottom strip of Nb (typical thickness: 0.5 µm, typical width: 1 mm) is sputtered through a mechanical mask onto a substrate. Then, the multilayer sample (typical thickness: 0.5 – 1.0 µm) is sputtered in situ through a second mask and, finally, the top Nb strip is laid down through a third mask. To measure the CPP resistance, the sample is cooled down to 4.2 K, where the Nb is superconducting, and the current is passed through one leg of the lower superconducting strip along the path of least resistance, through the small square (area: 1.2 mm^2) where the Nb strips overlap, and out of one leg of the top Nb strip (e.g., via the path e–h). Due to the superconductivity of the Nb contacts, a uniform perpendicular current–density distribution over the square area is realized. The sample's length/width ratio

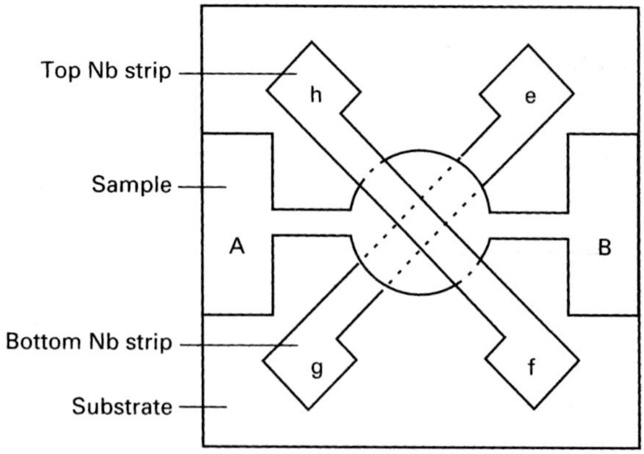

Fig. 5.1. Sample geometry. The CPP–MR is measured with the current injected at e and removed at h; the voltage is measured between g and f. The CIP–MR is measured by applying current between A and B and measuring the voltage between A and B [39]

of $\approx (1\,\mu\text{m})/(1\,\text{mm}) = 10^{-3}$ makes fringing currents negligible. The voltage is measured along the two remaining other legs of the Nb strips (g and f), and because there is, of course, no voltage drop across the Nb strips in their superconducting state, one can measure the perpendicular resistance of the multilayer only, which is of the order of $0.1 - 1\,\mu\Omega$. For typical measuring currents of 50 mA, the voltage to be measured is of the order of 10 nV, which is very hard to detect with a conventional voltmeter. Therefore, a SQUID-based measuring technique was developed, in which the sample is placed in one branch of a bridge configuration and the resistance is recorded by employing the SQUID as a very sensitive null detector. To protect the sensitive SQUID voltmeter from the applied magnetic field, it it necessary to have ferromagnetic shielding. Using Nb leads as sample contacts, it has been possible to measure MR in fields up to 2 T; for still higher fields it is possible to utilize a Nb–Ti alloy for the contacting material, which has proven to preserve its superconducting properties up to fields of 3.5 T.

With this technique MR of nearly any multilayer system can be investigated at 4.2 K. Moreover, the technique has the advantage of a relatively simple sample fabrication, allowing to study, within a reasonable time, systematically various multilayer systems. The following systems have been studied: Ag/Co [37–43], Cu/Co [40,43,44], Cu/NiFe [44], Ag/Ni [45], Co/Cu/NiFe/Cu [32,46], AgSn/Co [30,31,37,38] and AgMn/Co, CuMn/Co, AgPt/Co, CuPt/Co [30,31]. In the latter series a few atomic percent of the metals Mn or Pt were dissolved in the normal metals Cu and Ag, just like it was done with Sn for the AgSn/Co system. All samples were prepared in a cryopumped ultrahigh-vacuum-compatible system with a base pressure before sputtering of $< 10^{-8}$ mbar and a sputtering pressure of 2.5×10^{-3} mbar.

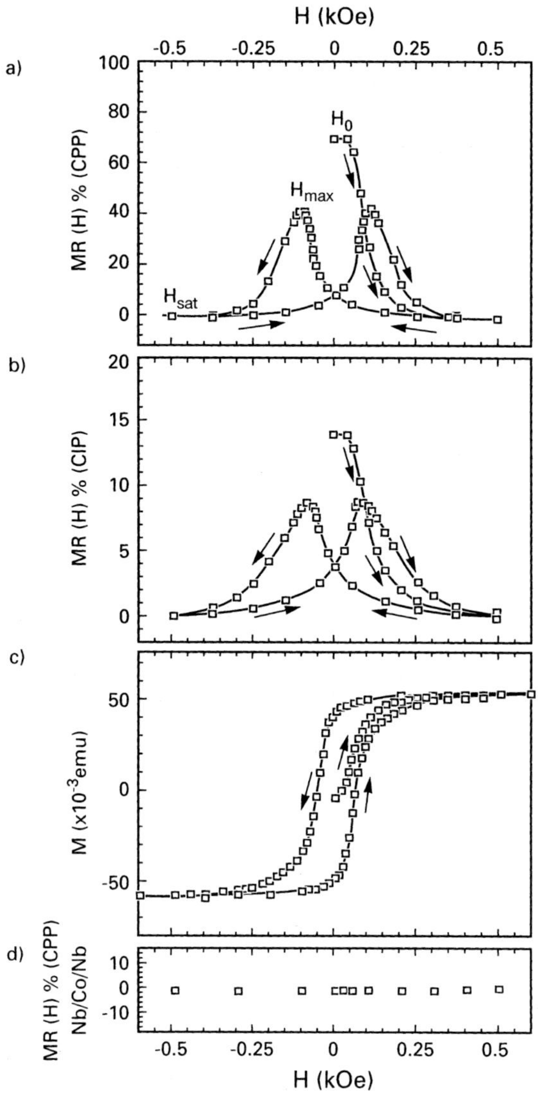

Fig. 5.2. (**a**) CPP–MR data for a 60 × (6 nm Ag + 6 nm Co) multilayer. The fields H_0, $H_{\rm max}$ and $H_{\rm sat}$ are defined in the text. (**b**) CIP–MR data for the same multilayer. (**c**) Magnetization for the same multilayer. (**d**) CPP–MR for a Nb/9 nm Co/Nb test sample [39]

The substrates were sapphire for the Ag-based systems and silicon, with or without a 5 nm buffer layer of Fe before growth of the multilayer, for the Cu-based systems. Samples showed predominantly a (111) texture. CIP resistance measurements were made on the thin horizontal strips of the sample, situated between the contacts A and B in Fig. 5.1, and are used for comparison with CPP–MR.

In addition, MBE-grown Cu/Co multilayers were investigated using the superconducting contacting technique by a group at the University of Leeds [47]. For these samples, the lower electrical contact was made of Nb, grown at 950°C on (1120) sapphire with a (110) bcc orientation. The subsequent Cu/Co multilayer was epitaxially grown at 20°C and has a (111) fcc orientation. Finally, a top Nb contact is grown at 20°C. The CPP–MR results of these MBE-grown layers are essentially in agreement with the results found for the sputtered samples.

The first experiments on CPP–MR were obtained for Ag/Co multilayers, characterized by relatively low saturation fields ($H_{\text{sat}} < 1\,\text{kG}$) for MR, enabling the application of relatively small magnetic fields [39]. In Fig. 5.2 a and b the CPP– and CIP–MR results are shown for a $60 \times (6\,\text{nm Ag} + 6\,\text{nm Co})$ multilayer. The field-dependent magnetoresistance ratio is defined as

$$MR(H) = \frac{R(H) - R(H_{\text{sat}})}{R(H_{\text{sat}})} . \tag{5.10}$$

CPP–MR is about a factor of 5 larger than CIP–MR. For both cases, MR is a maximum at H_0, a zero applied field for the as-prepared sample before exposure to any magnetic field cycle. Then, the field is increased until the resistance saturates at the field H_{sat}, which for the present sample is at 0.5 kG. When cycling the applied magnetic field to opposite polarity, a strong hysteretic behavior is present with a maximum in resistance at the fields H_{max}, which is appreciably smaller than the first resistance maximum at H_0. As is clear from the magnetization curves in Fig. 5.2 c, the total magnetization is $M = 0$ at both H_0 and H_{max}. However, the detailed relative orientations of the individual magnetizations of the Co layers is not exactly known; the MR experiments indicate that they, indeed, are different for H_0 and H_{max}. For the interpretation of the experiments, it was assumed that, at H_{max}, the magnetizations are randomly oriented in the layer plane, while, at H_0, the ordering is nearly antiparallel along a single axis. The latter is supported by the analysis of sample series with both strongly and weakly or uncoupled samples (i.e., with thin and thick normal-metal spacer layers). The maximum resistances observed at H_0 for hysteretic samples, indeed, show a continuous behavior as a function of normal-metal layer thickness to the maximum resistance of a truely antiferromagnetically coupled sample [40]. The resistance value measured at H_0 therefore seems the most useful choice for the calculation of the MR effect, although details of the magnetic structure of the sample are still unknown.

As this CPP measuring technique is inherently "two-terminal", it was checked whether the Nb/Co boundaries of the multilayers contributed noticeably to the CPP–MR by separately measuring MR of Nb/Co/Nb sandwiches in the CPP-geometry. As shown in Fig. 5.2 d, the resulting MR was of the order of the measuring uncertainty ±0.5%. The sandwich data reveal that the contributions to the multilayer CPP–MR are small and that isolated single Co layers do not produce significant CPP–MR.

The great strength of CPP–MR experiments with the superconducting contacting technique is the relatively simple sample preparation: a lot of samples can be made in a short time and, hence, it is well possible to vary parameters like the magnetic layer thickness t_{F}, the normal-metal layer thickness t_{N} or the number of magnetic/nonmagnetic bilayers in the sample N. Also switching to different material systems provides no problem. Systematic studies are possible and fundamental spin-dependent scattering parameters can be determined.

Theorectical Analysis of the Data. To analyze the experimental data, a "two-channel" series-resistor model was assumed, involving the field dependence of $AR_{\mathrm{T}}(H)$, where A is the experimental cross section area and $R_{\mathrm{T}}(H)$ the field dependent total resistance of the sample. This quantity has the dimension of a resistivity times a thickness. There are basically two contributions to the field-dependence of $AR_{\mathrm{T}}(H)$: one proportional to the magnetic-layer resistivity $\rho_{\mathrm{F}}(H)$ and one proportional to the ferromagnet normal-metal interface $R_{\mathrm{F/N}}(H)$. Based on the theoretical work of Zhang and Levy [22], the Michigan State University group showed that the two-channel model, in which electrons were divided into spin-up (+) and spin-down (−) channels which carry the current independently (i.e., no spin-flip scattering is present), could well explain their data [37,38]. This two-channel model was later given a microscopic basis and justified by Camblong [23] and by Valet and Fert [28,6] (the latter model was explicitly discussed in Sect. 5.1.1.

The resistivity of a ferromagnetic layer in the case that the electron spin and the local magnetization are parallel to each other, is defined as $\rho_{\mathrm{F}}^{\uparrow} = 2\rho_{\mathrm{F}}/(1+\beta)$. $\rho_{\mathrm{F}}^{\downarrow} = 2\rho_{\mathrm{F}}/(1-\beta)$ is the resistivity when spin and local magnetizations are antiparallel, where the notation introduced by Valet and Fert [6] is adopted. ρ_{F} is the experimentally measured bulk resistivity of a ferromagnetic Co film. Just like for the bulk of the magnetic layer, also for the ferromagnet/normal metal interface, spin-dependent resistances can be defined: $R_{\mathrm{F/N}}^{\uparrow} = 2R_{\mathrm{F/N}}/(1+\gamma)$ and $R_{\mathrm{F/N}}^{\downarrow} = 2R_{\mathrm{F/N}}/(1-\gamma)$. The important spin-orientation-dependent scattering parameters are then defined as $\alpha_{\mathrm{F}} = \rho_{\mathrm{F}}^{\downarrow}/\rho_{\mathrm{F}}^{\uparrow} = (1+\beta)/(1-\beta)$ and $\alpha_{\mathrm{F/N}} = R_{\mathrm{F/N}}^{\downarrow}/R_{\mathrm{F/N}}^{\uparrow} = (1+\gamma)/(1-\gamma)$. It is assumed that the injection of the electrons from the Nb superconductor is independent of the spin, and that spin-flip scattering is absent, i.e., an electron preserves the same spin when passing through the multilayer.

At H_0, the assumed antiparallel alignment of the magnetizations results in an AR_{T} value, which is half of the equal AR_{T} values for + spin and − spin channels, since the resistances of these channels add in parallel. As the number of bilayers N in a sample generally is rather large, one can neglect differences between N, $(N-1)$ and $(N+1)$ in writing the equations. For a *fixed total multilayer thickness* $t = N(t_{\mathrm{N}} + t_{\mathrm{F}})$ – experimentally a value

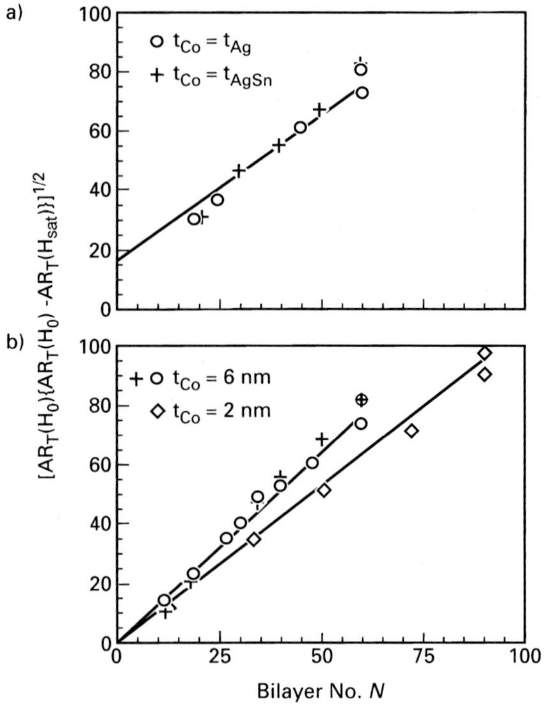

Fig. 5.3. (**a, b**) $A\sqrt{R_T(H_0)[R_T(H_0) - R_T(H_{sat})]}$ versus N for Co/Ag and Co/AgSn multilayers with a fixed total thickness of 0.72 µm and (**a**) $t_{Co} = t_{Ag} = t_{AgSn}$ and (**b**) a fixed Co thickness of 6 nm and 2 nm, respectively. Solid lines are best fits using the parameters of table 5.1 [43]

$t = 0.72\,\mu m$ is most often realized – the following equation linear in N is obtained for samples with a fixed Co thickness (t_F is constant):

$$AR_T(H_0) = 2AR_{S/F} + N(\rho_N t_N + \rho_F^* t_F + 2AR_{F/N}^*)\,, \tag{5.11}$$

with $\rho_F^* = \rho_F/(1-\beta^2)$ and $R_{F/N}^* = R_{F/N}/(1-\gamma^2)$. Note that the quantity $AR_{F/N}^*$ corresponds to the quantity r_b^* of the Valet–Fert theory.

Equation (5.11) predicts that a plot of $AR_T(H_0)$ versus N for a fixed Co thickness t_F should give a straight line with intercept $(2AR_{S/F} + \rho_N t)$ and slope $(\rho_F^* - \rho_N)t_F + 2AR_{F/N}^*$.

At H_{sat}, one adds the resistances of spin + and spin – electrons in parallel and gets

$$AR_T(H_{sat}) = AR_T(H_0) - N^2 \frac{[\beta\rho_F^* t_F + 2\gamma AR_{F/N}^*]}{[AR_T(H_0)]}\,. \tag{5.12}$$

Equation (5.12) contains a term in N^2, that deviates from zero when β or γ deviates from zero. For the cases of interest, $AR_T(H_0)$ will have a constant

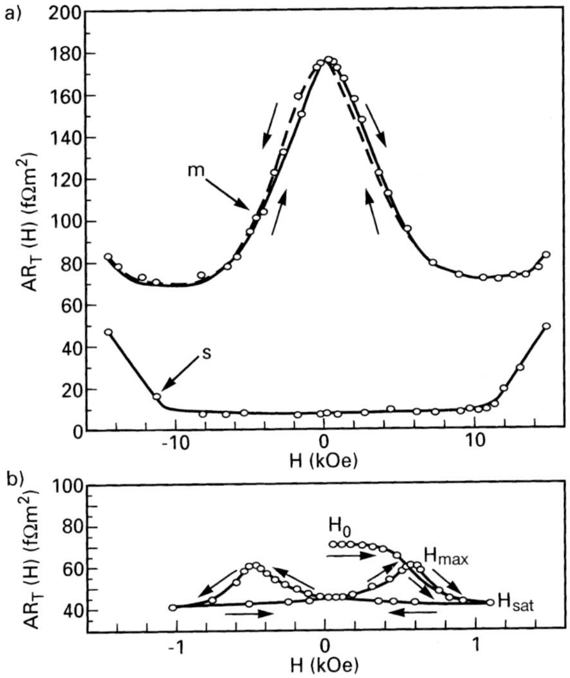

Fig. 5.4. (**a**) $AR_T(H)$ curve for an antiferromagnetically coupled Co/Cu multilayer ($t_{Co} = 1.5\,\text{nm}$, $t_{Cu} = 0.9\,\text{nm}$, with 5 nm Fe buffer layer, denoted by m). The $AR_T(H)$ curve for a Nb/Co/Nb sandwich ($t_{Co} = 9\,\text{nm}$) is denoted by s. (**b**) $AR_T(H)$ curve for an uncoupled Co/Cu multilayer ($t_{Co} = 1.5\,\text{nm}$, $t_{Cu} = 4.0\,\text{nm}$) [43]

part and a term which is proportional to N. Hence $AR_\text{T}(H_\text{sat})$ can be non-linear in N, depending on the relative magnitude of the different resistances in the problem. The constant part of $AR_\text{T}(H_0)$ contains the quantity ρ_N. For Ag/Co multilayers, the combination of a small Ag resistivity ($\approx 1\,\mu\Omega\,\text{cm}$) with a small value for $2AR_{\text{S/F}} \approx 6\,\text{f}\,\Omega\,\text{m}^2$ leaves $AR_\text{T}(H_0)$ dominated by a term proportional to N. The predictions of the two-channel model will then still yield linear plots in N for $AR_\text{T}(H_\text{sat})$. The situation is different for the AgSn/Co-system, characterized by a much higher resistivity of the AgSn ($\approx 18\,\mu\Omega\,\text{cm}$), which makes the constant term in $AR_\text{T}(H_0)$ more dominant. This situation leads to a curvature in $AR_\text{T}(H_\text{sat})$ versus N.

From (5.12) one finds that a very useful way of plotting the results is according to

$$A\sqrt{R_\text{T}(H_0)[R_\text{T}(H_0) - R_\text{T}(H_\text{sat})]} = N(\beta\rho_\text{F}^* t_\text{F} + 2\gamma AR_{\text{F/N}}^*)\,, \tag{5.13}$$

which is the expression to be used when t_F is fixed. In the case that $t_\text{F} = t_\text{N}$, (5.13) becomes

$$A\sqrt{R_\text{T}(H_0)[R_\text{T}(H_0) - R_\text{T}(H_\text{sat})]} = \frac{(\beta\rho_\text{F}^* t)}{2} + N[2\gamma AR_{\text{F/N}}^*]\,. \tag{5.14}$$

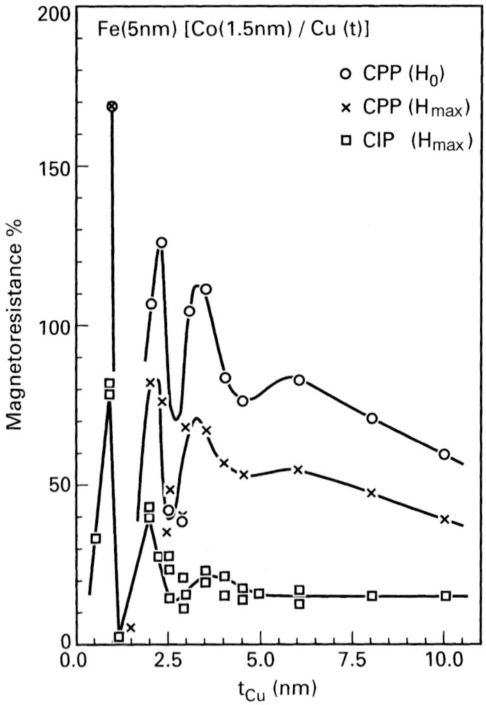

Fig. 5.5. CPP–MR at the fields H_0 and $H_{\max}$ and CIP–MR at the field H_0 versus Cu-layer thickness t_{Cu} for Co(1.5 nm)/Cu(t_{Cu}) multilayers, grown on 5 nm Fe buffer layers [44]

These equations are the same as those derived from the microscopic theory, i.e., (5.9). Lee et al. [37] and Pratt et al. [38] showed that (5.13) and (5.14) are especially useful for testing the model, since simple straight lines are predicted for the dependence on N. The other remarkable fact is that in the quantity on the right-hand side, there is no term proportional to ρ_{N} present. This means that the square-root function only contains information about the magnetic layers and interfaces, and is not influenced by the properties of the normal-metal layer. This was, indeed, confirmed by experiments ([37,38]) Fig. 5.3 a, where data are plotted for Ag/Co and AgSn/Co multilayers, for the case that $t_{\mathrm{Co}} = t_{\mathrm{Ag}}$ and $t_{\mathrm{Co}} = t_{\mathrm{AgSn}}$, respectively. The data indeed lie on a single straight line, though the values of ρ_{N} are markedly different for both data sets. Figure 5.3 b is a plot for two data sets with a constant Co thickness according to (5.13), showing the linear behavior in N and the passing through the origin, just as predicted by the model. In conclusion, the two-channel model has proven to be adequate in describing the experimental CPP–MR data, and thereby has led to considerable insight into the important spin-dependent scattering parameters of the multilayer.

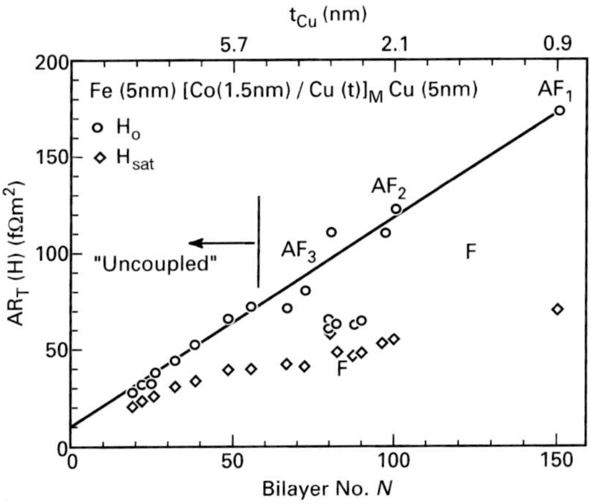

Fig. 5.6. $AR_{\mathrm{T}}(H_0)$ and $AR_{\mathrm{T}}(H_{\mathrm{sat}})$ versus bilayer number N for Co(1.5 nm)/Cu(t_{Cu}) multilayers, grown on 5 nm Fe buffer layers with a fixed total thickness. The data demonstrate the continuous behavior from uncoupled (= small$N$) to antiferromagnetically (AF) coupled (= large$N$) multilayers [44]

As emphasized before, a lot of different multilayer systems were studied with the superconducting contacting technique. Among them is the Co/Cu(111) system. Figure 5.4 a, (*upper curve*) exhibits the highest CPP–MR for a magnetic multilayer reported yet: for a $150 \times (1.5\,\mathrm{nm\,Co} + 0.9\,\mathrm{nm\,Cu})$ multilayer grown on a 5 nmFe buffer layer [48] a CPP–MR of 170% was reported [43]. For this multilayer, the small Cu thickness leads to strong antiferromagnetic coupling between adjacent Co layers at $H = 0$, by which hysteresis due to field cycling is reduced to a minimum. The lower curve of Fig. 5.4 a is for a Nb/Co/Nb sandwich and serves as a control experiment. It can be seen that the Nb/Co boundary resistance only gives a contribution of about $5\,\mathrm{f}\Omega\,\mathrm{m}^2$ to the measured resistance. The upturn in $AR_{\mathrm{T}}(H)$ for the samples somewhat above 10 kOe is due to magnetic flux penetration in the Nb superconducting contact leads to the sample and limits the applicable field range of the experiment. However, using superconducting NbTi contact electrodes it is possible to extend the experimental field range to about 35 kOe [49]. Figure 5.4 b illustrates CPP–MR measured for a $130 \times (1.5\,\mathrm{nmCo} + 4.0\,\mathrm{nmCu})$ multilayer and has a similar behavior as seen before for the uncoupled Ag/Co layers. It should be noted that the field scale in b is a factor ten smaller than in a. Figure 5.5 exhibits MR of a series of multilayers, grown on a 5 nm buffer layer with a fixed Co thickness $t_{\mathrm{Co}} = 1.5\,\mathrm{nm}$ and as a function of Cu thickness t_{Cu} [44]. Data are shown for CPP–MR, first by taking the resistance at H_0 as the maximum value in the expression for the magnetoresistance according to (5.10) and secondly by taking the value at H_{max}. For comparison CIP–MR data, taking the maximum resistance at H_{max}, are presented. As already observed for the Ag/Co system, also for Cu/Co CPP–MR is larger

Table 5.1. Fitting parameters for the two channel model applied to different multilayer systems

	Ag/Co	AgSn/Co	Cu/Co	Cu/NiFe
ρ_N [μΩ cm]	1.0 ± 0.3	18.5 ± 1.0	0.78 ± 0.02	
ρ_F^* [μΩ cm]	10.7 ± 1.0	7	7.5 ± 0.4	
$AR^*_{F/N}$ [fΩ cm^2]	0.56 ± 0.03	0.67 ± 0.06	0.50 ± 0.03	
β	0.48 ± 0.05	0.41 ± 0.12	0.47 ± 0.14	(0.52)
γ	0.85 ± 0.03	0.83 ± 0.06	0.72 ± 0.06	(0.69)
α_F	2.9 ± 0.4	2.4 ± 0.8	2.8 ± 0.8	(3.2)
$\alpha_{F/N}$	12 ± 4	11 ± 5	6.1 ± 1.5	(5.5)
Ref.	[38]	[38]	[44]	[44]

than CIP–MR. In the region $t_{Cu} < 6$ nm, the oscillations in MR are characteristic for a system, changing its coupling from strong antiparallel to parallel, and vice versa. As was already shown for the Ag/Co samples, data can be analyzed with the two-channel model. Therefore, samples with a fixed total layer thickness and with a constant Co thickness can be plotted versus N (Fig. 5.6). An important point of this figure is that antiferromagnetically coupled samples (denoted by AF_1, AF_2 and AF_3, for the first, second and third antiferromagnetic coupling peaks, respectively) are on the *same* line as the resistance data taken at H_0 for uncoupled samples. This behavior indeed suggests that in the uncoupled region the magnetizations at H_0 behave as if they were antiferromagnetically aligned.

A summary of the fitting parameters obtained by the Michigan State University group from a systematic comparison of the two-channel model with experimental data is presented in Table 5.1. ρ_N is around 1 μΩ cm for the noble metals Ag and Cu and, due to the alloying, ρ_N is considerably higher for the AgSn system. The resistivity of the Co can be determined as $\rho_{Co} = \rho^*_{Co}(1-\beta^2)$, so from the Ag/Co system, one finds that $\rho_{Co} = 8.2$ μΩ cm and from the Cu/Co system that $\rho_{Co} = 5.8$ μΩ cm: The difference may be due to different growth conditions for the two multilayer systems. For the AgSn/Co system, the value for ρ_F^* was just assumed in the fit. For the latter multilayer system, a few different parameter sets were able to fit the experimental data accurately [38]; only one set of these, consistent with the parameters of the Ag/Co, is given in Table 5.1. The other fits contain parameters which are considerably different from the Ag/Co reference samples. From the values of $AR^*_{F/N} \approx 0.6$ fΩ cm^2, we see that the total resistance of the multilayer is dominated by interface scattering: the resistance of one interface is of the order of a 10 nm thick bulk layer with a resistivity of 6 μΩ cm. The parameters β and γ, and α_F and $\alpha_{F/N}$ really are the important fitting parameters which fundamentally describe the spin-polarized transport in the multilayer. For bulk Co, values of about 3 are found for α_F, while, for interface, scattering the spin dependence is considerably larger, especially for the Ag-based systems. The observation that it is the interface which is the dominant source of spin-dependent scattering, is an important result of these experiments. The parameters quoted for the NiFe/Cu systems are placed in brackets, because they were determined from only a limited number of samples [44].

Not only "simple" multilayer systems (containing just one type of magnetic layer) were measured and analyzed with the two-current model, but also more complicated structures with two *different* magnetic materials in the multilayer unit cell (Co and $Ni_{84}Fe_{16}$) were investigated [46,32]. In such systems an antiparallel orientation between the magnetizations of neighboring magnetic layers can easily be obtained due to the difference in magnetic hardness of the constituents. Of course, (5.11) and (5.12) of the two-current model will become more complicated, as two kind of interface resistances and two different bulk resistivities of the magnetic layer now play a role. However, it is possible to extract the spin-dependent scattering parameters via constraint fits of the CPP–MR experimental data, by systematically varying the individual layer thicknesses and the number of bilayers. An important result is that the spin-dependent scattering parameters obtained from such a constraint fit are essentially the same as the values obtained from the fits of the "simple" multilayer data with the two-current model [46]. This analysis demonstrates that it is possible to *predict* the spin-dependent scattering properties and the magnetoresistance of complicated multilayer systems, if one knows the scattering parameters of the simpler system. This is a strong demonstration of the validity of the two-current model for the interpretation of CPP–MR experiments. The agreement also supports the argument that, in the H_0 state, resistances of the uncoupled Co/Cu and NiFe/Cu multilayers are close to those for perfectly antiparallel ferromagnetic layer alignment.

So far, in the analysis it was assumed that an electron conserves its spin, when moving through the multilayer, i.e., there is no spin-flip scattering. This assumption is valid at low temperatures for pure-metal systems, where the spin-flip diffusion lengths l_{sf} of a few 100 nm are realized [6], while individual layer thicknesses are only a few nm. It was shown that l_{sf} is the characteristic length scale of the CPP–MR experiment, just like the elastic scattering length λ is the length scale of CIP–MR [28,6,23,50]. More specifically, l_{sf} is the length scale for the relaxation of the spin accumulation, which is a consequence of a current applied perpendicular to the multilayer plane. When l_{sf} is large, conventional transport theory holds in the two-current model [6,22,23,28,50], and the analysis of experimental data is particularly simple, as was shown before. An interesting problem is what happens with CPP–MR, when l_{sf} becomes of the order of the multilayer period. Stated reversely, the question is if one can study l_{sf} by measuring CPP–MR for such systems. Experimentally

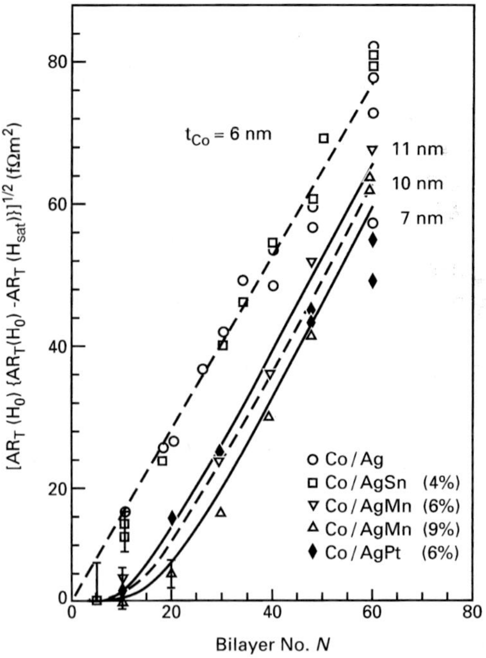

Fig. 5.7. $A\sqrt{R_T(H_0)[R_T(H_0) - R_T(H_{sat}]}$ versus N for Co/Ag, Co/AgSn (4 at.%), Co/AgMn (6 at.%), Co/AgMn (9 at.%) and Co/AgPt (6 at.%) multilayers with fixed $t_{Co} = 6$ nm. *Dashed* and *solid curves* were calculated with the Valet–Fert model. The *dashed line* through the origin is for $l_{sf} = \infty$. The best-fit values for l_{sf} for the other curves are indicated [30]

l_{sf} can be reduced by increasing the temperature, thereby enhancing the electron scattering by thermally excited magnons. This will be discussed in Sect. 5.1.1. Or, at a low temperature, a reduction of l_{sf} can be realized by adding a strong spin-orbit interaction via a scattering impurity such as Pt, or by adding a strong spin-spin interaction by introducing an impurity with a magnetic moment, such as Mn in Ag or in Cu. The Michigan State University group has reported a series of experiments where both types of reduction of l_{sf} in the normal-metal layers of the multilayer were applied; more specifically, multilayer systems with Co as the magnetic layer and AgSn (4 at.%), AgMn (6 at.%), AgMn (9 at.%), AgPt (6 at.%), CuMn (7 at.%) and CuPt (6 at.%) as the "normal"-metal layers were studied [30,31]. The influence of shortening l_{sf} on AR_T, when the magnetizations of the multilayer are both in the parallel and antiparallel states, was explicitly modeled by Valet and Fert [6,28]. Their general result was already given in (5.3)–(5.5). These equations contain two spin-flip diffusion lengths, one for the ferromagnetic metal l_{sf}^F and one for the normal metal l_{sf}^N. As the expected influence of shortening l_{sf}^N is larger than of shortening $l_{\mathrm{sf}}^{\mathrm{F}}$ [31], the experiments focused on modifying the normal-metal layer.

Quantitative conclusions on the influence of shortening $l_{\mathrm{sf}}^{\mathrm{N}}$ on CPP–MR can be obtained by studying complete sample series, as was done for the pure-metal multilayer systems. A constant total sample thickness of $t = 720\,\mathrm{nm}$ was chosen, and the number of bilayers N was varied for a fixed magnetic-layer thickness $t_{\mathrm{F}} = 6\,\mathrm{nm}$. To isolate the effect of a reduced $l_{\mathrm{sf}}^{\mathrm{N}}$, the square-root quantity of (5.13) is investigated. Square-root data for several Ag-based multilayer systems are shown in Fig. 5.7. The Co/Ag and Co/AgSn (4 at.%) data are consistent with a single straight line through the origin, as predicted by (5.13). In fact, this result was already indicated in Fig. 5.3 b and demonstrated that the normal-metal resistivity does not influence the square-root quantity, when $l_{\mathrm{sf}}^{\mathrm{N}}$ is sufficiently large. When $l_{\mathrm{sf}}^{\mathrm{N}}$ becomes of the order of the multilayer period, the square root no longer has the simple form of (5.13), but is considerably more complicated because the full equations of the Valet–Fert theory have to be used, i.e., (5.3)–(5.5). The result of such a calculation is that the square-root data are below the straight line through the origin, and that the deviation is larger for the smaller $l_{\mathrm{sf}}^{\mathrm{N}}$ values. The experimental data for the normal metal layers with magnetic Mn impurities and the strong spin-orbit Pt scatterers indeed are below the straight line, the relative deviation being larger for small N (i.e., large-normal layer thickness, when the total sample thickness remains fixed). The curves of Fig. 5.7 (solid lines for the Co/AgMn and dashed lines for the Co/AgPt) were calculated from the Valet–Fert model and are best fits, from which values for $l_{\mathrm{sf}}^{\mathrm{N}}$ can be deduced. The latter are listed in the figure and indeed are of the order of the multilayer period: They are in good agreement with independent estimates obtained from electron-spin resonance and weak-localization experiments [30]. One should note that these spin-flip diffusion lengths are of the order of the elastic scattering lengths in the multilayer, while the Valet–Fert theory was established in the limit $\lambda_{\mathrm{el}} \ll l_{\mathrm{sf}}^{\mathrm{N}}$. The good fitting of the results might indicate that theory has some broader applicability.

In summarizing, combining the superconducting contacting technique with the appropriate theory provides a tool for investigating various electron-scattering lengths in various material systems at low temperatures. Thereby this experimental method has led to a good insight into the origin of the spin-dependent scattering, leading to the GMR effect, and has allowed to determine quantitatively all relevant scattering parameters.

5.2 Perpendicular Magnetoresistance up to Room Temperature: Exploring Routes to Practical Devices

5.2.1 The Two-Channel Model at Higher Temperatures

The two-current model for the conduction in ferromagnets [7] and the corresponding models for CIP–MR include the three relaxation rates, $1/\tau_\uparrow$, $1/\tau_\downarrow$ and $1/\tau_{\uparrow\downarrow}$ (and the three associated characteristic lengths or mean free paths, $\lambda_\uparrow$, $\lambda_\downarrow$ and $\lambda_{\uparrow\downarrow}$). The rates $1/\tau_\uparrow$ and $1/\tau_\downarrow$ characterize the relaxation of the momentum to the lattice within each channel, while $1/\tau_{\uparrow\downarrow}$ determines the momentum transfer between the two channels (the so-called spin-mixing effect). However, for the low-temperature limit of the two-current model, the usual assumption is $1/\tau_{\uparrow\downarrow} = 0$ (no spin mixing). Consequently, there are three characteristic lengths, $\lambda_\uparrow$, $\lambda_\downarrow$ and $\lambda_{\uparrow\downarrow}$ at finite temperatures and only two, $\lambda_\uparrow$ and $\lambda_\downarrow$ in the low-temperature limit. In the models for CIP–MR, the assumption $1/\tau_{\uparrow\downarrow} = 0$ has generally been assumed valid at any temperature and only a few models introduce a non zero $1/\tau_{\uparrow\downarrow}$ at finite temperature to describe the temperature dependence of CIP–MR [51].

For CPP–MR the Valet–Fert model discussed so far has been worked out for the low-temperature limit and therefore assumes $1/\tau_{\uparrow\downarrow} = 0$. In addition, to describe the relaxation of the spin accumulations, it introduces another relaxation rate, the spin-flip relaxation rate $1/\tau_{\mathrm{sf}}$. The only scaling length appearing in the final expressions for CPP–MR is the spin-flip diffusion length l_{sf}, which is a function of the characteristic lengths $\lambda_\uparrow$, $\lambda_\downarrow$ and $\lambda_{\mathrm{sf}} = v_{\mathrm{Fermi}}\tau_{\mathrm{sf}}$. At finite temperatures, it is neccessary to introduce the spin-mixing effect (momentum transfer) into the model and, as described below, $\lambda_{\uparrow\downarrow}$ will also come into play with $\lambda_\uparrow$, $\lambda_\downarrow$ and λ_{sf} [33].

As discussed in [33], it is important to mention the difference between the relaxation rates $1/\tau_{\uparrow\downarrow}$ and $1/\tau_{\mathrm{sf}}$. The relaxation rate $1/\tau_{\uparrow\downarrow}$ characterizes the momentum transfer between the two spin channels by spin-flip scattering. *Only the scattering processes which conserve the momentum* contribute to $1/\tau_{\uparrow\downarrow}$. Especially when T increases, *spin-flip scattering by magnons* comes into play to give $1/\tau_{\uparrow\downarrow}$ a nonvanishing value. On the other hand, $1/\tau_{\mathrm{sf}}$ characterizes the relaxation of the out-of-equilibrium magnetization (or spin accumulation). *All the spin-flip scattering processes*, with and without momentum conservation, may contribute to $1/\tau_{\mathrm{sf}}$. At low temperatures, $1/\tau_{\mathrm{sf}}$ includes contributions from *spin-orbit scattering* by impurities or defects and, for non magnetic layers doped with paramagnetic impurities, contributions from *exchange scattering*. As T increases, $1/\tau_{\mathrm{sf}}$ can receive additional contribution from *electron-magnon scattering* in magnetically ordered layers.

The above discussion allows one to classify the effects which must be taken into account to describe the temperature dependence of CPP–MR:

a) At finite temperatures, the intrachannel relaxation rates $1/\tau_\uparrow$ and $1/\tau_\downarrow$ receive an additional contribution from thermally excited scatterings (phonons, magnons). This can be described in terms of an additional contribution to the resistivity of each channel, that is $\rho_\sigma(T) = \rho_\sigma + \rho_{i\sigma}(T)$, where ρ_σ is the low-temperature resistivity and $\rho_{i\sigma}(T)$ a temperature-dependent contribution, originating from electron-phonon or "intrachannel" electron-magnon scattering. This leads to temperature-dependent values of ρ_{F}^* and β: $\rho_{\mathrm{F}}^*(T) = [\rho_\uparrow(T) + \rho_\downarrow(T)]/4$ and $\beta(T) = [\rho_\uparrow(T) - \rho_\downarrow(T)]/[\rho_\uparrow(T) + \rho_\downarrow(T)]$.

b) Electron-magnon scattering not only contributes to $\rho_{i\uparrow}(T)$ and $\rho_{i\downarrow}(T)$, but also to the spin-mixing rate $1/\tau_{\uparrow\downarrow}$, or in terms of resistivity, to the spin-mixing resistivity $\rho_{\uparrow\downarrow}(T)$ [7].

c) Spin-flip electron-magnon scattering could also contribute to the spin-relaxation rate $1/\tau_{\mathrm{sf}}$ and therefore can shorten the spin-flip diffusion length in the magnetic layers.

While it is believed that the latter effect does not provide an additional efficient mechanism of spin-accumulation relaxation [33], the temperature dependence of CPP–MR should be predominantly due to the spin-mixing effect. In the limit of a long spin-flip diffusion length of the Valet–Fert model, the temperature-dependent version of (5.9) is given by [33]

$$\sqrt{(R^{(\mathrm{AP})} - R^{(\mathrm{P})})R^{(\mathrm{AP})}} = N\frac{\beta(T)t_{\mathrm{F}}\rho_{\mathrm{F}}^*(T) + 2\gamma r_{\mathrm{b}}^*}{1 + N\,t_{\mathrm{F}}\rho_{\uparrow\downarrow}(T)/R^{(\mathrm{AP})}}\,. \tag{5.15}$$

Compared with the low-temperature expression, the differences are: ρ_{F}^* and ρ_{N}^* are replaced by $\rho_{\mathrm{F}}^*(T) = \rho_{\mathrm{F}}^* + [\rho_{i\uparrow}(T) + \rho_{i\downarrow}(T)]/4$ and $\rho_{\mathrm{N}}^*(T) = \rho_{\mathrm{N}}^* + [\rho_{i\sigma}(T)]/2$, β is replaced by $\beta(T)$, and the reduction factor $[1 + N\,t_{\mathrm{F}}\rho_{\uparrow\downarrow}(T)/R^{(\mathrm{AP})}]$ expresses the spin-mixing effects.

The variation of $\sqrt{(R^{(\mathrm{AP})} - R^{(\mathrm{P})})R^{(\mathrm{AP})}}$ with T results from several competing effects. The increase of $\rho_{\mathrm{F}}^*(T)$ with T tends to increase it but, if the spin-asymmetry between $\rho_{i\uparrow}(T)$ and $\rho_{i\downarrow}(T)$ is less pronounced than that between the low-temperature resistivities $\rho_\uparrow$ and $\rho_\downarrow$, the resulting decrease of $\beta(T)$ tends toward the opposite direction. Also, the increase of $\rho_{\uparrow\downarrow}(T)$ with T in the denominator tends to decrease the square root. In addition, the variation is no longer linear in N since N is also involved in the denominator.

While (5.15) is explicitly derived for a multilayer in the CPP geometry and gives a description of the temperature dependence of CPP–MR in terms of spin-dependent bulk and interface contributions to the resistance, the original model of Fert and Campbell on the temperature dependence of the resistance of ferromagnetic alloys [7] has also been applied to fit CPP–MR experimental data [52]. This is possible because CPP–MR of a magnetic multilayer is self-averaging [22,6], i.e., for a given spin direction, the resistance can be written as a weighted average of the separate resistances of the constituents of the multilayer, see (5.7). This results in effective spin currents that are constant (but not necessarily equal) through the system, just as assumed in the alloy model (note that this is not the case for the CIP geometry). In the alloy model, the total resistivity as a function of temperature is written as

$$\rho(T) = \frac{\rho_\uparrow(T)\rho_\downarrow(T) + \rho_{\uparrow\downarrow}(T)(\rho_\uparrow(T) + \rho_\downarrow(T))}{\rho_\uparrow(T) + \rho_\downarrow(T) + 4\rho_{\uparrow\downarrow}(T)}\,, \tag{5.16}$$

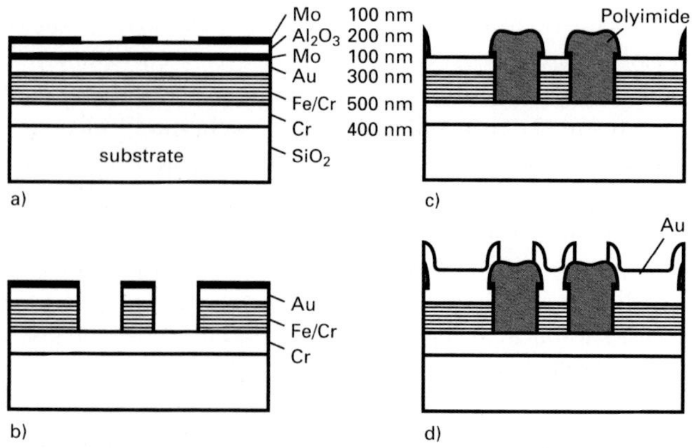

Fig. 5.8. Schematic diagram of different processing steps in the pillar structuring and contact fabrication [54,55]

with the resistivities as defined above. In the limit of small spin mixing (5.16) was used to extract the temperature dependence of the spin-dependent scattering of the two spin channels, however, without discriminating between the contributions of bulk and interface effects, as will be shown below.

5.2.2 Lithographically Defined Pillar Structures

Experimental Method. For studying the CPP–MR effect up to room temperature, other techniques that do not rely on superconducting properties are necessary. Therefore, another approach in the experimental study of the CPP–MR effect in magnetic multilayers relies on lithographic and microfabrication techniques for the preparation of the samples. If "pillar"-like samples with a sufficiently small cross-section could be made so that their resistance is in the mΩ-range or higher, they could be measured with a conventional voltmeter. If optical lithography is utilized, the minimum width of a pillar is of the order of a few micrometers, e-beam lithography can give rise to smaller sample dimensions, but it is somewhat more complicated. Because no longer ultra-low resistances need to be measured, it is possible to replace the superconducting contacts by normal-metal contacts, thereby opening the possibility to study the CPP–MR effect from 4 K up to room temperature. However, this experiment is not at all straightforward, because (i) any – even extremely small – spurious contact resistance easily is of the same order or larger than the perpendicular multilayer resistance to be measured, and (ii) the perpendicular resistance of a microstructured thin film generally is much smaller than the sheet resistance $R_\square$ of the contact leads, so that unwanted resistance contributions from the contacts easily contribute to the experimentally measured resistance. The solution for the first problem is to deposit the different layers of the structure in situ as for as possible and to take extreme care during contact fabrication. To overcome the second problem, a specific contact geometry needs to be chosen, which allows a proper measurement of the perpendicular resistance of a metal film without excessive spurious resistance contributions from the contact leads. Another complication is that, opposite to the experiment on samples with superconducting contacts, the perpendicular current density no longer is distributed uniformly over the cross-section of the pillar, when normal-metal contacts are used and when the height/width ratio of a pillar is much smaller than 1. For a correct interpretation of the experimental data and to understand the scaling of the measured perpendicular resistance with a pillar size, it is necessary to know the current distribution in the samples. The first measurements of the CPP–MR effect in microstructured samples were reported by the Philips–Eindhoven University group for Fe/Cr multilayers [53,54]. Subsequently, also Co/Cu multilayers [55,52] were studied using this technique. Later also CPP–MR experiments on NiFe/Cu/Co/Cu multilayers with different coercivities for the NiFe and Co layers were reported by a group at the Naval Research Laboratory [56,57]. The main advantage of experiments on microstructures is that the CPP–MR effect can be studied at all temperatures and that application of the room-temperature perpendicular magnetoresistance effect eventually becomes possible. In this section, we will discuss as an example the deposition, characterization and microstructuring of the Fe/Cr multilayers, as it has been developed at Philips, and only shortly mention other fabrication technologies for different multilayer systems. The processing of a Fe/Cr pillar sample is schematically shown in Fig. 5.8. First a 0.4 μm thick Cr base layer is rf-sputtered onto a SiO_2 substrate held at room temperature. Then, a $0.4-0.7$ μm thick Fe/Cr multilayer is deposited, using dc sputtering for the Fe and rf sputtering for the Cr, followed by a dc sputter deposition of a 0.3 μm thick Au layer. The central part of the structure consists typically of a $100 \times (3\text{nm Fe} + t_{\text{Cr}}\text{Cr})$multilayer with a variable Cr thickness t_{Cr}. All depositions were done during one vacuum run. The system pressure prior to deposition was 4×10^{-7} mbar, and the Ar-pressure during sputtering was 4×10^{-3} mbar. X-ray diffraction reveals predominantly (110) growth of the multilayer. For lithography the sample is covered by a 0.1 μm thick Mo layer, then by a 0.2 μm thick Al_2O_3 layer, followed by the deposition of another 0.1 μm thick Mo layer. The latter is structured using conventional lithography and wet etching (Fig. 5.8 a). It serves as a mask for the structuring of the Al_2O_3 layer, which is carried out in a CHF_3/Ar plasma (10 sccm CHF_3/40 sccm Ar, 25 μbar, 0.52 W/cm^2). The structured Al_2O_3 layer, in turn, is employed as a mask for the etching of the Au and the Fe/Cr multilayer; this is done in an HCl plasma (40 μbar, 2.1 W/cm^2) (Fig. 5.8 b). After removal of the remaining Al_2O_3, the sample is covered with an insulating polyimide layer. Contact holes in the polyimide are made using reactive ion etching, after which the remaining Mo in the hole is removed by wet etching (Fig. 5.8 c). Subsequently, the structure is covered

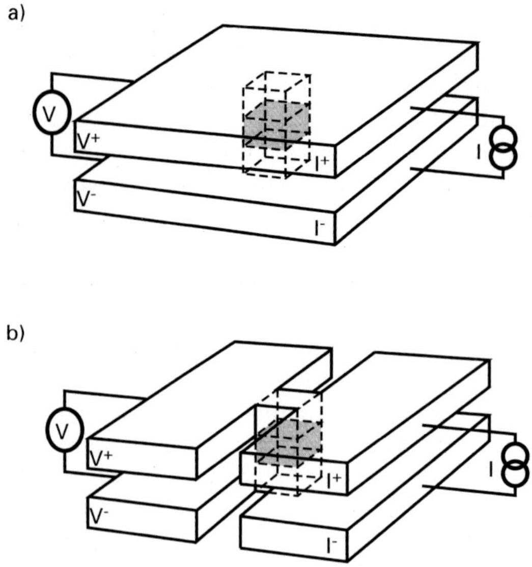

Fig. 5.9. Schematic drawing of (**a**) the experimental measurement geometry used for the study of CPP–MR in Co/Cu multilayers (geometry I) and (**b**) the experimental measurement geometry used for the study of CPP–MR in Fe/Cr multilayers (geometry II). The upper and lower current contacts are indicated by I^+ and I^-, respectively. Upper and lower voltage contacts are indicated by V^+ and V^-, respectively [55]

by a 1.9 μm thick Au film, in which the contact electrode pattern is made (Fig. 5.8 d). The pillar in the middle is the actual sample. Finally two Au contact leads are attached on top of it for the resistance measurement: also the side parts of the structure are contacted and serve as the lower contact leads for the pillar in the middle.

Similar CPP–MR structures in NiFe/Cu/Co/Cu multilayers were reported in [56,57]. The concept of the measurement geometry is the same, but there are, of course, differences in the lithographic processing technology. For example, the pillar structures are defined with Ar ion milling and Si_3N_4 instead of polyimide used for the insulating layer. Another difference is that for the NiFe/Cu/Co/Cu multilayers the terminating layer on top of the multilayer stack is Cu instead of Au. Cu may be somewhat more reactive than Au, and it can eventually give rise to unwanted contact resistances.

Figure 5.9 shows schematic drawings of the two different measuring geometries which were used in the experiments at Philips. Figure 5.9 a represents a structure showing an upper and a lower contact plane inbetween a microstructured multilayer pillar (shaded in grey). Current (I^+ and I^-) and voltage contacts (V^+ and V^-) are indicated. This experimental geometry was employed for the study of CPP–MR in the Co/Cu multilayers and will be called "geometry I" in the following. Figure 5.9 b differs from Fig. 5.9 a by a narrowing down of the upper and lower contact plane (experimentally this was done by "etching" trenches in the leads with a Nd-YAG laser). At the location of the pillar, the contacts have the same width as the pillar itself. This experimental geometry was used for the study of CPP–MR in the Fe/Cr multilayers, and will be called "geometry II" in the following. The reason for choosing different geometries for the two multilayer systems is the minimization of spurious contact-electrode contributions in the measurement of the perpendicular resistance, which sensitively depend on the relative values of the electrode and multilayer resistivities, as will be explained further. An alternative for the second geometry was presented by Vavra et al. [57], who have fabricated pillar structures with contact leads that intrinsically have the same width as the pillar itself, so that the necessity of postprocess "trimming" of the leads is avoided. For the largest Fe/Cr multilayer pillars fabricated on the substrate (width: $10-12\,\mu$m), the trench terminates exactly at the edges of the pillar, so that the only connection between the two parts of the structure is established through the pillar. This trench plays a crucial role in the measurement of the Fe/Cr pillars, because it strongly reduces the spurious spreading-resistance effect of the contact leads (particularly of the lower Cr electrode, with relatively high resistivity), which is necessary for a proper measurement of the very small perpendicular resistance (typically a few mΩ). For Fe/Cr the resistance of the structure without a trench is dominated by resistance contributions from the electrodes, as will be discussed below.

The fabrication of the Co/Cu multilayer pillars is similar to that of the Fe/Cr samples. However, it is well known that the growth of AF coupled Co/Cu multilayers with large MR is very sensitive to, for instance, the choice of the substrate and the use of a buffer layer [58,48]. In [55,52] Co/Cu(100) oriented layers were investigated. The multilayer stacks were prepared on single crystalline Si(100) substrates in a multichamber MBE system. The first stage of the multilayer stack involved the deposition of a 300 nm Cu layer, later to be employed as the lower electrical contact. Subsequent deposition of the Co/Cu multilayer with an identical Cu rate and a Co deposition rate of 0.05 nm/s, maintained the (100) orientation established for the Cu base layer. The multilayer stack was completed with a deposition of 300 nm of Au (deposition rate: 0.03 nm/s), to be used as the top electrical contact. The microstructuring process of the Co/Cu layers is similar to that of Fe/Cr. The only difference is that, during the pillar etching in the HCl plasma, only about 75% of the Co/Cu multilayer is etched through, so that the lower part of the multilayer becomes part of the lower contact electrode. The reason for this is the extremely high etching rate of pure Cu in the HCl plasma, which makes the etching down to the last Co/Cu bilayer extremely critical.

Typically, a few hundred pillars with a cross-section S ranging between $6\,\mu\mathrm{m}^2$ and $130\,\mu\mathrm{m}^2$ can be fabricated on one substrate for each multilayer

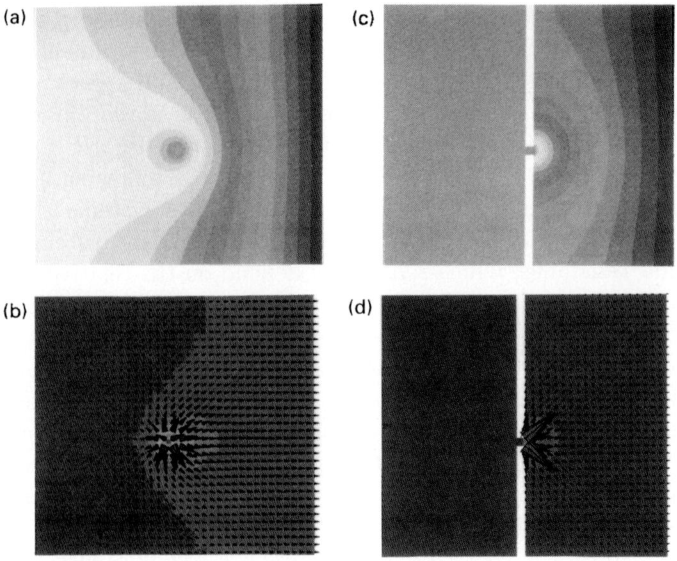

Fig. 5.10. (**a**) Finite element calculation of the equipotential line distribution in the bottom electrode of a structure without trench (geometry I). (**b**) Finite-element calculation of the electrical field modulus and vector plot of the electrical field in the bottom electrode corresponding to a structure without trench. (**c**) and (**d**) show the same as (**a**) and (**b**), but for a structure with laser-etched trenches (geometry II). For each simulation the current drain is on the right and voltage probes are situated on the left [60,55]

composition. A lot of identical pillar structures of the same cross-section and with the same layer thicknesses can be measured to check the reproducibility. However, to analyze the experimental data and to understand the scaling of the measured resistance with the pillar cross-section, knowledge of the current distribution for the specific measurement geometry is necessary.

The main difference between geometry I and geometry II can easily be clarified by using a 3-dimensional finite-element simulation package [59]. More specifically, electrostatic calculations, based on the finite-element method, were done for a structure consisting of two $140 \times 140\,\mu m^2$ square films with a thickness of $1.9\,\mu m$ and $0.3\,\mu m$, respectively, centrally connected by a $4 \times 4\,\mu m^2$ pillar with a height of $1\,\mu m$ [60,55]. In one case the contact planes were not trenched (geometry I), in the other case they contained trenches (geometry II). If one applies a current to the contacts situated on the right of the contact planes, one finds the equipotential lines in the $0.3\,\mu m$ thick bottom electrode of geometry I as shown in Fig. 5.10 a. Figure 5.10 b exhibits the corresponding vector plot of the electrical field. The direction of the arrows corresponds to the direction of the field, and their thickness to the field magnitude. It is clear that, between the voltage probes situated on the right

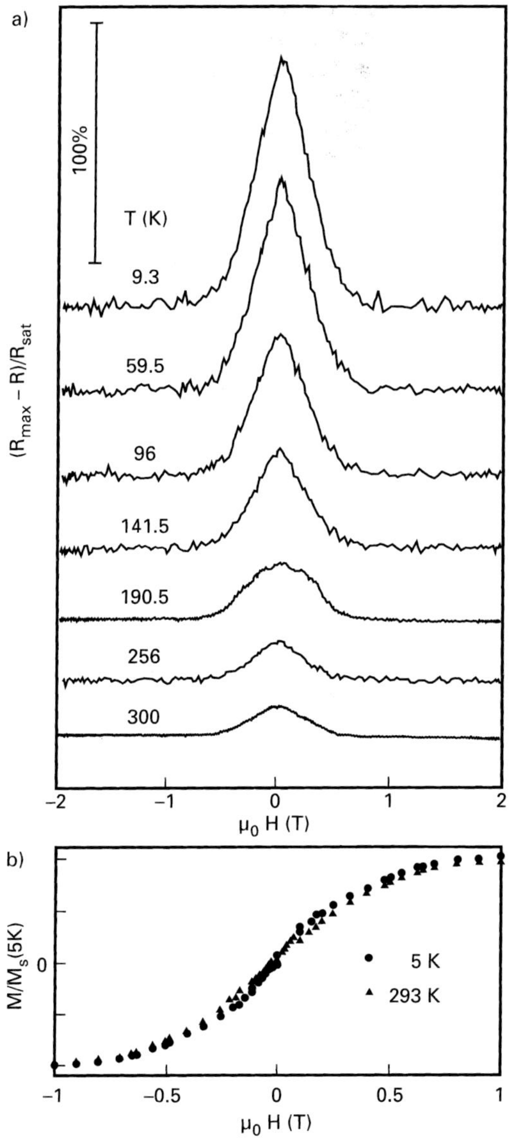

Fig. 5.11. (**a**) Giant magnetoresistance effect of a microstructured pillar in a 100×(3 nm Fe+1 nm Cr) multilayer with the current perpendicular to the multilayer plane (CPP) as a function of temperature. (**b**) Magnetization loops at 5 K and at 293 K showing the very similar AF coupling behavior at both temperatures [53]

of the planes, a large spurious voltage contribution from the leads is present. For the simulated structure, one finds, using contour integration, that only 3 % of the totally measured voltage actually corresponds to a voltage drop over the perpendicular resistance. Corresponding voltage and electrical field plots of the lower contact electrode for the structure *with* trenches are given in Fig. 5.10 c and d, respectively. In Fig. 5.10 c there is no voltage drop at all in the left voltage-sensing electrode which, of course, is equivalent to the locally vanishing electrical field in Fig. 5.10 d. In the simulations the resistivity of the contact electrodes and the perpendicular pillar resistivity were taken equal. This corresponds more or less to the Fe/Cr case. For Co/Cu multilayers, grown on a highly conducting Cu lower electrode, the spurious influence of the contacts is much smaller, and the geometry I could be used in the experiment.

The choice of geometry I or geometry II has a strong influence on the perpendicular current-distribution pattern in the pillar itself. Of course, this distribution will determine the value of the experimentally measured resistance and its scaling behavior. The exact current distribution in the pillar can directly be determined from finite-element calculations. On the other hand, it is also possible to calculate analytically the current distribution in the two experimental geometries [61]. The results of such calculations were shown to be consistent with the finite elememt results.

Similar considerations on the effect of the current density on the measurement of the perpendicular resistance were given in [56,57]. Here, it was demonstrated that the ratio of the perpendicular and the in-plane resistivities is of particular importance and that a different temperature dependence of these resistivities can give rise to current-spreading effects, leading to a *decrease* of the measured resistance with increasing temperature. As a concluding remark, one can say that knowledge of the current-distribution pattern in an experimental microstructure with normal contacts is necessary to deduce the proper values for the perpendicular resistivity. Care has to be taken not to mix perpendicular with in-plane resistance contributions, in order not to get a wrong interpretation of the experimental results.

Experimental Results. Figure 5.11 a displays CPP–MR curves at different temperatures for a pillar with a cross-section of $S = 90\,\mu m^2$ structured in a $100 \times (3\,nm\,Fe + 1\,nm\,Cr)$ multilayer: For clarity, the curves are shifted vertically [53,54]. Measurements were done using the experimental geometry II. Data have been corrected for a contact resistance which was of the same order of magnitude as the perpendicular multilayer resistance itself. At 9.3 K a MR effect of 108% was observed, more than four times higher than the corresponding CIP–MR effect in identically prepared unstructured films [20,62]. The MR effect is weakly temperature-dependent below about $T = 60\,K$; above that temperature the decrease is much stronger. At room temperature a 14% CPP–MR remains. The saturation field $B_S (\equiv \mu_0 H_{sat})$ is defined by the crossing of the low-field resistance decrease with the horizontal line of constant resistance at higher fields. One finds that $B_S = 0.54\,T$, nearly independent of temperature. This is comparable to the $B_S = 0.51\,T$ value found for the corresponding CIP-sample [20]; the small difference can be explained

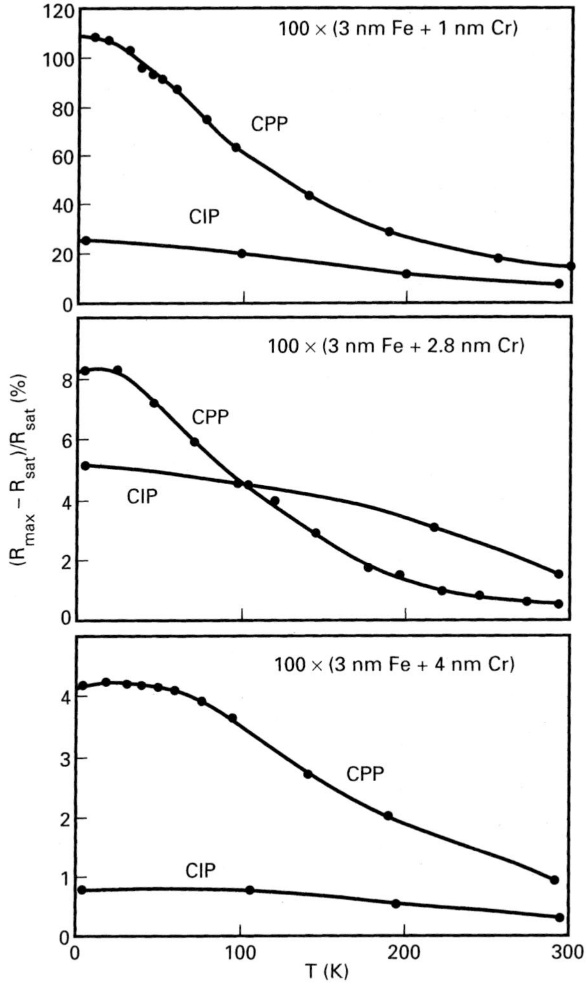

Fig. 5.12. Temperature dependence of the CPP magnetoresistance of pillar structures made in 100 × (3 nm Fe+t_{Cr} Cr) multilayers with (**a**) t_{Cr} = 1 nm, (**b**) $t_{\mathrm{Cr}} = 2.8$ nm and (**c**) $t_{\mathrm{Cr}} = 4$ nm. Corresponding CIP–MR data are given for comparison. The solid curves are a guide to the eye [53]

by a slightly different demagnetization factor. This B_{S}-value corresponds to an interlayer exchange-coupling energy per unit surface of $1.3\,\mathrm{mJ/m^2}$ [20]. Hence, the interlayer-coupling energy is orders of magnitude larger than the thermal energy for the microstructures. This implies that the strong temperature dependence of CPP–MR cannot be attributed to incomplete AF coupling at higher temperatures. Additional evidence for this is presented in Fig. 5.11 b, where magnetization loops at 5 K and at 293 K are plotted for the 100×(3 nm Fe + 1 nm Cr) multilayer, as measured by a SQUID magnetometer. All magnetization values are normalized by the saturation magnetization M_s at 5 K. The similarity of the two magnetization curves is striking. Another observation is the absence of hysteresis in the magnetoresistance. For the corresponding CIP sample, a small but clear hysteretic effect was found [20,62]. One should note that in the very weakly or uncoupled Ag/Co and Cu/Co CPP samples of [39] the magnetoresistive hysteresis is very important. It seems that both strong AF coupling and microstructuring, which leads to a reduction in the number of magnetic domains, are important for obtaining nonhysteretic MR characteristics. In Fig. 5.12 the temperature dependence of CPP–MR is shown for multilayers with different t_{Cr}. Comparison is made with the corresponding CIP–MR data [63]. Figure 5.12 a is for $t_{Cr} = 1$ nm and corresponds to the measurements presented in Fig. 5.11 a. The enlarged CPP–MR effect and its strong decrease with temperature is immediately clear. This decrease is also pronounced for the sample of Fig. 5.12 b ($t_{Cr} = 2.8$ nm, $S = 20\,\mu m^2$). Above about 100 K, CPP–MR is smaller than CIP–MR. This could be an intrinsic effect, but another possibility is a somewhat different contact resistance for this sample. One should note that the perpendicular multilayer resistance itself only is of the order of a few mΩ, a value which is easily obtainable by a spurious contact resistance. In Fig. 5.12 c, a sample with $t_{Cr} = 4$ nm and $S = 6\,\mu m^2$ is shown. For this Cr thickness, CIP and magnetization data indicate weak AF coupling. One observes that CPP–MR is enhanced with respect to CIP–MR, and that the temperature dependence is markedly stronger.

The low-temperature experimental data can be compared with the microscopic model of Valet and Fert [6,28], which like the work of [23] justified the original phenomenological approach of [37,38,43]. Pratt et al. [43,38], and Valet and Fert argued that the most appropriate experimental quantity to compare with the model is the "square-root" expression (5.9) or (5.13). Equation (5.9) is valid when the spin-flip diffusion length l_{sf} is much larger than the individual layer thicknesses in the structure, a condition which is generally well fulfilled at low temperatures. At 4 K, one finds that $\sqrt{(R_{max} - R_{sat})R_{max}S^2} \simeq 65\,m\Omega\mu m^2$ for the sample with $t_{Cr} = 1$ nm, while for the other two samples this value is reduced to a number of about $40\,m\Omega\mu m^2$ (R_{max} is the maximum resistance at zero field and R_{sat} the value at saturation). This decrease could indicate the limiting role of l_{sf}, when increasing the Cr thickness, but more probably is an artificial contact resistance effect. Due to the fact that spurious contact-resistances are of the same order of magnitude as the perpendicular multilayer resistance itself and due to the complicated microfabrication procedure, it is very hard to fabricate many kinds of samples in a reproducible way by microlithography. This is in contrast to the superconducting contact technique, where complete sample series with varying layer thicknesses can easily be made and measured. Therefore, one can only check if the scattering parameters derived for a certain microlithographic sample are consistent with the scattering parameters

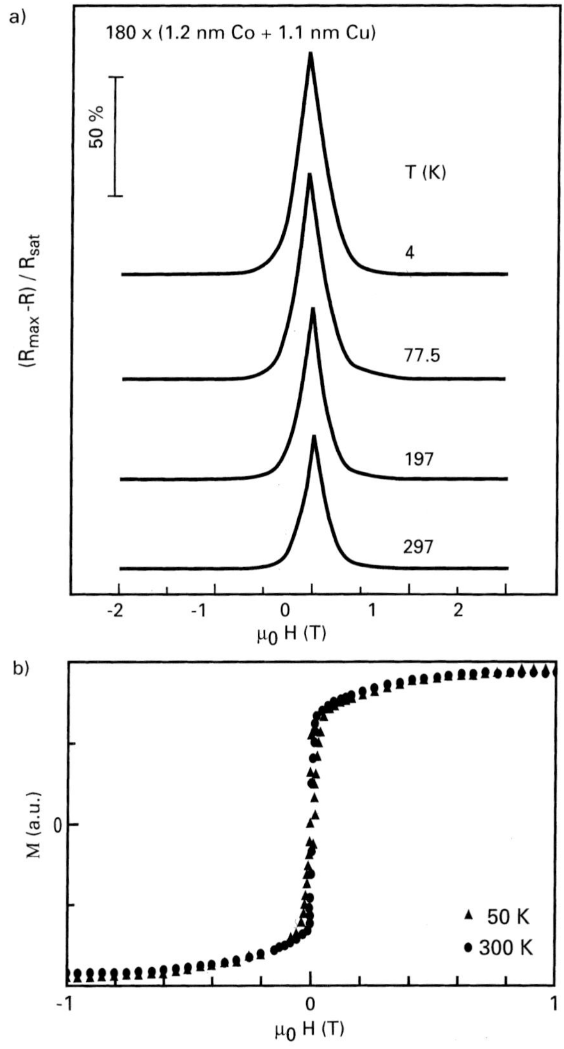

Fig. 5.13. (**a**) Giant magnetoresistance of a microstructured pillar in a 180 × (1.2 nm Co+1.1 nm Cu) multilayer at different temperatures. (**b**) Magnetization curves at 50 K and 300 K for the multilayer of (**a**) [52]

derived for the superconducting contact samples. For example, to estimate the magnitude of the Fe/Cr interface resistance $AR^*_{\mathrm{F/N}}$, we can use (5.9) and take $\beta = 0.33$ and $\gamma = 0.78$; these values correspond to a ratio of spin-up to spin-down resistivity of $1+\beta/1-\beta \simeq 2$ for Fe bulk scattering and a ratio of spin-up to spin-down interface resistance of $1+\gamma/1-\gamma \simeq 8$ for interface scattering, as obtained from a theoretical fit of the corresponding CIP–MR data [20,64]. With $N = 100$, $t_{\mathrm{Fe}} = 3\,\mathrm{nm}$ and $\rho_{\mathrm{Fe}} = 5 - 10\,\mu\Omega\,\mathrm{cm}$

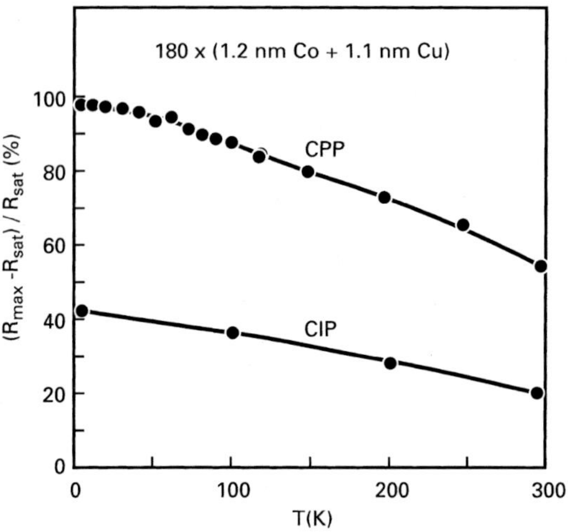

Fig. 5.14. Temperature dependence of the CPP magnetoresistance of a Co/Cu(100) pillar structure made from a 180 × (1.2 nm Co+1.1 nm Cu) multilayer. Corresponding CIP–MR data are given for comparison. The solid curves are guides to the eye [55,52]

one finds that $AR^*_{\mathrm{F/N}} = 0.2 - 0.4\,\mathrm{f\Omega\,m^2}$ for $t_{\mathrm{Cr}} = 1\,\mathrm{nm}$. For the other Cr thicknesses $AR^*_{\mathrm{F/N}}$ is somewhat reduced. These values are close to the value of $AR^*_{\mathrm{F/N}}$ found for the Michigan State University Cu/Co and Ag/Co multilayers (Table 5.1). Although comparison with theory is most convenient at low temperatures, where spin-flip scattering is minimum and the simple two-current model is well applicable, the temperature dependence of the effect is particularly interesting because it provides important new information on the temperature dependence of the spin-dependent scattering. Due to the increasing importance of electron-phonon and electron-magnon scattering, the two spin currents are equalized (as in the CIP geometry) and also the reduction of l_{sf} in the limit can lead to a decreased CPP–MR effect. An analysis of the temperature dependence of CPP–MR, based on the two-current model for magnetic alloys [7], will be presented after the discussion of the Co/Cu CPP–MR data.

In Fig. 5.13 a results are presented on the temperature dependence of CPP–MR of a 180 × (1.2 nm Co+1.1 nm Cu) multilayer, measured in the experimental geometry I [65,52]. At low temperature one finds MR of 90% and, at room temperature, MR of 56%. It should be noted that these are as-measured values and that no evidence for a substantial serial-contact resistance was found. AF coupling is incomplete, as shown by magnetization experiments (Fig. 5.13 b. Besides the similarity of the two curves at 50 K and at 300 K, one observes that about 2/3 of the magnetization signal around the origin reflects ferromagnetic coupling, and only 1/3 of the sample is AF coupled and, hence, contributes to the measured CPP–MR. This is also consistent with a CIP–MR reference experiment on a multilayer, grown at the same time on a 30 nm thick Cu buffer layer on a Si(100) substrate. In Fig. 5.14 CPP–MR is shown as a function of temperature together with the CIP reference sample. The CIP–MR is 43% at 4 K and 21% at room temperature. This last value is one third of the maximum MR (65%), obtained for a completely AF coupled (100)-oriented Co/Cu multilayer [21]. The problem of growing a thick, perfectly AF-coupled sample is related to the extremely narrow width of the first AF-coupling peak of Co/Cu(100): it has a maximum at 1.0 nm Cu and a width of only 0.3 nm [66]. Hence, it is very sensitive to even a minor variation of the Cu rate during the evaporation time (3 hours) in the MBE system. Resistance data like those of Fig. 5.13 a were obtained for a large number of pillar structures with variable pillar cross section, all situated in the first AF-coupling peak of Co/Cu(100), and are used to deduce the minimum and maximum perpendicular resistivities of the CPP–MR effect, with the current-density model [61]. One finds that, at 4.2 K, absolute resistivities are about a factor of 2.5 higher than the "intrinsic" resistivity of the multilayer, which is estimated from experiments by Pratt et al., made on a comparable sample with superconducting contacts [43]. The origin for this difference most probably is a small contamination during the Au top-contact fabrication on the microstructured sample.

Theoretical Analysis of the Temperature Dependence. To analyze the temperature dependence of CPP–MR one can apply (5.16) and compare it with the experimentally determined resistivities at zero field and at saturation of the resistance. The important spin-dependent scattering parameters are defined as $\alpha \equiv \rho_{0\downarrow}/\rho_{0\uparrow}$ and $\mu \equiv \rho_{i\downarrow}(T)/\rho_{i\uparrow}(T)$, α describing the overall spin-dependent scattering at low temperatures – and hence contains contributions both from bulk scattering and from interface scattering. μ is a newly defined parameter which accounts for the overall spin-dependent scattering of the temperature-dependent part of the resistivity. In the limit $\rho_{i\uparrow}, \rho_{i\downarrow}, \rho_{\uparrow\downarrow} \ll \rho_{0\uparrow}, \rho_{0\downarrow}$, (5.16) can be written as:

$$\rho(T) - \rho_0 = \left(1 + \frac{(\alpha - \mu)^2}{(1+\alpha)^2 \mu}\right) \rho_i(T) + \frac{(\alpha - 1)^2}{(\alpha + 1)^2} \rho_{\uparrow\downarrow}(T)\,, \tag{5.17}$$

with ${\rho_0}^{-1} \equiv {\rho_{0\uparrow}}^{-1} + {\rho_{0\downarrow}}^{-1}$ and ${\rho_i}^{-1} \equiv {\rho_{i\uparrow}}^{-1} + {\rho_{i\downarrow}}^{-1}$. One should bear in mind that the validity of (5.17) is roughly restricted to temperatures below about 100 K, where the limiting conditions leading to the equation are fulfilled. In the case that the magnetizations in the multilayer are aligned parallel, the experimental minimum resistivity can, in principle, be directly compared with (5.17). If the maximum resistivity corresponds to a completely AF-

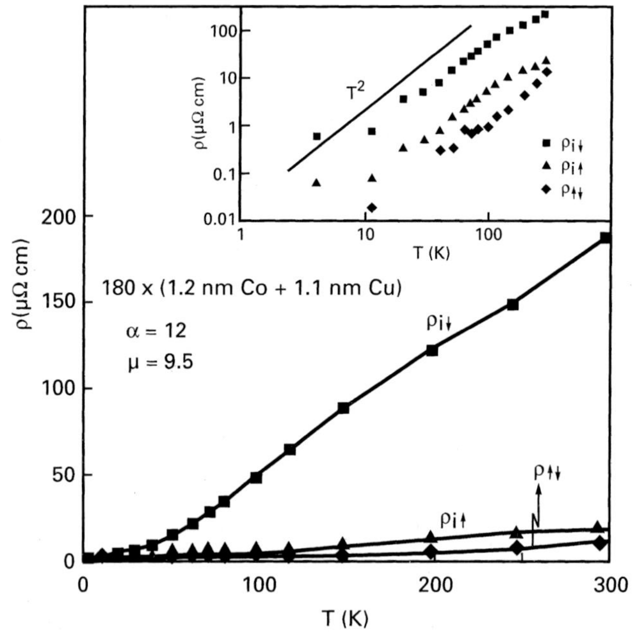

Fig. 5.15. Temperature-dependent resistivities ($\rho_{i\downarrow}(T)$, $\rho_{i\uparrow}(T)$ and $\rho_{\uparrow\downarrow}(T)$), for a Co/Cu multilayer, determined from a comparison of experimental CPP–MR data with the Fert–Campbell theory. Solid curves are a guide to the eye. *Inset*: Temperature-dependent resistivities on a double-logarithmic scale. The temperature dependence is consistent with a T^2 behavior [52]

coupled situation, where the two spin channels have the same resistivity, $\alpha^{\mathrm{AF}} = \mu^{\mathrm{AF}} = 1$ and (5.17) is replaced by

$$[\rho(T) - \rho_0]^{\mathrm{AF}} = [\rho_i(T)]^{\mathrm{AF}} = \frac{(1+\mu)}{4\mu}\rho_{i\downarrow}(T)\,, \tag{5.18}$$

with $\rho_{i\downarrow}(T)$ is the resistivity of the spin-down channel, and μ the scattering-asymmetry parameter in the *ferro*magnetic alignment configuration of the multilayer. One uses the experimental minimum and maximum resistivities for comparison with (5.17) and (5.18) respectively, with the condition that $\mathrm{d}\rho_{\uparrow\downarrow}(T)/\mathrm{d}T = 0$ at the lowest temperatures. The result is displayed in Fig. 5.15. Prior to the analysis, the value for α is determined from a comparison of CPP–MR at 4.2 K with the (self-averaged) magnetoresistance model [6,22,28], where the magnetoresistance is given by $(1-\alpha)^2/(4\alpha)$. One finds that $\alpha = 12$ for a perfectly AF-coupled sample, indicating the strong spin-dependent scattering at low temperatures. Very striking is the large value for $\mu = 9.5$, which means that also the temperature-dependent part of the resistivity is strongly spin-dependent. This number is in agreement with the CIP–MR result of [51], and not inconsistent with experiments on Co bulk alloys [67]. When plotting the results on a double-logarithmic scale, one obtains that the temperature dependence of the three resistivities below 100 K can be described by a T^2 law (see inset of Fig. 5.15). This seems reasonable for $\rho_{i\uparrow}(T)$ and $\rho_{i\downarrow}(T)$, which are determined by a combination of both electron-phonon and intra-channel electron-magnon scattering processes [7]. These can give rise to such a power-law behavior. $\rho_{\uparrow\downarrow}$ is the smallest resistivity. The spin-mixing scattering rate ${\tau_{\uparrow\downarrow}}^{-1}$, originating from elastic inter-channel electron-magnon scattering, has been microscopically modeled by Fert for low temperatures [68]. The result of this calculation is in good agreement with the observed magnitude and temperature dependence of $\rho_{\uparrow\downarrow}$ [52]. In addition, Fe/Cr CPP–MR data were analyzed using the "magnetic-alloy" model, although the analysis there is more complicated due to the presence of a serial pillar contact resistance. Nevertheless, if the same analysis is applied, one finds a much smaller value of $\mu = 2.2$. This explains why CPP–MR of Fe/Cr strongly decreases with increasing temperature: the thermally induced scattering is much less spin-dependent than for Co/Cu.

It may be clear from the previous discussion that the superconducting contact samples are much easier to fabricate and to vary than the lithographic samples. Therefore, for scientific studies at low temperatures, the superconducting contact technique clearly is the more flexible one. However, the superconducting contacting technique is limited to 4.2 K, while with the microfabrication technique a temperature range of 4.2 – 300 K or higher is possible. As a result, only the microfabrication technique has application potential, although for real applications a lot more is required than experiments under laboratory environment. One of the problems is that the perpendicular sample resistance is still very low. This problem could be solved by placing several microstructured pillars in series, but such a technique could easily lead to additional problems with the contact resistance. In fact there has been an attempt to realize such samples, by using microstructured Cu/Ni multilayers and (superconducting) Nb contacts [69], but no CPP–MR was observed, probably due to the choice of the multilayer system with intrinsicly very low magnetoresistance. In the following other and simpler fabrication techniques will be outlined, that *do* allow systematic studies of CPP–MR from 4.2 K up to 300 K.

5.2.3 Electrodeposited Multilayer Pillars

Experimental Method. Electrodeposition of a metal takes place by reduction of its ions in an electrolyte at the cathode of an electrochemical cell. Since different metals have different reduction potentials, it is possible to alter the composition of the alloy deposited from an electrolyte containing more than one species of ions by altering the cathode potential. Hence, it is possible to electrodeposit a two-component metallic superlattice on a conducting substrate attached to the cathode by switching between two cathode potentials and using coulometric control [70]. Giant magnetoresistance in such

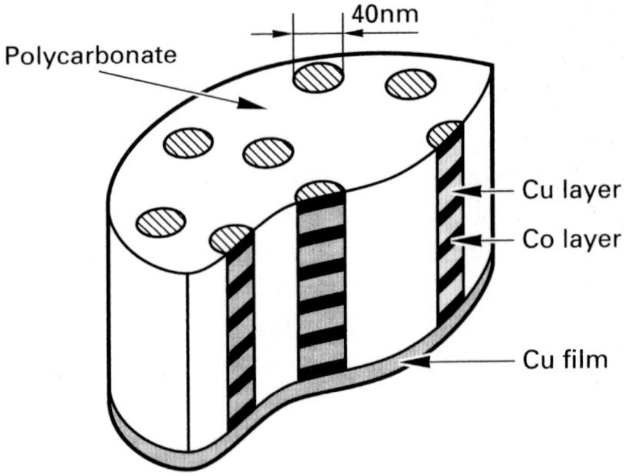

Fig. 5.16. Schematic representation of the array of nanowires in an insulating polymer matrix [77]

electrolytically grown superlattices was for the first time reported by Alper et al. [71–73] for CoNiCu/Cu multilayers in the CIP geometry, followed by similar experiments by other groups [74,75]. The electrochemical cell contains an electrolyte with Co^{2+}, Ni^{2+} and Cu^{2+} ions, the concentration of the Cu^{2+} ions being much smaller than that of the magnetic species. Cathode potentials of -0.15 and $-1.8\,V$, measured relative to a saturated calomel reference electrode, were used. At $-0.15\,V$ only the Cu, the most noble of the metals present in the electrolyte, will be deposited, while at $-1.8\,V$ all three metals are deposited to give a ferromagnetic CoNiCu alloy. The cathode potential is set by means of a potentiostat controlled by a computer, which also monitors the charge passing between anode and cathode. When this charge reaches the value corresponding to the desired thickness of the superlattice component being deposited, the computer switches the cathode potential to start depositing the other superlattice component. Since the film always remains within the same electrolyte, there is no exposure to the atmosphere, as would be the case if two separate electrolytes were used.

Another application of electrolythic growth was demonstrated by Whitney et al. [76], who showed that arrays of metallic Co and Ni wires of nanometer size can be electrodeposited into a template of nanometer-sized pores created by nuclear track etching in a polycarbonate membrane. The basic process for fabricating such a template is that the rapid passage of heavy ions, such as those emitted from a radioactive transuranium isotope through a dielectric material creates latent nuclear damage tracks. A suitable etchant can preferentially dissolve the damaged zones with a very high selectivity compared to the undamaged regions for mica and some polymers, resulting in etched pores that can have extremely small lateral dimensions of a few tens of a nanometer through the thickness of the dielectric. Commercially available membranes are approximately 6 μm thick and have pores with a

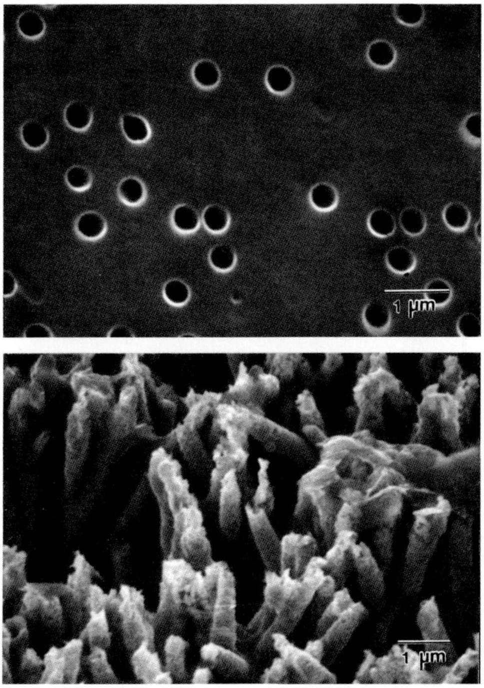

Fig. 5.17. Top-view scanning electron-microscope image of a membrane with 400 nm diameter pores (*upper image*). Scanning electron-microscope image of 400 nm multilayered nanowires with the membrane removed (*lower image*) [78]

diameter ranging between 30 and 200 nm. The pore density is of the order of $\approx 6 \times 10^8\,\mathrm{cm}^{-2}$. For the growth of the wires in the pores, a Cu film was first sputter-deposited on one side of the porous membrane, providing the conducting cathode in the electrolyte. The electrolyte is confined to the bare side of the membrane so that deposition was initiated onto the Cu film from within the pores. In this way Ni and Co wires with a diameter of only 60 nm and a length of 6 µm could be grown. It is immediately clear that if multilayer "wires" of sufficiently good quality and with a very high length/diameter ratio could be grown within the pores, and if appropriate electrical contacting can be achieved, this will be very advantageous for the study of the CPP–MR effect. Evidently, the role of eventual contact resistances would be minimum and the perpendicular current distribution would be simply uniform over the wire cross-section. An example of such experimental geometry is schematically presented in Fig. 5.16. Figure 5.17 a is a scanning electron microscope micrograph of the top of a bare polycarbonate membrane with 400 nm pores [78]. A scanning electron-microscope micrograph of the 400 nm nanowire arrays, after the polycarbonate membrane has been etched away, can be seen in Fig. 5.17 b. CPP–MR in such wires was studied by several groups [77,79,78]

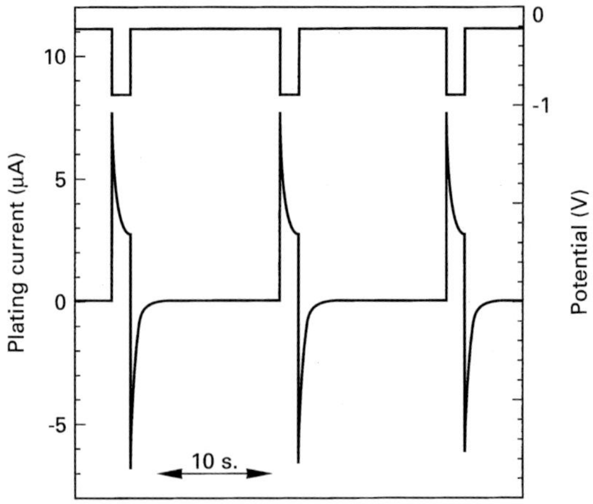

Fig. 5.18. Time traces of the plating current during deposition of 10 nm Cu-rich layers and 10 nm Co-rich layers at the cathodic potentials −0.2 and −0.9 V, respectively; the Cu deposition current at −0.2 V is approximately 0.2 μA [77]

using very similar ways of electrolytic growth. However, caution has to be exercised with regard to resistance and magnetoresistance measurements. The contacting method used in [78] is to cover the top membrane surface after the completion of the deposition of the Co/Cu multilayer wires with Cu to electrically connect all of them. Using a simple photoresist technique, the Cu layers on top and below the membrane are etched away except that a thin strip ($\simeq$ 100 μm) remains; the two remaining Cu strips on the top and the bottom surface of the membrane are perpendicular to each other, so that a limited number of wires can be selected for the electrical measurements.

A very peculiar feature of the electrolytic multilayer deposition process is illustrated in Fig. 5.18 for the deposition of Co/Cu nanowires [77]. A square-waveform potential is applied to the cathode, by using −0.2 V for the deposition of the Cu layers and −0.9 V for the Co layers. These voltages for the Co/Cu system are somewhat different for the different research groups, which is due to the different ion concentrations used in the electrolyte. The plating current obviously does not have the square waveform of the voltage excitation, indicating the importance of the electrodeposition process on the local ion concentration, diffusion, etc. Especially the negative current peak at the end of the −0.9 V pulse is a manifestation of strong nonequilibrium effects in the voltage-transient regime. It indicates that Co *dissolves* at the beginning of the Cu deposition. This feature of the electrodeposition process can give rise to a selective removal of Co atoms, and eventually can lead to pinholes between the layers. This may explain why it is such a problem to grow multilayers with small individual normal metal-layer thicknesses (around 1-2 nm) and with no ferromagnetic coupling between the individual ferromagnetic layers.

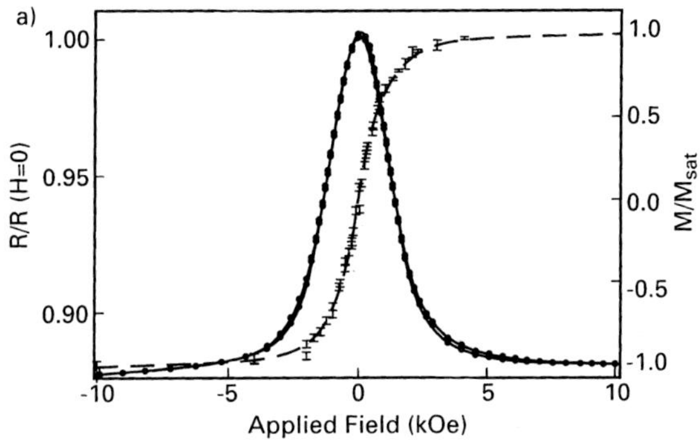

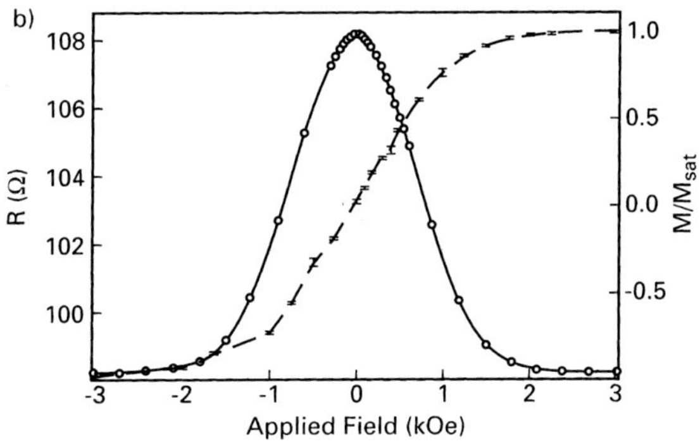

Fig. 5.19. (**a**) Room-temperature resistance ratio versus magnetic field and magnetization curve for a 5 nm Cu/5 nm Co electrolytically grown multilayer. (**b**) Resistance versus magnetic field and magnetization curve for a 5 nm Cu/5 nm$Ni_{80}Fe_{20}$ electrolytically grown multilayer [79]

Experimental Results. We show here experimental results from Blondel et al. [79], which are very similar to the measurements of other groups [77,78]. Resistance measurements were done on a batch of membranes, having an average pore diameter of 80 nm and a length of 6 μm. The samples were contacted by using Au as the lower contact and two gold wires glued with silver epoxy on a small area (typically 1 mm^2) on the other side of the membrane as the upper contact. Measurements of large resistance values with a reproducibility of the relative magnetoresistance R/R(0) within 20 % were obtained. Samples were prepared by taking identical Cu and ferromagnet thicknesses, set in the range from 2 nm to 20 nm by the duration of the voltage pulses. The magnetoresistance was found to be maximum at a thickness of 10 nm. Magnetization and magnetoresistance curves at room temperature are shown in Fig. 5.19 a for a 5 nm Cu/5 nm Co multilayer and in b for a 5 nm Cu/5 nm $Ni_{80}Fe_{20}$ multilayer.

The applied field is in the multilayer plane. The magnetization curves reveal that the ferromagnetic material is difficult to saturate, even for the $Ni_{80}Fe_{20}$-based multilayer. The reason is the rather large demagnetizing field, occuring in all directions. If the applied field is perpendicular to the wires and parallel to the wires, a demagnetizing field of a few thousand Oe was observed.

The room-temperature CPP–MR (using the definition of (5.10)) is 13% for the sample of Fig. 5.19 a and about 10% for the sample of Fig. 5.19 b. CPP–MR increased to 22% at 20 K for the sample of Fig. 5.19 a, but this value is rather far below the values reached in [43], where CPP–MR of 170% was reported at 4.2 K for a 0.9 nm Cu/1.5 nm Co) multilayer, and in [55], where 95% was reached at 4.2 K and 60% at room temperature for a 1.1 nm/1.2 nm Co) multilayer. Of course, the electrolytically grown samples are characterized by much larger individual layer thicknesses (5 nm), which also for sputtered samples would lead to lower magnetoresistance values than if they would be around 1 nm. The difference between the electrolytically grown and the sputter- or MBE-deposited multilayers may rest in the properties of the normal metal/ferromagnet interface. It was shown in Sect. 5.1.2 that the interface is of crucial importance for obtaining large magnetoresistance effects. Interfaces may have an optimum roughness for a maximum in spin-dependent scattering. Redissolution effects in the electrolytic deposition sequences could lead to less suitable interfaces and hence a lowered magnetoresistance. It is clear that this subject needs further study for obtaining more definite conclusions.

Important is that electrodeposition in nanoporous membranes of a multicomponent wire is a new technique enabling control of the sample morphology down to the 10 nm length scale, opening the way to new types of ferromagnetic composite materials. The electrodepostion presents some advantages and useful pecularities: an easy production of a large number of layers (reaching 1500 in the experiments), the possibility of controlling the interface chemistry (for example, by changing the transition between the two deposition potentials), the possibility of controlling the coercive field either by varying the thickness or by changing the pore diameter, the possibility to enforce antiparallel alignement by depositing alternatively two different widths of the magnetic layers. These issues clearly may lead to new developments in the field of CPP–MR.

5.2.4 Structures on Grooved Substrates

Experimental Method. An alternative approach to the CPP–MR effect is not to use a micro- or nanofabrication technique for structuring the multilayer itself, but to apply structuring technology to the substrate on which the multilayer has to be deposited. By combining the specially created texture of the substrate with an oblique evaporation of the multilayer, one can realize a CPP–MR-like geometry. This can be experimentally advantageous if one can get rid of a complex multilayer patterning technology and replace it with a

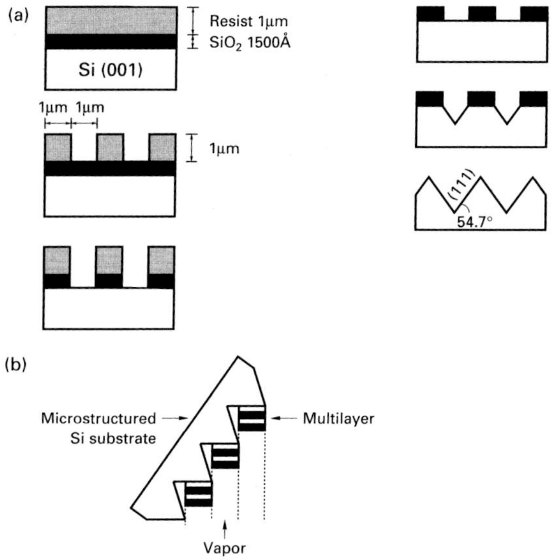

Fig. 5.20. (**a**) Process of V-groove fabrication on a Si(100) wafer. (**b**) Cross-sectional illustration of multilayers deposited on a V-grooved substrate [80]

much more standard substrate patterning technique, provided, of course, that the quality of the multilayer remains the same. Such a technique was introduced by Shinjo et al. [80], who made quasi-one-dimensional multilayer wires by evaporation at an angle on grooved substrates. The substrate fabrication technique is illustrated for Si(001)-oriented substrates [81] in Fig. 5.20 a. The surface of a flat Si(001) wafer is covered by 150 nm SiO_2 and subsequently by a 1 μm thick layer of photoresist. The resist is exposed through a mask, and a stripe pattern of resist is realized on the surface. Then the SiO_2 layer is partly removed by etching in a HF solution and the remaining resist is removed with acetone. Finally, a pattern of SiO_2 with a periodicity of 2 μm remains on the substrate. By wet anisotropic etching in a KOH solution, a V-shaped groove pattern is realized due to the strong difference in etching rates between a (100) and (111) plane of the Si. Multilayers of metallic elements are then deposited on one side of the (111) planes of the V-grooved substrate, as illustrated in Fig. 5.20 b. In that way quasi-one-dimensional multilayer wires are realized [80], but, of course, the substrate patterning technique also can be used to study the CPP–MR effect.

The latter experiment was done by Gijs et al. [82,83] following a very similar substrate patterning approach. By using holographic laser-interference lithography and anisotropic etching techniques, a V-groove pattern was realized into the surface of semi-insulating InP substrates. A pattern periodicity of 0.2 μm was achieved. Subsequently, Co/Cu multilayers were evaporated at an angle with the substrate normal, giving rise to a CPP-like measuring con-

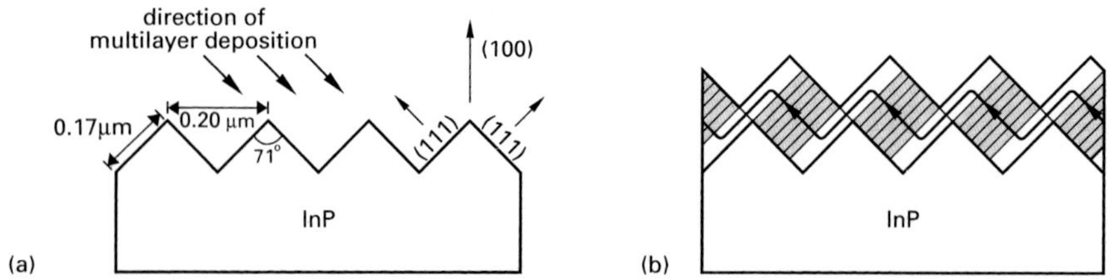

Fig. 5.21. Schematic diagram of **a** the grooved InP-substrate surface and **b** a multilayer evaporated under an angle with an indication of the current flow for the CPP-like case [83]

figuration. The grooved substrate is schematically shown in Fig. 5.21. First a semi-insulating (100) oriented InP substrate is covered with a thin (100 nm) photoresist layer (HPR 204). The resist is then exposed to a line pattern (period: 201.4 nm), produced by a single-frequency holographically interfering Ar-laser beam (Spectra Physics 2045E). The photoresist is developed and the InP layer is anisotropically etched in a $60\,H_2O : 30\,HBr(47\%) : 0.1\,Br_2$ solution. The result is the grooved substrate shown in Fig. 5.21 a. By evaporating a multilayer at an angle with the substrate normal onto *just one of the two (111) side planes* (with a width of 172 nm) of the groove, the structure of Fig. 5.21 b can be grown. Before and after the multilayer deposition, a well-conducting layer (Au or Cu) is deposited, which redistributes the current I, so that predominantly a CPP-measurement configuration is obtained, as schematically shown in the figure.

The multilayer stacks are prepared in a multichamber MBE system (VG Semicon V80M). All depositions are carried out at room temperature, at a pressure of better than 10^{-10} mbar. Characterization of the samples by X-ray diffraction is not easy due to the intrinsic slope of the multilayers with respect to the substrate plane. However, Co/Cu multilayers are grown on both Au buffer layers (which is known to result in (111) growth) and on Cu buffer layers (which results in (100) or mixed (100)-(111) growth) and no important differences in magnetoresistance are observed for the chosen layer thicknesses. Figure 5.22 reproduces scanning electron-microscopy pictures of the top and of the cross section of a complete multilayer structure (20 nm Cu + 32 × [1.5 nm Co + 6.0 nm Cu] + 20 nm Cu). It is obvious that the real structure is much more rounded than the schematic drawing in Fig. 5.21 b. A consequence is that the total layer thickness has to be chosen large enough to get a sufficiently low contact resistance between "neighboring" multilayer structures. Another consequence is that the nonmagnetic layer thickness (Cu in this case) should be chosen large enough to prevent significant ferromagnetic coupling between the magnetic layers via pinholes or thickness fluctuations.

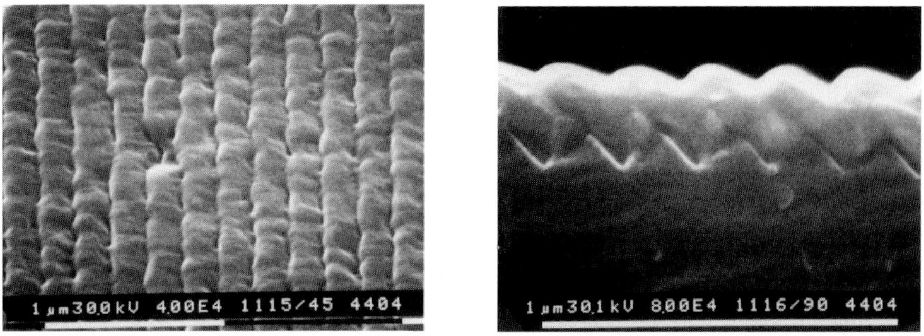

Fig. 5.22. Scanning electron-microscopy micrograph of the top (**a**) and of the cross section (**b**) of the complete multilayer structure grown on the grooved substrate [83]

Experimental Results. For a correct interpretation of the experiment it is important to realize a uniform current distribution perpendicular or parallel to the direction of the grooves, not to get a mixed CIP-CPP geometry. To prevent the latter the length/width ratio of the samples in the measuring direction is taken to be at least 4. The actual measuring configuration may probably be not "purely" CPP since, depending on the position in the multilayer, the current may have an angle of nonnormal incidence with the multilayer plane. In Fig. 5.23 CPP-like and CIP-like MR curves are depicted for a (20 nm Cu + 32 × [1.5 nm Co + 6.0 nm Cu] + 20 nm Cu) sample, measured as a function of temperature. For the CPP experiment the field was applied in the substrate plane in a direction parallel to the grooves and with the current perpendicular to the direction of the grooves (thus corresponding to the schematic current situation of Fig. 5.21 b. Due to the fact that the magnetic layers are uncoupled, a randomly oriented magnetization pattern is obtained around the resistance maximum and a hysteretic MR behavior is observed. The CPP–MR is about 37% at low temperatures and 17% at room temperature. These measurements are to be compared with the CIP–MR data (measured with the current parallel to the grooves). One finds that MR is considerably smaller than for the CPP case. Of course, the CIP resistance is shunted by the Cu layers on top and below the multilayer, giving rise to a reduction of the measured CIP–MR effect. Taking this into account, the magnitude of the CIP–MR is of the same order of magnitude as previous results on (111) oriented [84] or (100) oriented [21] Co/Cu multilayers. The CPP–MR effect at 4.2 K is in agreement with the literature results of [44]. This suggests that this CPP–MR measuring technique is indeed promising for the future. A similar behavior was observed for several other uncoupled samples (with a relatively high Cu thickness). One finds that, when reducing the Cu thickness, ferromagnetic coupling – probably via pinholes – can be present, giving rise to smaller MR values; this may be intrinsically related to the growth on a (111) substrate plane [84] or can be due to the round-off growth of the multilayer shown in Fig. 5.22 b. In general, the grooved substrate technique

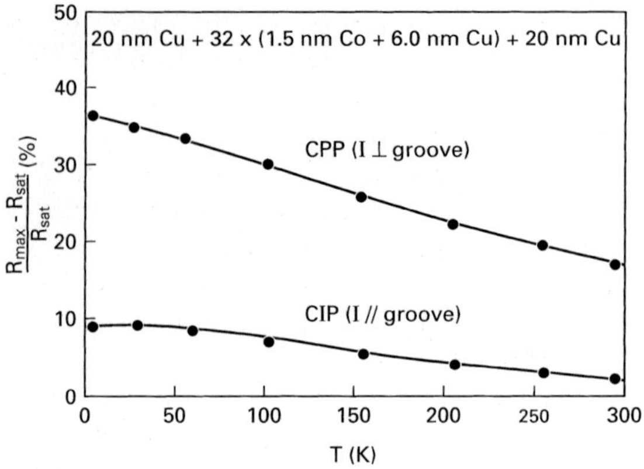

Fig. 5.23. Temperature dependence of the CPP and CIP magnetoresistance for a (20 nm Cu+32 × [1.5 nm Co + 6.0 nm Cu] + 20 nm Cu) multilayer measured in the CPP-like geometry ($I_{\perp groove}$) and in the CIP-like geometry ($I_{\parallel groove}$) [82,83]

seems, at least for the larger individual layer thicknesses, capable of realizing CPP samples of sufficiently high quality in a reproducible way, and hence, is very versatile for studying the physics of the CPP–MR effect, including its temperature dependence.

But the use of grooved substrates also can give rise to other interesting geometries [85]: a magnetic multilayer can be deposited at normal incidence to the plane of the grooved substrate and then a current can be applied, which has a certain angle with the deposited multilayer itself. In this way the magnetoresistance with the current under an angle to the plane (the so-called CAP–MR) can be measured. Such CAP experiment leads to magnetoresistance values in between the results of the CIP and CPP geometries [85].

5.3 Conclusion

In this chapter an overview is presented on the different techniques that are used to study the perpendicular GMR effect in magnetic multilayers. While, after the discovery of the giant magnetoresistance effect in 1988, there has been an explosion of experimental work on multilayers in the current-in-plane geometry, this has not been the case for the perpendicular geometry, although MR effects are appreciably higher here. The reason is that it takes much more experimental effort to get access to the CPP geometry.

In *a first experimental approach* superconducting contacts and sensitive SQUID-based techniques were used to measure the ultra-small perpendicular resistance of multilayers that were not microstructured. These techniques are only applicable at 4.2 K and have led to the determination of all relevant spin-dependent scattering parameters at low temperatures, thanks to the reproducibility of sample fabrication and measurement. Due to the use of superconducting contacts, the influence of contact resistances is minimized, the current distribution is uniform over the cross section of the structure, and the perpendicular resistance of the multilayer can be measured in a reliable way. This implies that individual layer thicknesses in the multilayer can be systematically varied, and influence of interfaces and the bulk can be unraveled. Also, the spin-flip diffusion length, the scaling length of CPP–MR, can be studied using these techniques. Therefore, the method is the preferable one for understanding the fundamental physics of the CPP geometry and for comparison of experiments with model calculations.

With *all other experimental techniques* developed it is possible to measure the CPP–MR effect at room temperature; hence, the application of the CPP–MR effect in principle is possible. In the microlithography approach "pillars" are structured in the multilayer, leading to perpendicular resistance values of the order of mΩ, which can be measured by conventional means. Normal contacts are used, and CPP–MR can be measured from 4.2 K up to 300 K and higher. However, spurious serial contact resistances can easily be of the same order of magnitude as the multilayer resistance itself and the perpendicular current distribution is nonuniform due the small height/width ratio of the pillars. It turns out that it is very hard to make reproducibly large sample series and to systematically study the origin of the spin-dependent scattering in the multilayers. In addition, the perpendicular resistances are still too low for realistic applications.

Some of these problems are avoided in the multilayer-wire electrodeposition technique. The wires have a very large height/width ratio wich strongly reduces the influence of contact resistances. In addition, the perpendicular current distribution is uniform over the cross section of the wire. However, published results so far have not led to the very large MR effects that were obtained with the superconducting-contact and microlithography technique. Very thin individual normal layer thicknesses with antiferromagnetic coupling have not been reported. Whether this is of fundamental origin or not is still an open question. The effect may be related to the redissolution effects occurring during electrolytic growth. Another problem, particularly for applications, may be the selection and contacting of the appropriate multilayer wire out of the multitude of wires grown in the porous membrane. Maybe a combination of electrolytic growth within well-controlled lithographically prepared "membranes" might solve the latter problem.

In using oblique evaporation of multilayers on prestructured substrates, one avoids the complex process of microfabrication of the multilayer and replaces it by a much simpler substrate-structuring technique followed by a standard evaporation. Large absolute resistance values can be obtained and problems with contact resistances are avoided. Multiple samples can be grown in a reproducible way and, hence, spin-dependent scattering parameters can be determined, not only at 4.2 K, but at all temperatures. Due to the rather simple fabrication process and the compatibility with lithographic processing technology, the grooved substrate technique seems also attractive for applications.

It is to be expected that the perpendicular magnetoresistance effect in magnetic multilayers will receive considerable attention in the near future. Although CPP–MR data at low temperature in combination with theoretical models have led to an understanding of the importance of bulk and interface contributions to the spin-dependent scattering, a lot more of the physics is to be investigated, such as the discrimination between spin-dependent scattering and potential steps at interfaces for the various multilayer systems. An even larger remaining challenge is to use the CPP–MR effect in realistic applications of magnetic field sensing.

Acknowledgements. I would like to express my gratitude to all my collaborators of Philips Research and Eindhoven University who have contributed to the work on the perpendicular magnetoresistance effect. Furthermore, I have benefited a lot from discussions with colleagues in the field, in particular with A. Fert, H. van Houten, W.J.M. de Jonge and especially G.E.W. Bauer.

References

1. M. N. Baibich, J. M. Broto, A. Fert, F. Nguyen Van Dau, F. Petroff, P. Etienne, G. Creuzet, A. Friederich, J. Chazelas: Phys. Rev. Lett. **61**, 2472 (1988)
2. A. Fert, P. Bruno: in *Ultrathin Magnetic Structures* Vol. II, ed. by B. Heinrich, A. Bland (Springer, Berlin, Heidelberg 1994), pp. 82–117
3. P. M. Levy: Solid State Physics: Advances in Research and Applications **47**, 367 (1994)
4. B. Dieny: J. Magn. Magn. Mater. **136**, 335 (1994)
5. M. A. M. Gijs, G. E. W. Bauer: Adv. Phys., **46**, 285 (1997)
6. T. Valet, A. Fert: Phys. Rev. B **48**, 7099 (1993)
7. A. Fert, I. A. Campbell: J. Phys. F **6**, 849 (1976)
8. S. Zhang, P. M. Levy, in: *Magnetic Ultrathin Films: Multilayers and Surfaces, Interfaces and Characterization* Mat. Res. Soc. Symp. Proc. **313**, 53 (1993)
9. W. H. Butler, J. M. McLaren, X.-G. Zhang, in: *Magnetic Ultrathin Films: Multilayers and Surfaces, Interfaces and Characterization* Mat. Res. Soc. Proc. **313**, 59 (1993)
10. K. M. Schep, P. J. Kelly, G. E. W. Bauer: Phys. Rev. Lett. **74**, 586 (1995)
11. R. E. Camley, J. Barnaś: Phys. Rev. Lett. **63**, 664 (1989)
12. A. Barthélémy, A. Fert: Phys. Rev. B **43**, 13124 (1991)
13. R. Q. Hood, L. M. Falicov. Phys. Rev. B **46**, 8287 (1992)
14. R. Q. Hood, L. M. Falicov, D. R. Penn: Phys. Rev. B **49**, 368 (1994)
15. P. B. Visscher, H. Zhang: J. Magn. Magn. Mater. **121**, 449 (1993)
16. V. V. Ustinov, E. A. Kravtsov: J. Phys.: Condens. Matter. **7**, 3471 (1995)
17. P. M. Levy, S. Zhang, A. Fert: Phys. Rev. Lett. **65**, 1643 (1990)
18. A. Vedyayev, B. Dieny, N. Ryzhanova: Europhys. Lett. **19**, 329 (1992)

19. Y. Asano, A. Oguri, S. Maekawa: Phys. Rev. B **48**, 6192 (1993)
20. M. A. M. Gijs, M. Okada: Phys. Rev. B **46**, 2908 (1992)
21. S. K. J. Lenczowski, M. A. M. Gijs, J. B. Giesbers, R. J. M. van de Veerdonk, W. J. M. de Jonge: Phys. Rev. B **50**, 9982 (1994)
22. S. Zhang, P. M. Levy: J. Appl. Phys. **69**, 4786 (1991)
23. H. E. Camblong, S. Zhang, P. M. Levy: Phys. Rev. B **47**, 4735 (1993)
24. G. E. W. Bauer: Phys. Rev. Lett. **69**, 1676 (1992), *erratum* **70**, 1733 (1993)
25. A. Brataas, G. E. W. Bauer: Europhys. Lett. **2**, 117 (1994)
26. T. Oguchi: J. Magn. Magn. **126**, 519, (1993)
27. I. Mertig, P. Zahn, M. Richter, M. Eschrig, R. Zeller, P. H. Dederichs: J. Magn. Magn. Mat. **151**, 363 (1995)
28. T. Valet, A. Fert: J. Magn. Magn. Mater. **121**, 378 (1993)
29. A. Fert, T. Valet, J. Barnaś: J. App. Phys. **75**, 6693 (1994)
30. Q. Yang, P. Holody, S.-F. Lee, L. L. Henry, R. Loloee, P. A. Schroeder, W. P. Pratt, Jr., J. Bass: Phys. Rev. Lett. **72**, 3274 (1994)
31. J. Bass, Q. Yang, S.-F. Lee, P. Holody, R. Loloee, P. A. Schroeder, W. P. Pratt, Jr.: J. Appl. Phys. **75**, 6699 (1994)
32. J. Bass, P. A. Schroeder, W. P. Pratt, Jr.,S.-F. Lee, Q. Yang, P. Holody, L. L. Henry, R. Loloee: Mat. Sce. Engin. B **31**, 77 (1995)
33. A. Fert, J.-L. Duvail, T. Valet: Phys. Rev. B **52**, 6513 (1995)
34. P. Monod, S. Schultz: J. Physique (Paris) **75**, 393 (1982)
35. P. C. van Son, H. van Kempen, P. Wyder: Phys. Rev. Lett. **58**, 2271 (1987), *comment and reply ibid.* **60** 377 (1988)
36. M. Johnson, R. H. Silsbee: Phys. Rev. Lett. **60**, 377 (1988)
37. S.-F. Lee, W. P. Pratt, Jr., Q. Yang, P. Holody, R. Loloee, P. A. Schroeder, J. Bass: J. Magn. Magn. Mater. **118**, L1 (1993)
38. W. P. Pratt, Jr., S.-F. Lee, Q. Yang, P. Holody, R. Loloee, P. A. Schroeder, J. Bass: J. Appl. Phys. **73**, 5326 (1993)
39. W. P. Pratt, Jr., S.-F. Lee, J. M. Slaughter, R. Loloee, P. A. Schroeder, J. Bass: Phys. Rev. Lett. **66**, 3060 (1991)
40. P. A. Schroeder, J. Bass, P. Holody, S.-F. Lee, R. Loloee, W. P. Pratt, Jr., Q. Yang: in *Magnetism and Structure in Systems of Reduced Dimensions* NATO ASI, Ser. B, Vol. 309 ed. by R. Farrow, B. Dieny, A. Fert, M. Donath, B. Nersmeier (Plenum, New York, 1993) pp. 129–142
41. J. Slaughter, J. Bass, W. P. Pratt, Jr., P. A. Schroeder, H. Sato: Jpn. J. Appl. Phys. Suppl. **26**, 1451 (1987)
42. S.-F. Lee, W. P. Pratt, Jr., R. Loloee, P. A. Schroeder, J. Bass: Phys. Rev. B **46**, 548, (1992)
43. W. P. Pratt, Jr., S.-F. Lee, P. Holody, Q. Yang, R. Loloee, J. Bass, P. A. Schroeder: J. Magn. Magn. Mater. **126**, 406 (1993)
44. P. A. Schroeder, J. Bass, P. Holody, S.-F. Lee, R. Loloee, W. P. Pratt, Jr., Q. Yang, in: *Magnetic Ultrathin Films: Multilayers and Surfaces, Interafaces and Characterization* Mat. Res. Soc. Symp. Proc. **313**, 47 (1993)
45. Q. Yang, S.-F. Lee, P. Holody, R. Loloee, P. A. Schroeder, W. P. Pratt, Jr., J. Bass: Physica B **194-196**, 327 (1994)
46. Q. Yang, P. Holody, R. Loloee, L. L. Henry, W. P. Pratt, Jr., P. A. Schroeder, J. Bass: Phys. Rev. B **51**, 3226 (1995)
47. N. J. List, W. P. Pratt, Jr., M. A. Howson, J. Xu, M. J. Walker, D. Greig: In: 14th Int. Colloquium on Magnetic Films and Surfaces, Düsseldorf, Germany (Heinrich Heine Universität, 1994) Digest pp. 802–803

48. S. S. P. Parkin, Z. G. Li, D. J. Smith: Appl. Phys. Lett. **58**, 2710 (1991)
49. W. P. Pratt, Jr.: personal commun. (1994)
50. S. Zhang, P. M. Levy: Phys. Rev. B **47**, 6776 (1993)
51. J. L. Duvail, A. Fert, L. G. Pereira, D. K. Lottis: J. Appl. Phys. **75**, 7070 (1994)
52. M. A. M. Gijs, S. K. J. Lenczowski, R. J. M. van de Veerdonk, J. B. Giesbers, M. T. Johnson, J. B. F. aan de Stegge: Phys. Rev. B **50**, 16733 (1994)
53. M. A. M. Gijs, S. K. J. Lenczowski, J. B. Giesbers: Phys. Rev. Lett. **70**, 3343 (1993)
54. M. A. M. Gijs, S. K. J. Lenczowski, J. B. Giesbers, in: *Magnetic Ultrathin Films: Multilayers and Surfaces, Interafaces and Characterization* Mat. Res. Soc. Symp. Proc. **313** 11 (1993)
55. M. A. M. Gijs, J. B. Giesbers, M. T. Johnson, J. B. F. aan de Stegge, H. H. J. M. Janssen, S. K. J. Lenczowski, R. J. M. van de Veerdonk, W. J. M. de Jonge: J. Appl. Phys. **75** 6709 (1994)
56. W. Vavra, S. F. Cheng, A. Fink, J. J. Krebs, G. A. Prinz: In: 14th Int. Colloquium on Magnetic Films and Surfaces, Düsseldorf, Germany, (Heinrich Heine Universität, Düsseldorf, 1994) Digest pp. 50–51
57. W. Vavra, S. F. Cheng, A. Fink, J. J. Krebs, G. A. Prinz: Appl. Phys. Lett. **66**, 2579 (1995)
58. S. S. P. Parkin, R. Bhadra, K. P. Roche: Phys. Rev. Lett. **66**, 2152 (1991)
59. *TOSCA Reference manual* (Vector Fields Ltd, Oxford, UK, 1992)
60. M. A. M. Gijs, J. B. Giesbers, S. K. J. Lenczowski, H. H. J. M. Janssen: Appl. Phys. Lett. **63**, 111 (1993)
61. S. K. J. Lenczowski, R. J. M. van de Veerdonk, M. A. M. Gijs, J. B. Giesbers, H. H. J. M. Janssen: Appl. Phys. Lett. **75**, 5154 (1994)
62. M. A. M. Gijs, M. Okada: J. Magn. Magn. Mater. **121**, 371 (1993)
63. M. A. M. Gijs, M. Okada: J. Magn. Magn. Mater. **113**, 105 (1992)
64. G. E. W. Bauer, M. A. M. Gijs, S. K. J. Lenczowski, J. B. Giesbers: J. Magn. Magn. Mater. **126**, 454 (1993)
65. M. A. M. Gijs, S. K. J. Lenczowski, J. B. Giesbers, R. J. M. van de Veerdonk, M. T. Johnson, J. B. F. aan de Stegge: Physica Scripta **T55**, 189 (1994)
66. M. T. Johnson, S. T. Purcell, N. W. E. McGee, R. Coehoorn, J. B. F. aan de Stegge, W. Hoving: Phys. Rev. Lett. **68**, 2688 (1992)
67. B. Loegel, F. Gautier: Phys. Chem. Sol. **32**, 2723 (1971)
68. A. Fert: J. Phys. C. **2**, 1784 (1969)
69. R. J. Highmore, M. G. Blamire, R. E. Somekh, J. E. Evetts: J. Magn. Magn. Mater. **104-107**, 1777 (1992)
70. D. S. Lashmore, R. Oberle, M. P. Dariel, L. H. Bennett, L. Schwartzendruber, in: *Multicomponent Ultrafine Mcrostructures* Mat. Res. Soc. Symp. Proc. **132**, 219 (1988)
71. M. Alper, K. Attenborough, R. Hart, S. J. Lane, D. S. Lashmore, C. Younes, W. Schwarzacher: Appl. Phys. Lett. **63**, 2144 (1993)
72. M. Alper, P. S. Aplin, K. Attenborough, D. J. Dingley, R. Hart, S. J. Lane, D. S. Lashmore, W. Schwarzacher: J. Magn. Magn. Mater. **126**, 8 (1993)
73. M. Alper, K. Attenborough, V. Baryshev, R. Hart, D. S. Lashmore, W. Schwarzacher: J. Appl. Phys. **75**, 6543 (1994)
74. S. Z. Hua, D. S. Lashmore, L. Salamanca-Riba, W. Schwarzacher, L. J. Swartzendruber, R. D. McMichael, L. H. Bennett, R. Hart: J. Appl. Phys. **76**, 6519 (1994)

75. S. K. J. Lenczowski, C. Schönenberger, M. A. M. Gijs, W. J. M. de Jonge: J. Magn. Magn. Mater. **148**, 455 (1995)
76. T. M. Whitney, J. S. Jiang, P. C. Searson, C. L. Chien: Science **261**, 1316 (1993
77. L. Piraux, J. M. George, J. F. Despres, C. Leroy, E. Ferain, R. Legras, K. Ounadjela, A. Fert: Appl. Phys. Lett. **65**, 2484 (1994)
78. K. Liu, K. Nagodawithana, P. C. Searson, C. L. Chien: Phys. Rev. B **51**, 7381 (1995)
79. A. Blondel, J. P. Meijer, B. Doudin, J.-P. Ansermet: Appl. Phys. Lett. **65**, 3019 (1994)
80. T. Shinjo, T. Ono, H. Yamamoto: 14th Int. Colloquium on Magnetic Films and Surfaces, Düsseldorf, Germany (Heinrich Heine Universität) 1994, pp. 245–246
81. E. Bassous: IEEE Trans. **ED-25**, 1178 (1978)
82. M. A. M. Gijs, M. T. Johnson, A. Reinders, P. E. Huisman, R. J. M. van de Veerdonk ,S. K. J. Lenczowski, R. M. J. van Gansewinkel: Appl. Phys. Lett. **66**, 1839 (1995)
83. M. A. M. Gijs, S. K. J. Lenczowski, J. B. Giesbers, R. J. M. van de Veerdonk, M. T. Johnson, R. M. Jungblut, A. Reinders, R. M. J. van Gansewinkel: J. Magn. Magn. Mater. **151**, 333 (1995)
84. G. R. Harp, S. S. P. Parkin, R. F. C. Farrow, R. F. Marks, M. F. Toney, Q. H. Lam, T. A. Rabedeau, R. J. Savoy: Phys. Rev. B **47**, 8721 (1993)
85. T. Ono, T. Shinjo: J. Phys. Sec. Jpn. **64**, 363 (1995)

6 Physics of and Methods for Studying Metallic Multilayers with Interlayer Exchange Coupling and GMR Response

H. A. M. van den Berg

6.1 Developments in the Field of Magnetoresistive Sensors

The magnetoresistive effect is a well-known physical mechanism for measuring magnetic fields [1] and has found a growing range of applications, particularly over the last twenty years.

The effect has been applied for direct field measurements, such as in disk drives in computer systems [2,4–8], for tape heads in consumer products, such as DCC (digital compact cassettes) [9–11], magnetometers [12–14] and compass systems [15]. Further, it can also be used for measuring currents in cables [16].

Position detection constitutes another field of application. In this case, a permanent magnetization pattern is attached to the object that has to be de tected. A sensor detects the change in the field as a result of object displacement [17–23]. In other cases, the permanent magnet is rigidly mounted to the sensor. Then, the object conducts the flux of the source and its motion causes the gap in the flux conducting circuit to vary and the sensor field to be modulated. A large number of physical quantities that can be correlated to positions and orientations are thus accessible, i.e., speed, acceleration, force, rotational speed, torque, etc.

It is useful to distinguish between analog and digital signals. In general, the requirements for the latter category are less stringent since only the passage of a specific reference field needs to be detected. In principle, its accuracy can be improved by increasing the steepness of the signal as a function of time. Examples are the recording read heads [2–8] and the digitized position sensors with soft-magnetic gear wheels and racks. In its most simple form, the magnetization is switched between two saturated states and the sensor response is not affected significantly by the domain effects. In analog systems, the field at the sensor has to be resolved with full accuracy at each instant of time, and limitation to a simple discrimination between two states with the field either above or below a certain threshold level is not possible.

Up to now, the anisotropic magnetoresistive effect (AMR) is the galvanomagnetic mechanism that has mainly been exploited in devices that are mass produced. The anisotropic magnetoresistivity ϱ depends on the angle Θ between the magnetization $\boldsymbol{M}$ and the current density $\boldsymbol{J}$ and obeys:

$$\varrho(\Theta) = \varrho_{\perp} + \Delta\varrho_{\mathrm{amr}} \cos^2\Theta\,, \tag{6.1a}$$

$$\Delta\varrho_{\mathrm{amr}} = \varrho_{\parallel} - \varrho_{\perp}\,, \tag{6.1b}$$

where $\Delta\varrho_{\mathrm{amr}}$ is generally positive. $\varrho_{\parallel}$ and $\rho_{\perp}$ are the resistivity parallel and normal to $\boldsymbol{J}$, respectively. At room temperature, $\Delta\varrho/\varrho_{\parallel}$ is typically of the order of a few percent in disordered alloys.

The AMR effect has a large number of unique and advantageous properties. First, the effect is also significant in thin ferromagnetic films so that the impedance of the sensors can be given suitable values for electronic processing at small lateral dimensions. This has the advantage of miniaturization and high spatial resolution. In addition, a large number of sensors can be accommodated on one single substrate so that the AMR sensors are well suited to mass production with the use of planar technology. The impedance levels of the sensor elements are relatively low so that they are immune to electromagnetic noise and interference. Since the magnetic field can bridge empty spaces, the sensors are, in principle, contactless and do not exhibit mechanical wear. As a further consequence, the sensors can be encapsulated in many cases and can operate in industrial environments without special precautions. The detectors have no moving parts and require no maintenance. A further important feature characteristic of the AMR effect and of the small sensor-film thickness is the useful frequency range, which starts at 0 Hz and extends far into the MHz range. Finally, AMR sensors have the reputation of being highly sensitive, an aspect that is commented on below.

Magnetic fields are vectorial physical quantities that are defined by their direction in space and by their magnitude. The physical effects and the magnetic sensors are classified according to these field aspects. In addition, one distinguishes a third category with a sensitivity to the field component along only one reference direction. Thus, one defines direction, component and amplitude sensors or effects. According to (6.1), the AMR effect responds to the direction of $\boldsymbol{M}$ and it bears the character of a direction effect. A sensor that is kept in the saturated state by an appropriate field is also of the direction type. In order to obtain component or amplitude sensors, special precautions have to be taken that define the desired relationship between the field and $\boldsymbol{M}$. In other words, the high sensitivity of the AMR effect is rather confusing and it actually reflects the existence of a group of magnetic materials, the soft ones, and circumstances, e.g., vast thin films, that allow $\boldsymbol{M}$ to be rotated by small fields. The additional sensor relationship between $\boldsymbol{M}$ and $\boldsymbol{H}$ may have a detrimental effect on the reproducibility and tolerances of the sensors so that the AMR effect appears to be destined for the employement as a direction sensor. This application seems to be most suitable because the sensor response is hardly affected by domain effects.

It should be kept in mind that the relative change in the resistivity is a few percent only. This fact leads to a very strong temperature dependence

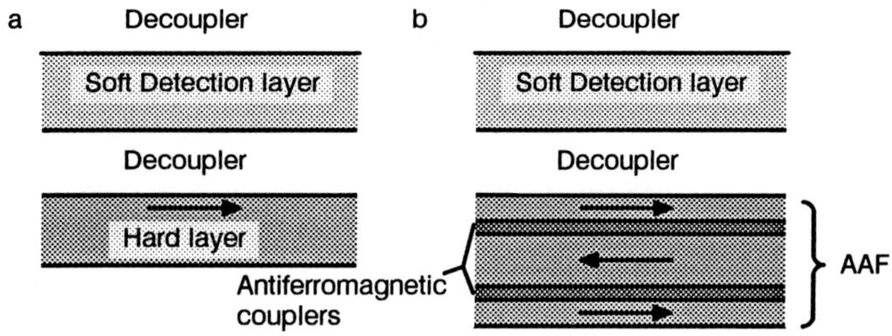

Fig. 6.1. a, b The principle of a hard-soft system. Archetype (**a**) with an artificial antiferromagnetic subsystem (AAF) (**b**) instead of a simple hard layer

since $\varrho_{\|}$ increases with T due to the increment in the frequency of electron-scattering events at phonons and magnons. This causes a change of typically about 0.3% per °C and a temperature shift of a few degrees already causes a ϱ change comparable to the maximal AMR span. These temperature effects are particularly disastrous for the analog devices. Wheatstone bridges [18] and electronic temperature compensation are the costly remedies against this deficiency of AMR. In principle, bridges with constant current supply should fully cancel the temperature variations of ϱ. However, a small bridge asymmetry always occurs due to material defects or manufacturing faults. Temperature drifts of typically several tens of μV still remain. For further balancing, additional resistors are added to the bridge. These are either thermistors or have to be laser trimmed.

It was emphasized already that analog applications, in particular, should benefit significantly from an increment in $\Delta\varrho/\varrho$ by an order of magnitude. For the recording heads also, there is a compelling need for sensors with improved performance due to the higher storage densities in media of the next generation. As a consequence, the magnetic flux emanating from the storage carriers and the dimensions of the sensors will be smaller, which place higher demands on sensor sensitivity.

The GMR effect discovered recently is a potential candidate for sensor applications that have much higher $\Delta\varrho/\varrho$. The phenomenon was discovered in multilayers in which adjacent metallic ferromagnetic layers are separated by a metallic interlayer that bears no permanent net magnetic moment. Neighboring magnetic layers have opposite magnetic moments in the ground state at zero external field due to a strong antiferromagnetic coupling across the interlayer. By applying a strong external magnetic field, typically of the order of 10 kOe, the magnetizations are forced to the parallel state and relative resistivity changes of up to 80% were achieved at ambient temperature. Thus, the sensitivity was very poor notwithstanding the high $\Delta\varrho/\varrho$. By taking thicker interlayers, the interlayer-coupling strength can be dramatically reduced and the sensitivity can be improved significantly at the cost of some loss in $\Delta\varrho/\varrho$. Various schemes have been proposed that exploit the GMR effect for sensor purposes. Most of these are based on the so-called hard-soft systems in which there are two types of magnetic layers, namely hard and soft ones (see Fig. 6.1 a). Each soft layer is only bounded by hard layer(s) and vice versa. In principle, only the soft layer is affected by the field H_m to be measured, and the magnetization in the hard layer should be fixed so that the relative orientation of the magnetization in both types of layers is changed by the field H_m. In reality, it has turned out to be difficult to combine sufficient hard and soft properties in one stack. Therefore, other schemes have been proposed.In Fig. 6.1 b one of these is shown.

Each hard layer is replaced by a subsystem with three magnetic layers, the so-called artificial antiferromagnet (AAF), that are strongly antiferromagneti cally coupled across metallic interlayers. The frictional torque against rotation of the AAF magnetization is proportional to the total volume of the three magnetic films of the AAF, while the torque exerted by the external field is proportional to the relatively small net magnetic moment of such a configuration with oppositely directed magnetizations. Such a system can therefore withstand much higher fields than a simple hard magnetic layer. The operational characteristics strongly depend on the phenomena that are the subject of the present study, namely the interlayer coupling and the magnetoresistive scattering properties of systems with a number of different magnetic and nonmagnetic layers. In the following, a brief introduction to the physics of both the interlayer (de)coupling and the giant magnetoresistive effect will be given. Subsequently, a number of experimental procedures and difficulties that arise when measuring these parameters will be discussed.

6.2 Physics of the Interlayer Coupling

The GMR effect has already reached a certain state of maturity and its feasibility with sensors has already been convincingly demonstrated. Its physics has close resemblance to the AMR effect. It was discovered by Grünberg et al. [24,25] in 1988. They observed that the resistive effect of their MBE-grown Fe(120 Å)/Cr(10 Å)/Fe(120 Å) magnetic bilayers on a GaAs(110) substrate exhibited astonishingly high changes upon saturation ($\Delta\varrho/\varrho = 1.5\%$) when compared to a pure Fe single layer with a thickness of 250 Å (0.2%). Furthermore, they observed that the effect does not depend on the direction of the current with respect to the external field so that the AMR mechanism could be excluded as a possible origin of the phenomenon. From the angular spectrum of light scattered at the spin waves in these magnetic bilayers, they concluded that both magnetic layers are antiferromagnetically coupled across the Cr interlayer [26–28]. The magnetizations in both layers are opposite in the ground state at zero field. The new magnetoresistive mechanism occurs when the magnetizations in the adjacent layers are forced to rotate from antiparallel alignment to a parallel one by the field H_0.

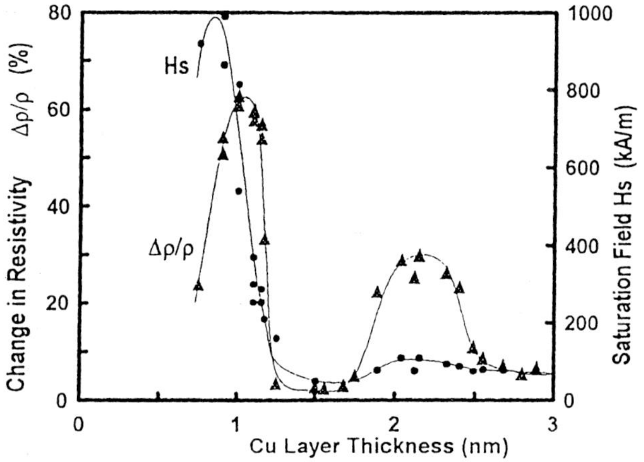

Fig. 6.2. The saturation field H_s and $\Delta\varrho/\varrho$ as a function of the interlayer thickness x of glass/Fe(6 nm)/[Co 1 nm/Cu x Å]N superlattices prepared by DC magnetron sputtering with total thickness of about 100 nm

The real breakthrough with this discovery happened when the group of Fert in Paris [29,30] reported on the giant changes in the magnetoresistivity in (Fe 30 Å/Cr 9 Å) superlattices with 60 periods deposited by MBE on GaAs(100). An $\Delta\varrho/\varrho$ of 80% was achieved at 4.2 K, at which ϱ was measured in the saturated state. Again, the opposite alignment of neighboring magnetic layers was observed in the ground state. The coupling between adjacent layers was strong and fields up to 20 kOe were required to overcome the coupling. Fields of this order of magnitude are not really attractive for sensor applications.

The nature of the interlayer coupling was initially not well understood. Parkin [31–34] demonstrated that the antiferromagnetic coupling across nonmagnetic interlayers is a very general phenomenon and is not typical of Cr only. Further, he discovered that the coupling is an oscillating function of the interlayer thickness and changes from ferro- into antiferromagnetic and vice versa. The changing sign of the coupling reflects itself both in the hysteresis curves and in the GMR signal. Figure 6.2 shows both the saturation field H_s and $\Delta\varrho/\varrho$ for glass/Fe(60 Å)/[Co 10 Å/Cu x Å]50 superlattices, prepared by DC magnetron sputtering [35]. Two maxima in $\Delta\varrho/\varrho$ can be recognized at Cu thicknesses of 10 and 21 Å. From the plot, it follows that a high GMR signal and high H_s are ostensibly correlated and that significant signals can only be expected in the antiferromagnetic thickness regimes. In other words, the sensitivity $d(\Delta\varrho/\varrho)/dH$ of the effect should then be intrinsically small and the usefulness of the GMR effect to sensors would be questionable. We shall, therefore, delve deeper into the physics of the interlayer coupling in the next section.

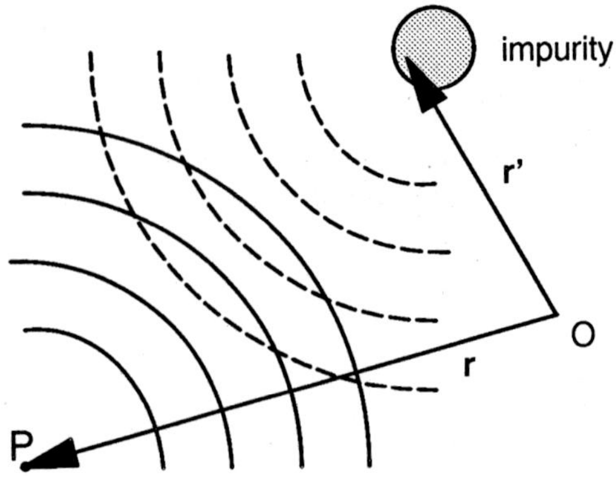

Fig. 6.3. Schematic representation of the interference of the waves emanated from point P and the scattered waves at the impurity leading to an spatial oscillating electron density (see also Fig. 6.4 **a**)

6.2.1 The Interlayer Coupling in Metallic Superlattices

Barthélémy et al. [30] concluded from the monotonic decline of the antiferromagnetic coupling strength at increasing Cr thickness in the Fe/Cr superlattices cited above that this interaction cannot originate in the RKKY coupling [36,37]. In principle,their argument holds. However, Parkin [31–34] convincingly demonstrated that the interlayer coupling does oscillate, although the thickness period of oscillation is much larger than the lattice-plane distance, as predicted by the RKKY theory.

The RKKY Coupling. The basic ideas of the RKKY theory are only roughly outlined here refering to the original works by Ruderman, Kittel [36], Kasuya and Yosida [37] for further details. The consequences of this mechanism on the interlayer coupling in multilayers can be found in the papers by Bruno et al. [38,39].

First, a nonmagnetic impurity in a nonmagnetic metallic host is studied, where the latter is consider as a free electron medium. The impurity potential is screened by the electrons so that the effective impurity potential V is much smaller. The effective potential causes a change in the electron density Δn, which is function of the distance to the impurity. Δn is given by

$$\Delta n(\boldsymbol{r}) = \int \chi_0(\boldsymbol{r}, \boldsymbol{r}')V(\boldsymbol{r}')\mathrm{d}(\boldsymbol{r}')\,, \tag{6.2}$$

where $\chi_0(\boldsymbol{r}, \boldsymbol{r}')$ is the susceptibility of the free electrons, and the integration is over the impurity volume, which is considered to be confined to one lattice cell. $\chi_0(\boldsymbol{r}, \boldsymbol{r}')$ gives the change in the electron density at point $\boldsymbol{r}$ due to a unit uDelta-potential jump at $\boldsymbol{r}'$). For a weak point-defect potential, which only slightly modifies the free electron density, it can be derived that the

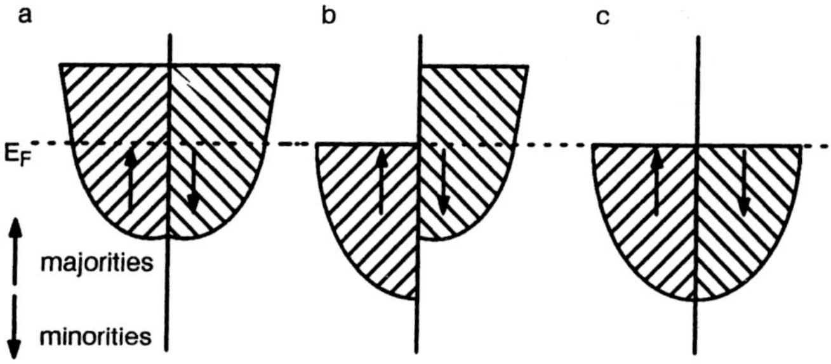

Fig. 6.4. a–c Schematic representation of the matching of the d bands of the magnetic element in the middle (**b**) and elements in columns at the left (**a**) and right (**c**) of it in the periodic table

susceptibility χ_0 of the free electrons, when the spherical quantum mechanical electron waves emanating from point P (see Fig. 6.3) are isotropically scattered at a point defect, obeys

$$\chi_0(\boldsymbol{r}-\boldsymbol{r}') = -\frac{2m}{\hbar^2 4\pi \mid \boldsymbol{r}-\boldsymbol{r}' \mid^2} \int_0^{k_F} \sin(2k \mid \boldsymbol{r}-\boldsymbol{r}' \mid) \, k \, \mathrm{d}k \,,$$

$$\text{with} \quad k = \sqrt{\frac{2m}{\hbar^2}E} \,, \tag{6.3}$$

where m, k_F are the electron mass and the wave vector $\mid \boldsymbol{r} \mid$ at the Fermi sphere and the energy of state $\boldsymbol{k}$, respectively. By integrating (6.3), one obtains

$$\chi_0(R) = \frac{m}{\hbar^2(2\pi)^3 R^4}\{2k_F R\cos(2k_F R) - \sin(2k_F R)\} \,, \tag{6.4}$$

where $R = \mid \boldsymbol{r}-\boldsymbol{r}' \mid$. The oscillation in χ_0 reflects the standing-wave pattern originating in the interference of the spherical waves with wave vectors between 0 and k_F. Note that the oscillation frequency in (6.4) is determined by the wave vector k_F at the Fermi surface and the other frequency components are mutually cancelling. One should bear in mind that the above results apply to weak coupling only. In the general case, the impurity potentials are rather strong so that the above formulae are approximations only. A more rigid non linear theory gives rise to deviations in both the magnitude and the phase of χ_0.

From (6.2), it is obvious that (6.4) gives the modulation of the uniform electron density in the case of a point defect. A superlattice contains a large number of parallel planes at which a potential jump is present. Let us take a closer look at the spatial variation in the free electron density near the origin of a Cartesian coordinate system in the case of a plane monolayer of magnetic atoms arranged in a plane perpendicular to the z axis. One shall ignore the fact that the scattering centers are located at a regular lattice in crystalline materials so that one can simply integrate (6.2). The change in the potential

$V(\boldsymbol{r}')$ is given by $V\delta(z)'$ with the δ–function $\delta(z)'$. The susceptibility χ_0 is given by

$$\chi_2(z) = \frac{4mk_{\mathrm{F}}^2}{\hbar^2(2\pi)^2} \int_{2k_{\mathrm{F}}Z}^{\infty} \frac{u\cos u - \sin u}{u^3} du \,,$$

at which $u = 2k_{\mathrm{F}}\boldsymbol{r}'$, which gives for $2k_{\mathrm{F}}z \gg 1$

$$\chi_2(z) \simeq -\frac{m}{\hbar^2(2\pi)^2} \frac{\sin(2k_{\mathrm{F}}z)}{z^2} \,. \tag{6.5}$$

It will be shown that the coupling strength is proportional to χ_2, so that the last expression shows that the strength is inversely proportional to the square of the distance to the monolayer at large z.

With regard to magnetic impurities in a nonmagnetic host, it is well known that the 3d band of the magnetic transition elements Ni, Co and Fe is split up into a majority band with electron spins parallel to the magnetization and a minority band with opposite spin direction. The majority band of the strong ferromagnetic elements Co and Ni lies completely below the Fermi level E_{F}, while the minority band intersects E_{F}.

For elements of columns to the left of the magnetic element in the periodic table, the minority band resembles the d band of the interlayer and the potential jumps of the majorities tends to be larger than for the minorities. In this case, the scattering potential could be approximated by

$$V^- \simeq V_{\mathrm{a}} \,, \tag{6.6a}$$
$$V^+ = V_{\mathrm{a}} + v_{\mathrm{m}} M_{\mathrm{s}} \,, \tag{6.6b}$$

where V^+, V^- and M_{s} refer to the scattering potentials of majorities and minorities and to the saturation magnetization, respectively. The opposite applies to interlayer elements to the right of the magnetic element, so that roughly

$$V^+ \simeq V_{\mathrm{a}} \,, \tag{6.6c}$$
$$V^- = V_{\mathrm{a}} + v_{\mathrm{m}} M_{\mathrm{s}} \,. \tag{6.6d}$$

According to (6.2), there will also be a difference in the electron densities of both spin directions outside the impurity. In the case of a ferromagnetic monolayer, the spatial variation in the electron density and in the difference between these densities are given by

$$\Delta n(z) = \Delta n^+(z) + \Delta n^-(z) = \chi_2(z) V_{\mathrm{a}} \,, \tag{6.7a}$$
$$|\, \boldsymbol{m}(z) \,| = \Delta n^+(z) - \Delta n^-(z) = -\chi_2(z) v_{\mathrm{m}} M_{\mathrm{s}} \,. \tag{6.7b}$$

The symbol $|\, \boldsymbol{m} \,|$ at the left in (6.7b) refers to the magnetic polarization (in the number of Bohr magnetons) in the nonmagnetic medium induced by the magnetic medium. Note that the magnetization distribution in the nonmagnetic medium is completely covered by the susceptibility χ_2.

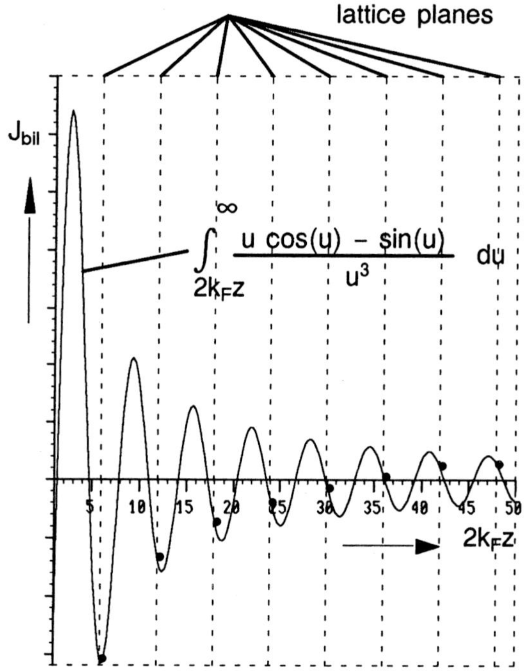

Fig. 6.5. The aliasing effect. $\chi_2(u)$ according to (6.5) is represented by the solid line. The heavy points are the intersections with the lattice planes and represent the real oscillations in $J_{\rm bil}$

Finally, one arrives at the RKKY coupling. For this, a magnetic atom is positioned in the nonmagnetic medium in point P at distance z from the magnetic monolayer with $z = 0$. For this atom at the origin, there is a potential increase $\Delta V^{+,-}$ for both spin directions, where the $^+$ and $^-$ signs refer to the spin directions in the magnetic monolayer. The interaction energy ΔE is given by

$$\Delta E = \int \{V^+(\boldsymbol{r}')n^+(z') + V^-(\boldsymbol{r}')n^-(z')\}\,\mathrm{d}\boldsymbol{r}' - \int 2V_{\rm a}(\boldsymbol{r}')n^0\,\mathrm{d}\boldsymbol{r}' , \qquad (6.8)$$

in which $n^+(z')$ and $n^-(z')$ follow from (6.7) and V^+ and V^- from (6.6) with $\Delta n^{+,-} = n^{+,-} - n^0$. n^0 is the free electron density in each spin channel. For the interaction energy, one obtains

$$\Delta E = \int \{V_{\rm a}(\boldsymbol{r}')\Delta n(z)' - v_{\rm m}(\boldsymbol{r}')\boldsymbol{m}(\boldsymbol{r}')\cdot \boldsymbol{m}_1(z')\}\mathrm{d}\boldsymbol{r}' ,$$

in which the first term describes the chemical interaction and the second one the magnetic exchange energy, which can be rewritten as

$$\Delta E_{\rm bil} = \boldsymbol{m}\cdot \boldsymbol{M}_1 \int v_{\rm m}(\boldsymbol{r}')\chi_2(z')v_{\rm m}\,\mathrm{d}\boldsymbol{r}' , \qquad (6.9)$$

Table 6.1. Comparison of the theoretical period length of the interlayer exchange coupling after Bruno's modified RKKY theory and experimental data

Period length after RKKY theory			Experimental period length		
Spacer material	Texture	Λ in ML	Multilayer	Λ in ML	Ref.
Ag	(001)	$\Lambda_1 = 2.4$	Fe/Ag/Fe	$\Lambda_1 = 2.4$	
		$\Lambda_2 = 5.6$		$\Lambda_2 = 5.6$	
Au	(001)	$\Lambda_1 = 2.6$	Fe/Au/Fe	$\Lambda_1 = 2$	[47]
		$\Lambda_2 = 8.6$		$\Lambda_2 = 7.5$	
Cu	(001)	$\Lambda_1 = 2.6$	Fe/Cu/Fe	$\Lambda_1 = 6$	[48]
		$\Lambda_2 = 5.9$	Co/Cu/Co	$\Lambda_1 = 2.6$	[45]
				$\Lambda_2 = 8$	
	(111)	$\Lambda_1 = 4.5$	Ge/Cu/Fe	$\Lambda_1 = 6$	[92]
			Co/Cu/Co	$\Lambda_1 = 5 - 6$	[34–170]

where $\boldsymbol{m}$ is the magnetic moment of the impurity in P. The interlayer-exchange density between the two plane parallel magnetic monolayers separated by a nonmagnetic slap with thickness D follows from the translation invariance in the lateral direction. By taking again the potentials for both spin directions of the second monolayer evenly distributed in lateral direction, the integration yields

$$\Delta E_{\mathrm{bil}} = C\ \chi_2(D)\boldsymbol{M}_1 \cdot \boldsymbol{M}_2 = -J_{\mathrm{bil}}\hat{\boldsymbol{\imath}}_1 \cdot \hat{\boldsymbol{\imath}}_2 \,, \tag{6.10}$$

where $\boldsymbol{M}_1$ and $\boldsymbol{M}_2$ are the saturation magnetizations of both monolayers. The subscript $_{\mathrm{bil}}$ refers to bilinear coupling.

For the present discussion, we assume that even the monolayers have uniform permanent moments parallel to the unit vectors $\hat{\boldsymbol{\imath}}_1$ and $\hat{\boldsymbol{\imath}}_2$. J_{bil} is the so-called bilinear coupling constant, which oscillates as a function of D due to χ_2. This can be understood as follows. One monolayer, let us say layer 1, probes via its conduction band the difference in the electron density of both spin orientations, induced by monolayer 2. The energy per majority electron of layer 1 is lower than per minority electron. The majorities of 1 will therefore occupy the spin state with the highest electron density. The sign of $\Delta n^+ - \Delta n^-$ oscillates as a function of D and so does J_{bil}. The oscillation period λ_{R} is equal to π/k_{F} (6.5), being of the order of the lattice-plane distance of the interlayer. The oscillation period actually observed is much longer (typically 10 Å) so that additional arguments are required.

A number of reseachers [40–42] independently gave the following explana tion, called the aliasing effect, to this dilemma. So far, it was assumed $\chi_2(D)$

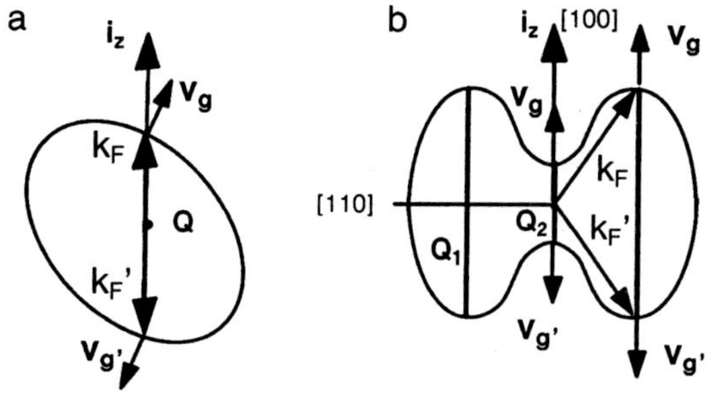

Fig. 6.6. a, b Construction scheme for the period length of the oscillation in the interlayer coupling based on the Fermi surface of the interlayer. (**a**) Elliptical Fermi surface for the direction R, (**b**) The multiplicity in the oscillation period in the case of a dog-bone type Fermi surface

to be a continuous function. However, the interlayer is a solid material and the atoms are arranged at discrete lattice planes. Therefore, the thickness D and along with it $\chi_2(D)$ can only take discrete values. The consequences of this observation are elucidated in Fig. 6.5. The lattice planes, indicated by the dotted lines, intersect the curve at the heavy points. At these planes, the monolayer 1 probes the electron-density variations Δn, induced by monolayer 2. It is obvious that the period length of the oscillations in $J_{\rm bil}$ is much larger than the one in χ_2. Coehoorn [40] gave the following expression for this period length Λ of the coupling oscillations:

$$\Lambda = \frac{1}{\mid 1/\lambda - n/d \mid}, \tag{6.11}$$

where d is the the atomic-plane distance of the interlayer parallel to the magnetic monolayers, λ is $\pi/k_{\rm F}$. The nonnegative integer n is such that Λ is larger or equal to 2d. It should be noted that Λ should depend strongly on the interatomic layer distance d and that, in principle, it might vary by orders. Tables 6.1 and 6.2 show that actually the measured Λ is always of the order of 10 Å so that one has to conclude that the above picture is too simple.

The aliasing approach is a first attempt to account for the discreteness of the crystal lattices. The spatial distribution of the atoms reflect itself in the shape of the Fermi surface. The Fermi surfaces of crystalline materials are generally not spherical as for the free electrons. As a consequence, $k_{\rm F}$ is directional-dependent and the oscillation period depends on the direction [39,43], as has been experimentally verified [44,45]. According to Bruno [39], there generally exists more than one period for a given crystal direction. The periods in multilayers can be derived from the Fermi surface by using the following procedure. Fig. 6.6 a shows the elliptical Fermi surface of a hypothetical interlayer in a stack with stack normal $\boldsymbol{i}_z$. One now searches for pairs of wave vectors $\boldsymbol{k}_F$ and $\boldsymbol{k}_{F'}$ for which $\boldsymbol{Q} = \boldsymbol{i}_z - \boldsymbol{k}_{F'}$ is parallel to the stack normal and for which the group velocities $\boldsymbol{V}_g$ and $\boldsymbol{V}_{g'}$ are opposite. Only one such direction is found for each $\boldsymbol{i}_z$ in the case of an elliptical surface. Figure 6.6 b gives the famous dog-bone Fermi surface, being typical for cubic lattices, as has been verified by de Haas-van Alphen and cyclotron-resonance measurements. Using the same construction method, one finds three solutions for the [100] $\boldsymbol{i}_z$ direction, two of which are equivalent. $\boldsymbol{Q}_1$ and $\boldsymbol{Q}_2$ represent a short and a long period, respectively. This multiperiodicity has been experimentally observed by several researchers [34,45–47,92,170]. Furthermore, this scheme provides an adequate tool to estimate the length of the long period, which agrees well with the data in Table 6.2. A comparison between measured period lengths and the predictions according to the modified RKKY theory by Bruno is shown in Table 6.1.

It is obvious that the period length cannot depend on the interlayer thickness only because there are reflections of the electron waves at each interface between different media, i.e., at transition from magnetic to nonmagnetic media and vice versa but also at eventual interfaces between two nonmagnetic media. As a consequence, the interference of these waves and thus the interlayer-coupling strength should depend on the magnetic layer thicknesses. Bruno [49] and Barnas [50,51] predicted that thickness variations of the magnetic layer cause oscillations in the magnitude and not so much in the sign of J_{bil}. This impact has also been experimentally confirmed [52–54].

Remarks on First-Principle Calculation of the Interlayer Coupling. As emphasized above, RKKY theory cannot provide a quantitative prediction of the coupling strength due to the ad hoc introduction of the scattering potentials in (6.6). Such estimates have to be based on first principle calculations, at which the surface-energy density E_{F} in the state with parallel magnetizations in the layers is compared with its counterpart E_{AF} with antiparallel $\boldsymbol{M}$'s, from which the coupling constant J can directly be derived by the relation $J = (E_{\text{AF}} - E_{\text{F}})/2$.

These calculations are hampered by the high accuracy required since the energy difference is orders of magnitude smaller than the total energy itself. A lot of these evaluations concern the Fe/Cr system [55–58]. Stoeffler et al. demonstrated that, in this system, the oscillation in the coupling originates in the antiferromagnetic interaction between neighboring Cr atoms, so that interlayers containing an odd or even number of atomic planes exhibit a ferromagnetic or antiferromagnetic coupling, respectively. The higher energy state is characterized by an ordering defect in the magnetic moments of the monolayers in the middle of the Cr layer and by a reduction of the Cr moments in this region in the so-called collinear state. They found that the Cr couples strongly antiferromagnetically to the Fe, which has been confirmed experimentally [59]. No long oscillation period was discovered and the strength of the calculated coupling was at least one order of magnitude too high. Stoeffler and coworker therefore used a direct-space method so that they could introduce interdiffusion near the Fe/Cr interface. This mixing resulted in a strong reduction in the coupling strength and an increase in the oscillation-period length [60].

A similar behavior was also found for the Fe/Mn system [60], in which Mn is also characterized by an intrinsic antiferromagnetic order. However, Mn can couple both ferromagnetically and antiferromagnetically to the Fe which gives an extra degree of freedom to prevent the coupling defect from occurring. Stoeffler et al. [59] introduced two additional ordering modes for Cr and Mn interlayers, to wit the helicoidal and the noncollinear states, which are favorable at interlayers thicker than about 10 ML in the frustrated state so to reduce the defect energy. In these states, the Cr moment is basically constant and the configuration could be evaluated with the micomagnetic theory, as presented in Sect. 6.3.2.

The coupling mechanism of Cr and Mn deviates strongly from the RKKY type. Also starting from first principles, Stoeffler studied the coupling across Ru and V and found a Friedel type of oscillation with the too short oscillation-period length of π/k_{F}.In a later paper on Co/Ru [61], he introduced atomic steps at the Co/Ru interfaces and found again a reduction in the coupling strength and an increase in the coupling period. It is striking that the ground state is antiferromagnetic at the thinnest Ru layers of one and two ML thickness.

The Biquadratic Coupling. A higher-order term, the so-called biquadratic coupling, of the form

$$\Delta E_{\mathrm{bq}} = -J_{\mathrm{biq}}(\hat{\boldsymbol{i}}_1 \cdot \hat{\boldsymbol{i}})^2 \tag{6.12}$$

has been discovered in a number of weakly coupled bilayers [48,62–64]. The electronic contribution to this effect was found to be too small [65,66] to account for the experimental data in the case of interlayers that couple by a RKKY type of interaction. Stoeffler et al. [67,68] discovered that an electronic origin of the biquadratic coupling arises in the case of interlayers with intrinsic antiferromagnetic arrangement, like Cr. As discussed in the previous section, the energy of the defect in the middle of the Cr layer in the frustrated situations can be reduced by the noncollinear or by helicoidal spin configurations in the Cr at Cr thicknesses larger than about 10 ML. The frustrated state is characterized by parallel and opposite orientation of the Fe magnetizations at Cr interlayers with even and odd numbers of ML, respectively. Later studies have revealed that these spin configurations are also possible in thinner Cr layers when the angle between the Fe magnetizations deviate from the situation of highest frustration. As a consequence, the coupling energy exhibits a significant biquadratic term, which agrees well with the experimental data. This mechanism does not apply to systems in which the coupling takes place by the polarization of conduction electrons. Therefore, other mechanisms such as surface roughness and loosy spins have been

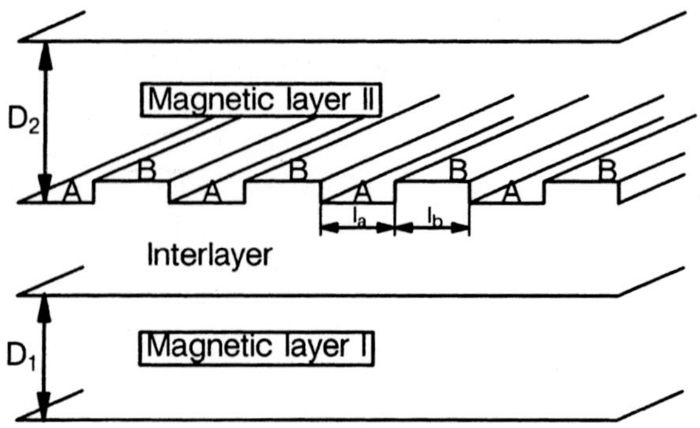

Fig. 6.7. Cross-section of a hypothetical stack with a stepped interface at one of the surfaces of the coupling layer. Regions A couple ferromagnetically and B antiferromagnetically

suggested by Slonczewski [69,70]. The ferromagnetic bridges [71] constitute a mechanism that is strongly related to the surface roughness mentioned above.

Let us elucidate the origin of biquadratic coupling due to the roughness of the interface. Figure 6.7 shows a stepwise varying interface at one side of the interlayer and an atomicly flat interface at the opposite side. For simplicity, it is assumed that the terraces are symmetrical and that the regions A and B have equal widths l_a and l_b, respectively. Furthermore, it is assumed that the regions A couple ferromagnetically via the interlayer to the bottom magnetic layer with coupling constant J_a, while the B terraces have antiferromagnetic coupling constant J_b. Let us first assume that the bulk-exchange constant is infinite. The magnetization in both layers is uniform with M directions $\bar{\Theta}_1$ and $\bar{\Theta}_2$. The interlayer exchange-energy surface-density is $E_\mathrm{int} = l_\mathrm{a}J_\mathrm{a}\cos(\bar{\Theta}_1 - \bar{\Theta}_2) + l_\mathrm{b}J_\mathrm{b}\cos(\bar{\Theta}_1\bar{\Theta}_2)$, which is zero and does not depend on $\bar{\Theta}_1$ and $\bar{\Theta}_2$ when $l_\mathrm{a}J_\mathrm{a} + l_\mathrm{b}J_\mathrm{b} = 0$. This changes when the bulk exchange constant is finite as follows from simple considerations. Let us assume that the lateral $\boldsymbol{M}$ variations are small due to the strong bulk-exchange constant and due to the high density of the steps. First, one takes the mean $\boldsymbol{M}$'s in the adjacent layers oppositely directed, i.e., $\bar{\Theta}_1 - \bar{\Theta}_2 = \pi$. $\boldsymbol{M}$ in the regions B are in the minimum energy state and the excursion from the mean will be very small here. The deflection in the ferromagnetic coupling regions A will be much larger. However, it is effectively suppressed by the strong lateral coherence. The energy reduction by this deflection is very small due to the quadratic character with respect to the deflection of the cosine function near $\bar{\Theta}_1 - \bar{\Theta}_2$ being zero or π. The situation is different when $\bar{\Theta}_1 - \bar{\Theta}_2 = \pi/2$, so that $\boldsymbol{M}$ in both the ferromagnetic and antiferromagnetic regions deflects from the mean at which the energy gain is proportional to the deflection angle. In other words, the perpendicular orientation of the mean $\boldsymbol{M}$'s is the state of minimum energy, which can be described by an energy term of the form $-J_\mathrm{biq}\cos^2(\bar{\Theta}_1 - \bar{\Theta}_2)$, where $J_\mathrm{biq} < 0$.

Let us consider the situation with $l_a J_a + l_b J_b \# 0$ in a more quantitative fashion. We confine ourselves to situations in which at least one of the coupling constants J_a, J_b is negative. The magnetization is uniform in the case of two positive constants.The magnetization angles Θ_i can be developed in Fourier series so that

$$\Theta_i = \sum_{n=0}^{\infty} \Theta_i^{(n)} \cos(nk_0 x) \quad \text{where} \quad k_0 = 2\pi/(l_a + l_b)\,, \tag{6.13}$$

where $x = 0$ is located at the middle of a terrace. By integrating the surface-energy densities due to the bulk- and the interlayer-exchange energy, one arrives at a mean surface-energy density of

$$\begin{aligned} E_\sigma = & \sum_{i=1}^{2} k_0^2 (d_i A_i)/2 \sum_{n=1}^{\infty} n^2 (\Theta_i^{(n)})^2 \\ & - \frac{J_a l_a + J_b l_b}{l_a + l_b} \cos(\Theta_1^{(0)} - \Theta_2^{(0)}) \\ & + \frac{J_a l_a + J_b l_b}{l_a + l_b} \frac{\cos(\Theta_1^{(0)} - \Theta_2^{(0)})}{4} \sum_{n=1}^{\infty} (\Theta_i^{(n)})^2 \\ & + (J_a - J_b) \frac{\cos(\Theta_1^{(0)} - \Theta_2^{(0)})}{8\pi} \sum_{n=1}^{\infty} \frac{(\Theta_i^{(n)})^2}{n} \sin \frac{n 2\pi l_a}{l_a + l_b} \\ & + (J_a - J_b) \frac{\cos(\Theta_1^{(0)} - \Theta_2^{(0)})}{4\pi} \\ & \times \sum_{n=1}^{\infty} \sum_{m=1 m\# n}^{\infty} \Theta^{(n)} \Theta^{(m)} \\ & \times \left\{ \frac{1}{n-m} \sin \frac{(n-m)\pi l_a}{l_a + l_b} + \frac{1}{n+m} \sin \frac{(n+m)\pi l_a}{l_a + l_b} \right\} \\ & + (J_a - J_b) \frac{\sin(\Theta_1^{(0)} - \Theta_2^{(0)})}{\pi} \sum_{n=1}^{\infty} \frac{\Theta^{(n)}}{n} \sin \frac{n \pi l_a}{l_a + l_b} \end{aligned} \tag{6.14}$$

where $\Theta^{(n)} = \Theta_1^{(n)} - \Theta_2^{(n)}$ and A_i and D_i are the exchange constant and the thickness of magnetic layer (i), respectively. From the second term in the above relationship, it is obvious that a bilinear coupling contribution occurs with constant $(J_a l_a + J_b l_b)/(l_a + l_b)$, being the weighted sum of both couplings constants. It is not self-evident that the remaining terms give rise to a biquadratic coupling term and have no significant contribution to the higher order terms. To evaluate the $\Theta_i^{(n)}$'s, one requires that the first derivatives of E_σ with respect to the $\Theta_i^{(n)}$'s are zero. This yields simple relations between the $\Theta_1^{(n)}$'s and $\Theta_2^{(n)}$'s, given by

$$D_1 A_1 \Theta_1^{(n)} + D_2 A_2 \Theta_2^{(n)} = 0, \quad \text{for} \quad n > 0\,, \tag{6.15}$$

and a second system of equations with a band type of matrix structure. Equation (6.15) shows that $\Theta_1^{(n)}$ and $\Theta_2^{(n)}$ have opposite signs for each n, and the Fourier component is smaller for the thicker and stiffer layer. The second system can be put into a relative simple relationship between the $\Theta^{(n)}$'s when $DA = D_2 A_2 = D_1 A_1$. The matrix exhibits a band type of structure and the $\Theta^{(n)}$'s tend to decrease by the second power of n due to the bulk-exchange stiffness. Therefore, as a first approximation, we confine ourselves to $\Theta^{(1)}$. Ignoring the higher-order terms in Θ, one gets:

$$\Theta^{(1)} = \frac{-(J_\mathrm{a} - J_\mathrm{b}) \frac{\sin(\Theta_1^{(0)} - \Theta_2^{(0)})}{\pi/2} \sin \frac{\pi l_\mathrm{a}}{l_\mathrm{a} + l_\mathrm{b}}}{2k_0^2(DA) + \left\{ \frac{J_\mathrm{a} l_\mathrm{a} + J_\mathrm{b} l_\mathrm{b}}{l_\mathrm{a} + l_\mathrm{b}} + \frac{J_\mathrm{a} - J_\mathrm{b}}{2\pi} \right\} \sin \frac{\pi l_\mathrm{a}}{l_\mathrm{a} + l_\mathrm{b}} \cos(\Theta_1^{(0)} - \Theta_2^{(0)})} . \tag{6.16}$$

It is important to have a closer look at the denominator of (6.16). The term with cosine-multiplication factor is typically significantly smaller than 5. For $A = 10^{-6}$ erg/cm, $D = 10^{-7}$ cm and $l_\mathrm{a} + l_\mathrm{b} = 5 \times 10^{-7}$ cm, $2\,k_0^2$ DA is larger than 10. Therefore, one can assume that the cosine term has small impact. By confining ourselves to the same first order term in the energy expression and by substituting (6.16) into (6.14), one gets for the biquadratic coupling constant

$$J_\mathrm{biq} = \frac{-(J_\mathrm{a} - J_\mathrm{b})^2 \sin^2(\pi l_\mathrm{a}/(l_\mathrm{a} + l_\mathrm{b}))}{\pi^2 k_0^2 (DA)} . \tag{6.17}$$

Note that the cosine term in (6.16) does not appear until the second order in J_biq, i.e.,J_biq is really a constant when the cosine term is small and that it does not vary when a hysteresis curve is traced. It is also obvious from (6.17) that J_biq is always negative. Note that the denominator is relatively large and that a significant J_biq only occurs when the sine term approaches one. The biquadratic coupling oscillates when a stack is growing in monolayer by monolayer mode. There is a minimum each time that a monolayer is completed, with an oscillation with a period of one lattice distance, like in Friedel oscillations. Note that the occurrence of these oscillations strongly depends on the structural condition of both interfaces and it would be the strongest when one of these interfaces is atomicly flat in all samples. $\mid J_\mathrm{a} - J_\mathrm{b} \mid$ is the largest at the zero-crossings of the bilinear constant $(J_\mathrm{a} l_\mathrm{a} + J_\mathrm{b} l_\mathrm{b})/(l_\mathrm{a} + l_\mathrm{b})$ and the biquadratic contribution becomes most apparent then.

It should be emphasized that the occurrence of a biquadratic coupling constant does not require that J_a and J_b have opposite signs. J_biq is also finite when J_a and J_b are both negative and different. In the latter case, the lateral coupling-strength variations cause spatial fluctuations in the angle of the $\boldsymbol{M}$'s of adjacent layers, which are again suppressed by the bulk-exchange forces.

The surface-energy density of a sandwich with equally thick magnetic layers can be put in the form

$$\begin{aligned} E_\sigma = & -H_0 M_\mathrm{s} D\cos(\Theta_1 - \Phi_{\mathrm{H}_0}) - H_0 M_\mathrm{s} D\cos(\Theta_1 - \Phi_{\mathrm{H}_0}) \\ & -J_\mathrm{bil}\cos(\Theta_1 - \Theta_2) - J_\mathrm{biq}\cos^2(\Theta_1 - \Theta_2) , \end{aligned} \tag{6.18}$$

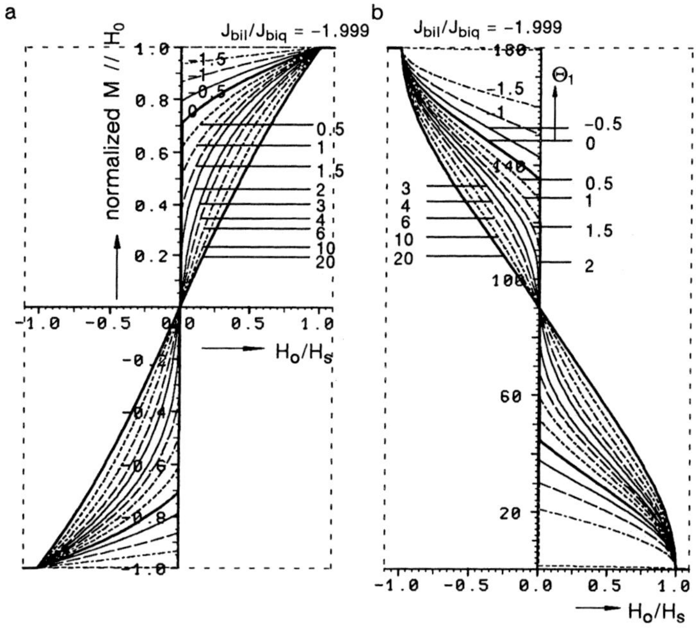

Fig. 6.8. a, b The normalized magnetization curve (**a**) and the magnetization angle Θ_1 of layer 1 (**b**) as a function of the normalized field H_0/H_s

in which Θ_1 and Θ_2 are the mean $\boldsymbol{M}$ directions. With $\Theta = \Theta_1 = -\Theta_2$, the following analytic solutions can be derived for the three ranges of J_{bil}:

$$\Theta = 0 \quad H_0 > 0\,, \tag{6.19a}$$

$$\Theta = \Pi \quad H_0 < 0\,, \tag{6.19b}$$

for $J_{bil} > 2 \mid J_{biq} \mid$, i.e., for the ferromagnetic case (I). Furthermore,

$$\cos(\Theta) = \left(-\frac{(J_{bil} - 2J_{biq})}{3J_{biq}}\right)^{1/2} \times \cos\left(\cos^{-1}\left(\frac{3DM_sH_0/2}{(J_{bil} - 2J_{biq})^{2/3}/(-3J_{biq})^{1/2}}\right)/3\right)\,, \tag{6.20a}$$

for $\mid H_0 \mid < 2(J_{bil} - 2J_{biq})^{3/2}/\{(-3J_{biq})^{1/2}3DM_s\}$, and

$$\cos(\Theta) = -\left(-\frac{(J_{bil} - 2J_{biq})}{12J_{biq}}\right)^{1/2} \times \left(\left(\frac{1-(1-\sin^2(\omega))^{1/2}}{\sin(\omega)}\right)^{1/3} + \left(\frac{1-(1-\sin^2(\omega))^{1/2}}{\sin(\omega)}\right)^{-1/3}\right)\,,$$

$$ \tag{6.20b} $$

with $\sin(\omega) = 2(J_{\rm bil} - 2J_{\rm biq})^{3/2}/\{(-3J_{\rm biq})^{1/2}3DM_{\rm s}H_0\}$, for $2(J_{\rm bil} - 2J_{\rm biq})^{3/2}/\{(-3J_{\rm biq})^{1/2}3DM_{\rm s}\} < \mid H_0 \mid < -(J_{\rm bil} + 2J_{\rm biq})/(2DM_{\rm s})$, and

$$ \begin{aligned} \Theta &= 0 \quad \text{for } H_0 > -2(J_{\rm bil} + 2J_{\rm biq})/(DM_{\rm s}) \,, \\ \Theta &= \pi \quad \text{for } H_0 < 2(J_{\rm bil} + 2J_{\rm biq})/(DM_{\rm s}) \,, \end{aligned} \tag{6.20c} $$

for $\mid J_{\rm bil} \mid < -2J_{\rm biq}$; i.e., for the remanent case (II). Finally,

$$ \begin{aligned} \cos(\Theta) = {} & \left(\frac{(J_{\rm bil} - 2J_{\rm biq})}{3J_{\rm biq}} \right)^{1/2} /2 \\ & \times \left(\left(\frac{-1}{\tan(\omega)} + \left(1 + \frac{1}{\tan^2(\omega)} \right)^{1/2} \right)^{1/3} \right. \\ & \left. - \left(\frac{-1}{\tan(\omega)} + \left(1 + \frac{1}{\tan^2(\omega)} \right)^{1/2} \right)^{-1/3} \right) , \end{aligned} \tag{6.21a} $$

with $\tan(\omega) = -2(2J_{\rm biq} - J_{\rm bil})^{3/2}/\{(-3J_{\rm biq})^{1/2}3DM_{\rm s}H_0\}$, for $\mid H_0 \mid < -(J_{\rm bil} + 2J_{\rm biq})/(2DM_{\rm s})$, and

$$ \begin{aligned} \Theta &= 0 \quad \text{for } H_0 > -2(J_{\rm bil} + 2J_{\rm biq})/(DM_{\rm s}) \,, \\ \Theta &= \pi \quad \text{for } H_0 < 2(J_{\rm bil} + 2J_{\rm biq})/(DM_{\rm s}) \,, \end{aligned} \tag{6.21b} $$

for $J_{\rm bil} < 2J_{\rm biq}$; i.e., for the reversible case (III). Note that a remanence occurs of

$$ \cos(\Theta) = ((J_{\rm bil}/J_{\rm biq} - 2)/2)^{1/2} $$

when $\mid J_{\rm bil} \mid > -2J_{\rm biq}$. The remanence requires $J_{\rm bil} > -2J_{\rm biq}$, i.e., the ferromagnetic case. Figure 6.8 provides an overview of the normalized M–H_0 and Θ–H_0 curves as a function of the coupling-constant ratio $J_{\rm bil}/J_{\rm biq}$. The biquadratic component leads to a curvature in the $M - H_0$ loops. The transition between the three regions is continuous.

6.2.2 Experimental Investigations Towards the Interlayer Coupling

A large effort has been put in the experimental investigation of the oscillating interlayer coupling. Parameters like the strength, phase and period length of interlayer-thickness have been studied for varying repetition numbers of the multilayers and for a large variety of material combinations. The coupling strength is measured by various methods. It was discovered by Brillouin scattering of light by spin waves [72], a method also used by other groups

Table 6.2. The interlayer-coupling constant J (in erg/cm^2) for a number of multilayers. The meanings of the abbreviations are as follows:

DC	DC magnetron sputtering	Sput	sputtering
e-be	e-beam deposition	IBS	ion beam sputtreing
Perm	permalloy	O	own data
T	temperature (K)	Prep	preparation

System	J	1st AF max Å	2nd AF max Å	T	Prep	Refs.
Co/Ag	–0.15	9	24	300	MBE	[81]
Co/Cu	–0.35	9.5	21	300	DC	[35]
	–0.2	9	20	4.5	DC	[82]
	–0.5	9	20	300	DC	[83]
	–0.7	9	20	300	MBE	[44]
	–0.3	9	20	300	IBS	O
Co/Os	–0.55	9	25	300	e–be	[84]
Co/Re	–0.29	5	–	300	DC	[85]
Co/Rh	–3.0	8	20	4	DC	[98]
	–11.0	< 8	20	4	MBE	
Co/Ru	–5	neg	17	4.5	DC	[86]
	–6	< 6	20	300	MBE	[87]
Fe/Ag	< -0.003			300	MBE	[88]
	< -0.0005			6	MBE	[89]
Fe/Al	–0.3	6	18	6	MBE	[47]
Fe/Au	–0.014	17	34	6	MBE	[47]
Fe/Cr	–0.6	9	24	300	DC	[90]
	–0.65	6	26	6	MBE	[89]
	–0.9	11	27	4.2	DC	[91]
Fe/Cu	–0.26	10	20	295	MBE	[88]
	–0.1	14	26	4.2	Sput	[92]
Fe/Mo	–0.07	12	22	4.2	DC	[93]
Ni/Ag	–0.02	11	29	4.2	Sput	[94]
Perm/Ag	–0.02	11		4.2	Sput	[95]
	–0.01	11		300	Sput	[95]
Perm/Cu	–0.035	9	19	300	DC	[96]
	–0.048	9	19	4.2	DC	[96]
	–0.05	8	–	300	DC	[97]
	–0.05	9	20	300	IBS	O
Perm/Ru	-0.1	6	18	300	DC	[78]

[73,74,88]. The same information, not spatially resolved however, can be extracted by ferromagnetic resonance [75–77]. Both methods provide the coupling both in the ferromagnetic and the antiferromagnetic coupling regime. However, they have the disadvantage of requiring high investments in equipment and the systems should be operated by highly skilled personnel. Neutron scattering, by means of which the period doubling of the chemical unit cell of the stack was demonstrated, is an even more complicated alternative [32,90]. Most frequently employed, the Kerr technique and the VSM are much simpler to handle. Furthermore, the magnetoresistivity measurement constitutes a powerful tool when the samples have significant GMR signal. These last methods, based on the B–H and $\Delta\varrho/\varrho - H$ curves, are simple, but provide the coupling constant in the AF regime only. Ferromagnetic coupling constants can only be determined with a trick, so-called spin engineering [78], which requires stacks with additional AF-coupled layers. In general, the constants are derived from superlattices with many identical periods containing one magnetic and one nonmagnetic layer. In samples with weak anisotropy, J can be derived from the in-plane saturation field H_s by using the simple relation

$$J = J_\mathrm{bil} + 2J_\mathrm{biq} = -D_\mathrm{m} M_\mathrm{s} H_\mathrm{s}/L \,, \quad \text{with} \quad L = 4 \tag{6.22}$$

where D_m and M_s are the thickness and saturation magnetization of the magnetic layer, respectively. In principle, the value of H_s should easily be measurable since it is theoretically signified by a kink in the curve. In reality, the approach to saturation is gradual, thus hampering the accurate determination of its value. In other cases, magnetic bilayers with two identical magnetic layers are used, for which the factor L in (6.22) should be 2. In the case of media with anisotropy, the multiphase magnetization approach [79,80] and measurements in a few directions of H_0 are required for attaining J. It will be returned to the analysis of stacks with anisotropic behavior. To derive both exchange constants, the details of the shape of the magnetization curves should be considered, as will be demonstrated in the sections on biquadratic coupling and surface anisotropy in Co/Ru systems.

In Table 6.2, a representative selection from the literature with data on J in the first AF coupling maximum, the interlayer thickness of the first and second AF peak, the measuring temperature and the method of preparation have been collected.

Table 6.2 demonstrates that the AF interlayer-exchange coupling is a rule rather than an exception. J varies from very weak coupling (in Fe/Ag, Fe/Au, Ni/Ag and Permalloy/Ag) to a strong one in the Co/Ru and Co/Rh system with $J \simeq -5\,\mathrm{erg/cm^2}$ and $-11\,\mathrm{erg/cm^2}$, respectively [87,99]. It should further be noted that the maximum in Co/Rh occurs at higher interlayer thickness that is about two times larger than in the Co/Ru, so that, according to the $1/(\text{interlayer thickness})^2$-correction of Parkin [33] (see also (6.5)), Co/Rh couples seven times stronger than Co/Ru. Note that Rh is located just below Co in the periodic table and that the electronic bandstructure of both elements

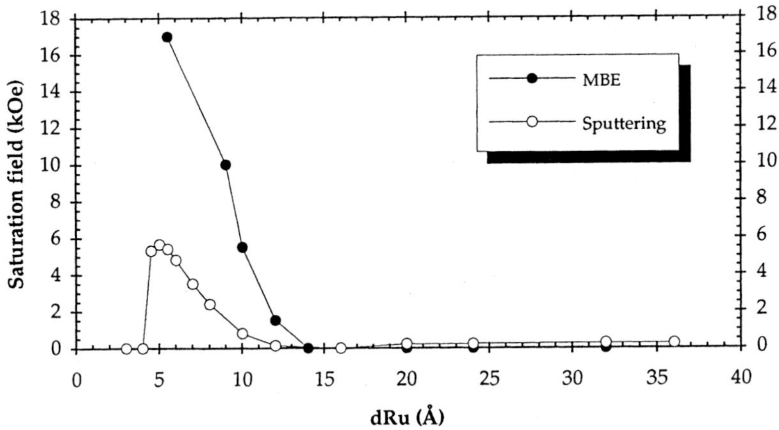

Fig. 6.9. The saturation field H_s as a function of the Ru thickness of sandwiches made by:
a) RF magnetron sputtering: glass/Ru50 Å/ [Co30 Å/Ru d_{Ru} Å]/Co30 Å/Ru50 Å
b) MBE: mica/Ru150 Å/[Co30 Å/Ru d_{Ru} Å]/Co30 Å/Ru100 Å

exhibit a close resemblance. It is astonishing that elements with a great similarity in the band structures give rise to the highest coupling known to date. One might anticipate that the potential jump ΔV and thus the coupling is small in such circumstances. it Parkin systematically screened the elements of the periodic table in the vicinity of the magnetic transition elements as interlayer materials [33]. He pointed out that the interlayers in a given row of the periodic table have increasing $\mid J \mid$ when going from left to right and that its value decreases when going downwards in a column.

Coupling of Co/Ru Sandwiches Prepared by MBE Versus Sputtering. It has experimentally been found that, in many cases, $\mid J \mid$ is larger for samples that have been grown by MBE rather than by sputtering. This phenomenon is thought to originate from the larger surface intermixing in the sputtered samples due to the substrate bombardment by neutral atoms. This gives rise to a random component in the atom distribution near the interface and to a destructive interference of the scattered electronic waves. The coupling is frustrated [39,99].

Figure 6.9 shows a comparison between H_s of the in-plane B–H curves of Co/Ru superlattices for both deposition methods. The MBE samples were grown epitaxially on mica. Both the Co and Ru have hexagonal crystal structure with sixfold symmetry axis parallel to the stack normal. The Co thickness of 30 Å is high enough to have a preferential in-plane orientation of $\boldsymbol{M}$ [114]. Notwithstanding the crystal anisotropy, J can directly be derived with (6.22) from the H_s values of the in-plane B–H curves since the high degree of isotropy in planes perpendicular to the hexagonal axis. The sputtered sam-

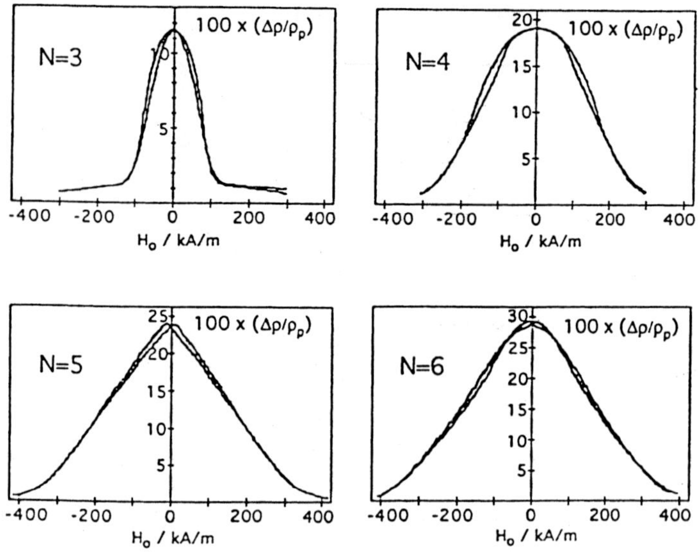

Fig. 6.10. The change in the magnetoresistivity $\Delta\varrho/\varrho$ as a function of the external field H_0 for Co/Cu stacks with varying number N of Co/Cu pairs. Stack geometry: glass/Fe60 Å/[Co10 Å/Cu10 Å]$\times N$

ples are poorly textured and do not exhibit a distinct anisotropy so that (6.22) can be used again. Thus, one can conclude that the coupling J in the sputtered samples is indeed weaker.

Variation of the Interlayer Coupling Strength in Co/Cu Multilayers. In (6.22), it is assumed that the magnetization in each layer does not vary across its thickness, which is a rather good approximation in lattices with thin magnetic layers, i.e., in the order of a few tens of Å. Secondly, (6.22) starts from a constant interlayer coupling in the superlattice across all the interlayers. This need not be true. Figure 6.10 shows the GMR curves of four stacks with different number of Co/Cu pairs. The curves show little hysteresis and are characteristic of AF coupling. The magnetization is in the plane at $H_0 = 0$ and the lateral anisotropy can be neglected because of the polycrystalline nature of the stacks. Therefore, (6.22) where $L = 4$ and 2 should give the upper and lower limits of the saturation field H_s, respectively, i.e., the upper boundary of H_s is two times the lower one. In Fig. 6.10, the H_s ratio of the stacks with $N = 6$ and $N = 3$ is about 3, which is too large. The actual span of H_s is even much larger as follows from the open squares in Fig. 6.11 with N ranging from 1 to 35. The span of 8 can also not be understood by taking account of the Fe buffer. Indeed, the thick ferromagnetic, rigid buffer may cause an increase of the span to 4 [101].

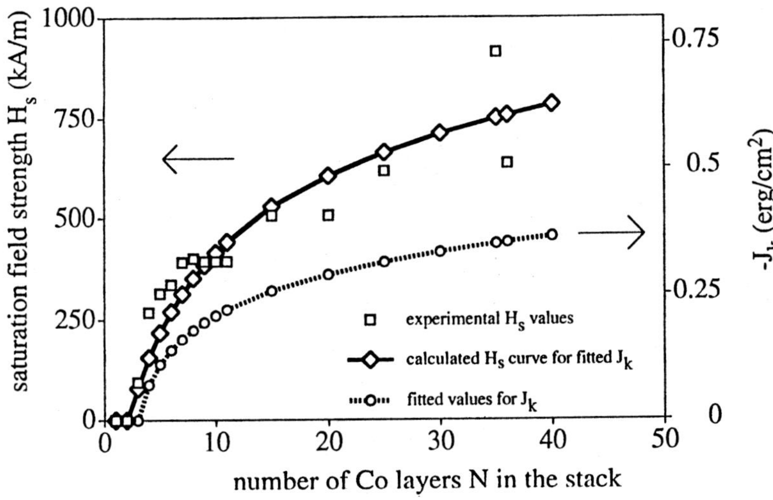

Fig. 6.11. The saturation field H_s as a function of the number of Co/Cu pairs: *open squares* experimental H_s's, *open lozenges* H_s's values for the fit, *open circles* evaluated J_k's

From these arguments, one has to conclude that the interlayer coupling has to vary in the stack. The Cu layers and the corresponding coupling constant J_k have been marked by the index k, with k monotonically increasing with the distance to the Fe buffer have been solved. The index k is 1 for the Cu layer closest to the substrate. The analysis is hampered by the presence of the thick Fe buffer layer, in which a nonuniform magnetization distribution cannot be excluded. The so-called atomic layer model (ALM) for evaluating the J_k's [102–104] has been adopted.

The principle of this method is demonstrated in Fig. 6.12. Each magnetic layer is decomposed in one or more plane parallel sheets parallel to the interfaces of the superlattice. Each sheet is uniformly magnetized so that the $\boldsymbol{M}$ distribution is one-dimensional with $\boldsymbol{M}$ varying along the z-axis. Each sheet interacts with its neighboring magnetic sheets by exchange interaction only. Bulk and interlayer exchange are equivalently treated and the bulk exchange is put into the form of (6.10) by introducing a J_b, given by

$$J_b = 2A/d_s \,, \tag{6.23}$$

where A is the bulk-exchange constant and d_s the distance between the centers of adjacent sheets. It should be noted that J_b is of the order of $100\,\mathrm{erg/cm^2}$ for a typical A of $10^{-6}\,\mathrm{erg/cm}$ and atomic plane distance of $2 \times 10^{-8}\,\mathrm{cm}$. This is much larger than the coupling constants in Table 6.2 and illustrates that the coherence inside the magnetic layers is much stronger than across the interlayers. The nonuniformity inside the magnetic layers does not become significant until the magnetic layer thickness D is of the order of

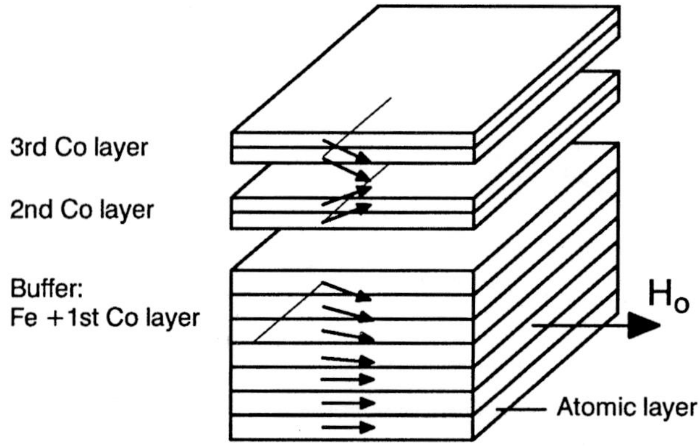

Fig. 6.12. The decomposition of the ferromagnetic portion of the stack into plane-parallel sheets that are mutually coupled by either bulk or interlayer coupling

magnitude of the domain wall. In this case, the helicoidal spin configuration is along the stack normal and its width is determined by the competition of the bulk-exchange stiffness, and the interlayer-exchange torques, which predominantly act on the outermost atomic planes of the magnetic layers and the magnetic field that affect all magnetic dipoles equally. In other words, the deflection from uniformity depends on H_0 and is the largest in the canted situation in which the magnetizations in the adjacent layers enclose angles of about 90°. The deflection from uniformity becomes significant for layer thicknesses of about $D_c = 0.2\,A/J(D_c = 20\,\text{Å}$ for $A = 10^{-6}\,\text{erg/cm}$ and $J = 1\,\text{erg/cm}^2)$.

Let us return to the atomic layer model. From the above arguments, one can see that it is necessary to decompose the 70 Å thick Fe/Co buffer. The surface-energy densities of the sheets, including the magnetic field, the magnetostatic and the anisotropy terms, are added and the total energy is minimized with respect to the $\boldsymbol{M}$ directions in the sheets. In the present case, the $\boldsymbol{M}$'s are in the plane when H_0 is in-plane and the equilibrium distribution is defined by a simple recurrence equation. This equation can conveniently be solved by a numerical scheme [101]. The procedure is as follows. One fits the B–H curve for the stack with $N = 2$ and finds J_1. With this J_1, one subsequently fits the curve for the stack with $N = 3$ and determine J_2. In this way, one scanns all the stacks consecutively and evaluates the J_k's given by the open circles in Fig. 6.11. J_k's vary from $J_2 = -0.07$ to $J_{35} = -0.35\,\text{erg/cm}^2$. It should be emphasized that the first Cu layer exhibits a weak ferromagnetic coupling and the stack with $N = 2$ gives no GMR signal.

The large variation in J_k in a specific stack may appear a little surprising. However, this phenomenon has been more frequently observed. It was also seen in Co/Pt superlattices, in which the coupling was also weakest near the substrate [105] and in Fe/Si multilayers [106]. Also, the coupling across the

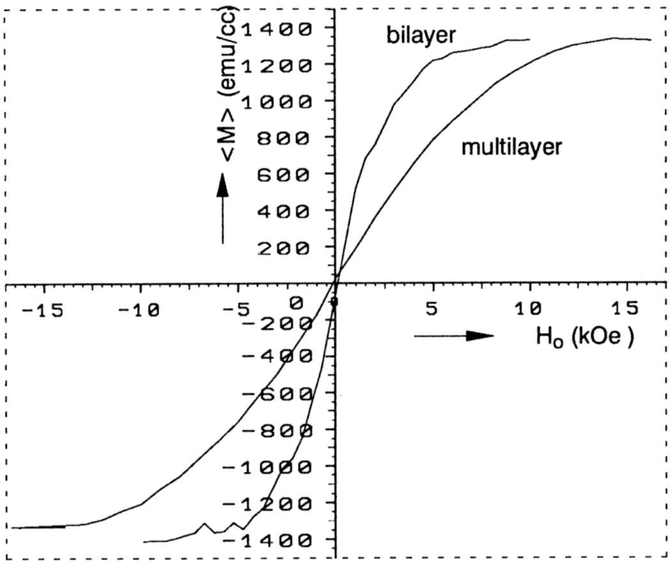

Fig. 6.13. The in-plane $M - H_0$ loops of two sputtered Co/Ru stacks with different number of periods.
a) Bilayer: glass/Ru50 Å/Co30 Å/Ru5 Å/Co30 Å/Ru20 Å
b) Multilayer: glass/Ru 50 Å/[Co30 Å/Ru5 Å] × 15/Ru50 Å

Cu layer in exchange-biased sensor systems based on permalloy always seems to be ferromagnetic. It might be speculated whether the variation has a fundamental origin. This is very unlikely because it must then be expected that the deviating coupling would be observed at both the top and bottom periods of the stacks. This is definitely not true. Figure 6.13 shows the M–H curves of a *DC* magnetron sputtered Co/Ru bilayer and of a multilayer with 15 periods. It is obvious that, in conformity with (6.22) that H_s of the multilayer is about two times higher than its counterpart. In this case, one has no significant dependence of J_k on k. Further, we achieved an AF-coupling across the 1st Cu layer by adapting deposition conditions in Fe/Co/Cu/Co stacks.The above facts rule out a fundamental origin and the J_k dependence presumably originates in a different conditioning of the interfaces due to changes during the film growth. A possible mechanism might be the so-called orange peel effect, in which the topography of the buffer is duplicated by the intermediate Cu layer into the bottom surface of the Co layer that is above it. This causes a ferromagnetic interlayer coupling as already was recognized by Néel a long time ago [107–109]. The effective interlayer coupling is the sum of the electronic and the magnetostatic terms which have opposite signs. The idea is that the roughness of the initial layers is smoothed out by covering the stack with Co/Cu periods. The smoother interfaces cause the antiferromagnetic electronic contribution to dominate stronger.

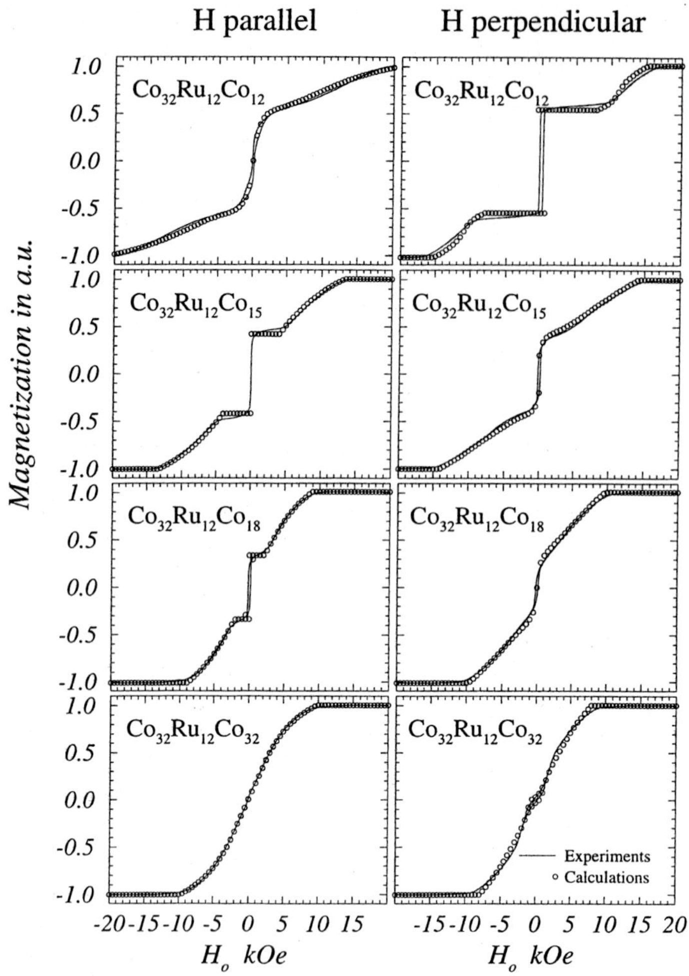

Fig. 6.14. The *M–H* loops for H_0 parallel and normal to the stack plane of Co/Ru bilayers with unequal thicknesses of both Co layers. Circles are the calculated moments for the experimental solid lines: mica/Ru160 Å/Co32 Å/Ru12 Å/CoY Å/Ru40 Å, where Y is 12, 15, 18 and 32 Å

Analysis of Co/Ru Sandwiches with Different Anisotropies. The Co/Ru bilayers exhibit another effect which is typical for thin layers, to wit surface anisotropy. Its impact is demonstrated by the *M–H* curves in Fig. 6.14 of MBE-grown asymmetrical bilayers (mica/Ru160 Å/Co32 Å/Ru12 Å/CoY Å/Ru40 Å with Y = 12, 15, 18 and 32). The samples have been epitaxially grown on mica substrates and have *hcp* crystalline structure with hexagonal axis perpendicular to the film plane [110–112]. The samples show no in-plane preferential directions because of the high symmetry of the per-

Table 6.3. The physical parameters as derived from the solid hysteresis curves in Fig. 6.14. J, interlayer-exchange constant; A, bulk-exchange constant; K_S, surface-anisotropy constant; K_V, bulk-anisotropy constant

Stack geometry	J	A	K_S	$K_V\,10^6$erg/cm^3	
	erg/cm^2	10^{-6}erg/cm	erg/cm^2	top layer	bottom layer
$Co_{32}Ru_{12}Co_{12}$	−1.4	0.45	0.43	4.4	10
$Co_{32}Ru_{12}Co_{15}$	−1.3	0.45	0.42	4.6	10
$Co_{32}Ru_{12}Co_{18}$	−1.0	0.40	0.42	5.6	10
$Co_{32}Ru_{12}Co_{32}$	−1.4	0.40	0.43	9.3	10

pendicular hexagonal axes. The Co layer nearest to the substrate has a fixed thickness of 32 Å. According to the critical thickness $D_c = 0.2\,\text{A/J}$, the impact of a twist in magnetization configuration, e.g. a wall, in the bottom layer should be expected in these stacks with strong interlayer coupling (see Table 6.2).

The M–H curves in the perpendicular and the in-plane direction are quite different and change dramatically with thickness Y of the top Co layer. The thinnest stack shows a preference for perpendicular $\boldsymbol{M}$ orientation, as can be deduced, among other things, from the higher in-plane saturation field H_s. The occurrence of this preference at the thinnest Co layer indicates the presence of surface anisotropy. For this thickness, the M–H curve in the perpendicular direction has a plateau as a consequence of the difference in the thickness and, as a result, in the total moments of both Co layers. This plateau is also observed in the in-plane curves of the stacks with a larger thickness of the top layer. Again, its origin is the asymmetry in the Co thicknesses and is not present in the symmetrical stack at the bottom of Fig. 6.14. Note that H_s of the symmetrical stack in Fig. 6.14 is much larger than its counterpart in Fig. 6.9 at $d_{Ru} = 12$ Å. This is caused by the much higher substrate temperature during the Co/Ru growth of the samples in Fig. 6.9.

Analysis of the hysteresis curves of stacks in which such a large number of parameters play a role requires a systematic approach. As a first step, a model was elaborated that takes account of the surface and bulk anisotropies in both layers and of the bulk- and interlayer-exchange couplings. Further, it incorporates the domain-wall configurations that match the contradicting requirements for $\boldsymbol{M}$ at the interfaces and in the volume of, in particular, thick magnetic layers [113–115]. The interface spins are different in that they are subjected to the interlayer coupling. Arbitrary direction of the external field can be used in the model so that a large number of different hysteresis curves for each stack can be analyzed. Thus, the freedom in the physical fit

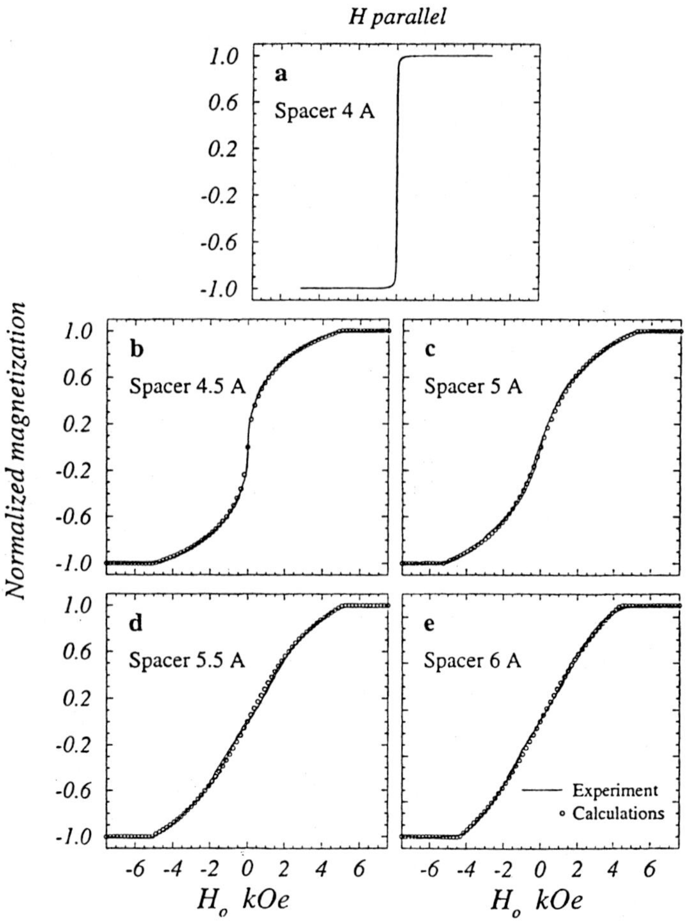

Fig. 6.15. The M–H loops for in-plane H_0 of Co/Ru bilayers with varying Ru thickness and equal thicknesses of both Co layers. Circles are the calculated moments for the experimental solid line. glass/Ru50 Å/Co30 Å/RuY Å/Co30 Å/Ru50 Å, where Y is 4, 4.5, 5, 5.5 and 6 Å

parameters discussed above can be systematically reduced by increasing the number of experimental hysteresis curves to be considered.

Table 6.3 presents the set of parameters that is achieved by matching the experimental solid curves of Fig. 6.14. The quality of the fits can be judged from the open circles. Compared to the value in Table 6.2, J is relatively small because the Ru thickness is rather large and on the flank of the $H_s - d_{Ru}$ curve in Fig. 6.9. It is therefore not surprising that there is some scattering in J. Most striking is the difference in the K_v's of the top and bottom layer. This can be attributed, first, to a magnetoelastic anisotropy contribution due to the lattice misfit of Ru and Co and, second, to a larger surface roughness

Table 6.4. The bilinear, biquadratic and effective coupling constants as derived from fitting the curves of Fig. 6.15 for the sputtered samples. Sputtered samples: glass/Ru50 Å/Co30 Å/Rud_{Ru}Å/Co30 Å/Ru50 Å. MBE samples: Ru150 Å/Co30 Å/Rud_{Ru}Å/Co30 Å/Ru100 Å (last two columns)

	Sputtered				MBE	
d_{Ru} Å	4.5	5	5.5	6	5	10
J_{bil} erg/cm^2	−0.5	−0.8	−0.9	−0.9	−1.8	−0.9
J_{biq} erg/cm^2	−0.18	−0.11	−0.05	−0.02	−0.3	−0.02
$J_{\mathrm{bil}}/J_{\mathrm{biq}}$	2.8	7.3	18	45	6	45
J_{bil} erg/cm^2	−0.86	−1.02	−1	−0.94	−2.4	−0.94

of the top Co layer [115]. The above discussion shows that consideration has to be given not only to eventual variations of J_k. Changes in the other physical parameters also influence the estimated values of J. It must be kept in mind that similar effects may also occur in polycrystalline samples. Mattheis et al. [116] investigated theoretically the course of H_{s} in multilayers with perpendicular anisotropy as a function of the number of periods in the stack.

Biquadratic Coupling in Co/Ru Sandwiches. In the foregoing section, we confined ourselves to sandwiches with relatively thick Ru layers so that the coupling constant is small compared to the value of J given in Table 6.2. The maximum in $-J$ is found at much smaller Ru thicknesses as is demonstrated by Fig. 6.9. At d_{Ru} smaller than about 4 Å, the coupling suddenly assumes a ferromagnetic character for both MBE-and sputter-grown samples. The magnetization curves exhibit a very typical course for the various d_{Ru} in this regime. Figure 6.15 illustrates this for sputtered samples. At 4 Å, one has the stepwise changing ferromagnetic curve, and the sandwich with $d_{\mathrm{Ru}} = 6$ Å exhibits a rather good linear behavior. In the intermediate range, one observes the curved M–H loops which are typical of a biquadratic component in the coupling, at which the strongest curvature is present at 4.5 Å (compare with Fig. 6.8 a).

These curves have been fitted by assuming isotropic materials, identical magnetic films and both bilinear and biquadratic coupling. The resulting curves are indicated by the dots in Fig. 6.15 and agree well with the solid experimental curves. The fitting parameters are summarized in Table 6.4.

Curved hysteresis loops can also be obtained by superposing linear B–H curves, i.e., by assuming lateral inhomogeneities and independent areas with different coupling constants or magnetic layer moments. How can one be sure that one is dealing with real biquadratic coupling? The sudden change within 2 Å from the pure ferromagnetic behavior at 4 Å to the dominantly antiferromagnetic one at 6 Å already indicates that the homogeneity of the samples

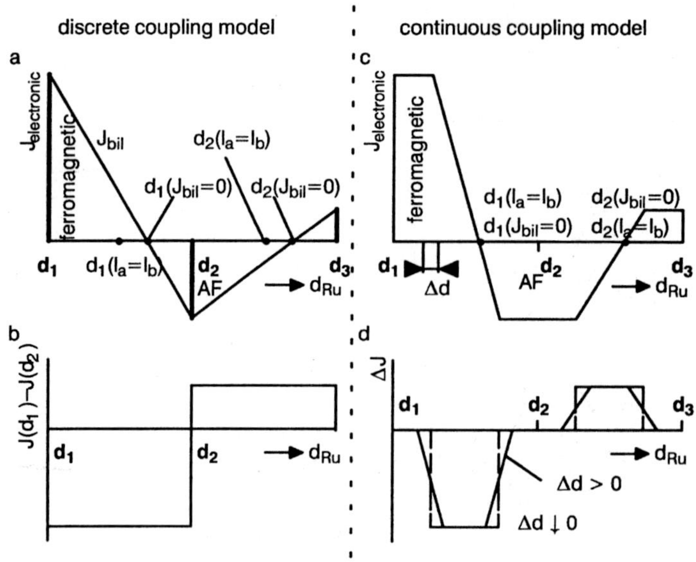

Fig. 6.16. Two possible explanations of the course of the thickness dependence of the bilinear and the biquadratic coupling constants as function of the Ru thickness

is excellent. Therefore, the question is whether the sizes of the regions with a particular type of coupling are smaller or larger than the lateral coherence length of $\boldsymbol{M}$. Let us assume that the dimensions of the coherent regions with the same coupling constant are large so that one has a superposition of curves of the various regions. In that case, Fig. 6.15 a with $d_{\mathrm{Ru}} = 4\,\text{Å}$ represents a situation with ferromagnetically coupled regions only. In Fig. 6.15 c with 5 ÅRu, i.e., with $\cong 0.5$ ML of additional Ru, only 50% of the ferromagnetically coupled region is covered by the added Ru so that one should expect that half of the stack has still to be ferromagnetically coupled. No remanence in Fig. 6.15 c can be observed, indicating that the remanence of the ferromagnetic portion is elevated by the strong coupling with the AF regions. In other words, the assumption of super-position of independent regions does not hold and the hypothesis of biquadratic coupling should apply.

There is one main reason for the short mean lateral coherence length of the spatial coupling fluctuations, namely the intermixing of Ru and Co near the interface. NMR spectra show no indication of a sharp interface for both the sputtered and the MBE-grown samples. The phase diagram of the Co-Ru system also demonstrates the very good intermiscibility over the entire composition range. Therefore, the interfacial composition fluctuations have a typical lateral coherence length of a few atomic distances, being small compared to $2\pi\,(\mathrm{DA})^{1/2}$ ($\simeq 200\,\text{Å}$ in (6.16 and 6.17)).

In all cases, the ratio $J_{\mathrm{bil}}/J_{\mathrm{biq}}$ is larger than 2, so that B–H curves without remanence are covered only. We see J_{bil} rise to thicknesses of about 6 Å. On the other hand, a continuous decline in $\mid J_{\mathrm{biq}} \mid$ at increasing d_{Ru} can be observed. How can this be plausibly explained?

The physical picture that should be used depends on the growth mode of the stack. First, one should consider the monolayer-by-monolayer growth and, assume for reasons of simplicity of the discussion that the bottom Co/Ru interface is atomically flat. There are one-monolayer thick terraces on its counterpart and, in principle, there are two discrete thicknesses of the interlayer. (In the case of two stepped interfaces, there are generally four interlayer distances.) As a consequence, one also has to deal simultaneously with two couplings constants. The left part of Fig. 6.16 a shows such a situation at small interlayer thicknesses with discrete couplings coefficients (indicated by the vertical heavy lines). It was assumed that the coupling is ferromagnetic at the smallest thicknesses. By increasing d_{Ru}, one increases the area with the AF-coupling J_{b}, i.e., l_{b} in (6.14), so that the effective bilinear coefficient $(J_{\mathrm{a}} + J_{\mathrm{b}} l_{\mathrm{b}}/l_{\mathrm{a}})/(1 + l_{\mathrm{b}}/l_{\mathrm{a}})$ crosses zero at $d_1(J_{\mathrm{bil}} = 0)$. Above this thickness, the strength of the AF-coupling constant $\mid J_{\mathrm{bil}} \mid$ increases till d_{Ru} is equal to d_2. This might explain the observation concerning J_{bil}. Let us now proceed to J_{biq} and consider (6.17).

The significant terms in (6.17) are $(J_{\mathrm{a}} - J_{\mathrm{b}})^2$ and $\sin^2(\pi/(1 + l_{\mathrm{b}}/l_{\mathrm{a}}))$. The first one is a constant up to d_2, as shown in Fig. 6.16 b. Note that $d_1(J_{\mathrm{bil}} = 0)$ is closer to d_2 than d_1 due to the tendency of declining electronic coupling strength at increasing interlayer thickness. (This need not always be true due to the sampling of the coupling function by the discrete atomic planes.) The maximum of $\sin^2(\pi/(1 + l_{\mathrm{b}}/l_{\mathrm{a}}))$ is at the thickness $d_1(l_{\mathrm{a}} = l_{\mathrm{b}}) = (d_1 + d_2)/2$, which is generally at the left of $d_1(J_{\mathrm{bil}} = 0)$. $\mid J_{\mathrm{biq}} \mid$ tends to decrease continuously with d_{Ru} above $d_1(l_{\mathrm{a}} = l_{\mathrm{b}})$ till d_2. This might explain the observation concerning J_{biq}.

There is one problem however. Co/Ru stacks do not grow layer by layer. As already discussed, the interfaces are intermixed so that one should expect the heavy lines in Fig. 6.16 a to be broadened. Fig. 6.16 c gives a hypothetical extreme course for the case that the original discrete course of the electronic coupling as a function of d_{Ru} is replaced by a continuous function with flat plateaus and straight slopes. The coupling variations ΔJ are represented by the thickness difference Δd (see Fig. 6.16 c). ΔJ is about the derivative of the curve in Fig. 6.16 c when Δd is infinitesimal small. In this case, it is given by the dashed curve in Fig. 6.16 d. Then, $d_1(l_{\mathrm{a}} = l_{\mathrm{b}})$ of the maximum in the $\sin^2(\pi/(1 + l_{\mathrm{b}}/l_{\mathrm{a}}))$ function and $d_1(J_{\mathrm{bil}} = 0)$ coincide. For $d_{\mathrm{Ru}} > d_1(J_{\mathrm{bil}} = 0)$, $\mid J_{\mathrm{bil}} \mid$ increases up to the maximum of the plateau, while $\mid J_{\mathrm{biq}} \mid$ decreases and drops to zero in the thickness regimes with $\Delta J = 0$, i.e., near the maxima of $\mid J_{\mathrm{electronic}} \mid$. The biquadratic contribution appears again at the second plateau in Fig. 6.16 d and might give rise to a local maximum in H_{s}, as frequently observed for Ru interlayers (see also Fig. 6.9).

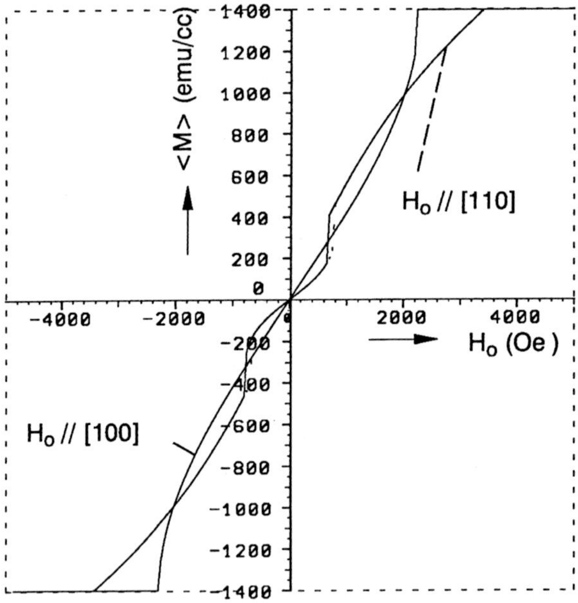

Fig. 6.17. The calculated in-plane M–H loops of a magnetic bilayer, in which both magnetic layers have equal thicknesses D (10^{-7} cm), identical M_s (1400 emu/cm^3), K_c (4×10^5 erg/cm^3) and coherent crystal structure with (100) texture. H_0 along [100] and [110] direction

Analysis of Stacks with In-Plane Anisotropy and Magnetizations. Up to now, we have confined ourselves to multilayers that are isotropic in the film plane. In this section, we shall look at a few aspects of anisotropy and provide some convenient formulae for obtaining the material parameters from the hysteresis curves. Magnetic bilayers with in-plane $\boldsymbol{M}$ and in-plane preferential anisotropy directions will be considered. For convenience, it is assumed that both magnetic layers have identical thickness D_m and equal magnetization M_s. Moreover, it is assumed that the crystal structures are coherent and the anisotropy strengths are equal. The interlayer coupling is considered to be purely bilinear. In Fig. 6.17, the M–H curves of a cubic stack with (100) texture are shown for the in-plane H_0 parallel to the [100] and the [110] direction. The curves are distinguished by both the saturation field H_s and the occurrence of jumps. The positions of these jumps cannot easily be employed for estimating J and K_c (K_c is the cubic anisotropy constant). The actual value of the jump field is determined by domain transformation processes, which are difficult to implement. These transformations involve nonuniform and nonlinear transition phases. Fortunately, the material parameters can easily be determined from the saturation field H_s by measuring the M–H curve in two nonequivalent in-plane directions. Two independent measurements are required to determine both K and J. The optimum choice

Table 6.5. The formulae for the interlayer-coupling constant J and the anisotropy constant K as a function of the saturation fields measured in two in-plane directions. H_s^n, H_s^p: Saturation fields normal and parallel to the uniaxial axis. H_s^{110}, H_s^{100}: Saturation fields along the [110] and [100] direction

Texture	Coupling constant J erg/cm^2	Anisotropy Constant K erg/cm^2
Uniaxial	$-(H_s^n + H_s^p)D_m M_s/4$	$(H_s^n + H_s^p)M_s/4$
Cubic (100)	$-(H_s^{110} + H_s^{100})D_m M_s/4$	$(H_s^{110} - H_s^{100})M_s/4$
Cubic (110)	$-(5H_s^{110} - 3H_s^{100})D_m M_s/4$	$(H_s^{110} - H_s^{100})M_s/4$

for these directions of the measurement, indicated by the superscripts of H_s, for a few symmetries of the anisotropy and the corresponding formulae are summarized in Table 6.5. The meaning of the superscript are given in the caption of the table.

Summary. A number of circumstances has been discussed, namely, the stacks with varying J_k, with inhomogeneous anisotropies and with in-plane preferential axes, in which determination of the interlayer-coupling constant from M–H or ϱ–H curves is problematic. In these cases, the problems have been overcome in two steps. First, the data from a number of M–H and/or ϱ–H curves have been combined and, second, numerical models to fit the measured data have been developed. In two cases, the curves for several external field directions for a given stack have been used, while curves from a large number of stacks form the basis of the analysis in the case of nonconstant J_k. In some cases, the saturation field H_s constitutes the only parameter used for the parameter fitting. However, in other situations, the shapes of the entire curves are considered. The study of the coupling and the anisotropies of the Co/Ru sandwiches exemplify this.

6.3 Physics of the GMR Effect

6.3.1 Theoretical Background of the GMR Effect

The origins of the GMR and AMR effect are not very distinct so that a number of basic physical insights for the understanding of GMR was proposed for decades. One of the basic ideas, to wit the decomposition of the conduction current into two distinct spin channels that can be treated as independent, harks back to the proposal of Mott [117] in the thirties. Let us first delve a little bit deeper into these classical ideas on the electric conduction in ferromagnetic media.

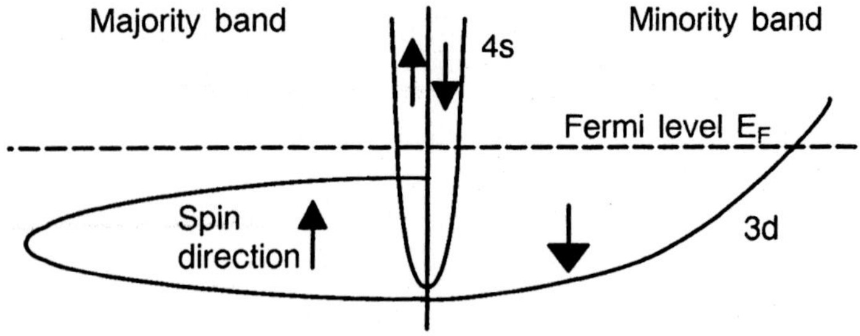

Fig. 6.18. A schematic representation of the band structure of the strongly ferromagnetic transition metals Ni and Co that are strong ferromagnetic

The Conduction in Two Spin Channels. Mott [117] stated that the conduction current in ferromagnetic metals can be decomposed into two carrier types, namely a majority and a minority current. He assumed that the spin of the charge carriers is preserved and that, even when electrons scatter at defects, no spin-flip processes occur. Such spin-flip transitions take place at collisions with magnons and their rate is small at temperatures substantially below the ordering temperature T_c [118] because of the low magnon density. Further, it was assumed that the conduction takes place almost exclusively by means of the s-electrons due to the large effective masses of the d-electrons. It should be emphasized that although this supposition is not completely justified [119], it does not change anything fundamental in the explanation. Figure 6.18 provides a rough schematic impression of a band structure typical for magnetic transition metals. The electrons are split up into majority spins with polarization parallel to the magnetization and minority electrons with opposite moments. The majority d-band lies below the Fermi level and is completely occupied at sufficiently low temperatures, while the minority band intersects the Fermi level E_F. This picture applies well to the strong ferromagnets Co and Ni and to a lesser extent to Fe with a density of state of the majorities at the Fermi level that is indeed smaller but still significant. Now, the total conductivity σ is the sum of the minority and majority contributions σ^- and σ^+, respectively. The conductivities satisfy expressions of the form:

$$\sigma^{+,-} = \frac{n_s e^2 \tau_s^{+,-}}{m_s^*} , \tag{6.24}$$

where n_s, τ_s and m_s^* are the density, relaxation time and the effective mass of the 4 s conduction electrons, respectively. The $^+$ and $^-$ sign indicate majority and minority carriers, respectively.

When minority carriers scatter, transitions can take place to the empty levels in the 4 s$^-$ and 3 d$^-$ band. In a first approximation, it is assumed that the associated scattering probabilities can be added, i.e.,

$$\varrho^- = \varrho_{ss}^- + \varrho_{sd}^- \cdot \tag{6.25}$$

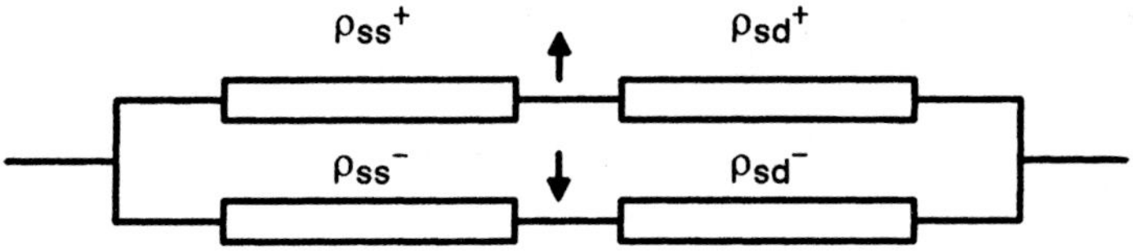

Fig. 6.19. The two current model
$\varrho_{\rm ss}^+$, $\varrho_{\rm sd}^+$: resistivity contribution of majorities to the s and d bands, respectively;
$\varrho_{\rm ss}^-$, $\varrho_{\rm sd}^-$: resistivity contribution of minorities to the s and d bands, respectively

Note that the density of states of the 3d$^-$ level is much larger than its counterpart 4s$^-$, so that $\varrho_{\rm sd}^-$ is significantly larger than $\varrho_{\rm ss}^-$. In other words, the lower branch in Fig. 6.19 is dominated by $\varrho_{\rm sd}^-$. A relation similar to (6.25) applies to the majorities, so that the total resistivity ϱ obeys

$$\varrho = \varrho^+\varrho^-/(\varrho^+ + \varrho^-)\,. \tag{6.26}$$

The 4s$^+$ majority carriers can only transfer to other 4s$^+$ states because the 3d$^+$ levels are completely filled up. Due to the symmetry of the s-band for both spin directions, it applies that $\varrho_{\rm ss}^- = \varrho_{\rm ss}^+$, so that it is obvious that most of the current is carried by 4s$^+$ electrons in the pure magnetic elements, i.e., $\alpha = \varrho^-/\varrho^+ > 1$. From these arguments, it follows that α is large for Co and Ni and small for Fe. In pure perfect metals, however, the scattering events occur at phonons and magnons at elevated temperatures. At temperatures well below $T_{\rm c}$, the scattering is mostly dominated by foreign atoms and defects.

Let us now proceed to disordered alloys with a relatively small portion of foreign atoms, which act as scattering centers. In this case, not only is the availability of states at the Fermi level to which scattering can take place relevant, but also the scattering potential of the foreign inclusions in a ferromagnetic host [120,121]. The scattering potentials are generally different for majority and minority conduction electrons. We return to Fig. 6.4 to elucidate this. It was discussed that the d-bands of elements at the left from the magnetic host in the periodic table resemble the minority d-band, and the majority bands are very different. Therefore, the majorities are scattered more strongly and $\alpha < 1$. The scattering occurs to virtual bound states near the defect. The opposite is true for foreign atoms at the right of the host in the table, for which α tends to be > 1.

If we depart now from the low temperature regime, processes at which momentum transfer between the channels takes place become relevant. These electron-magnon scattering events give rise to an increase of ϱ^{+-}, which is equal for both channels. Fert [122] derived the following expression:

$$\varrho = \frac{\varrho^+\varrho^- + \varrho^{+-}(\varrho^+ + \varrho^-)}{\varrho^+ + \varrho^- + 4\varrho^{+-}} \simeq \frac{\varrho^+\varrho^-}{\varrho^+ + \varrho^-} + \left(\frac{\alpha - 1}{\alpha + 1}\right)^2 \varrho^{+-}\,, \tag{6.27}$$

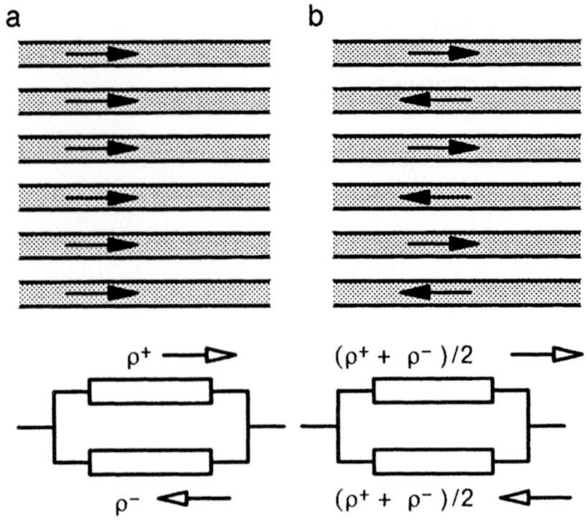

Fig. 6.20. a, b Cross-section of a superlattice with magnetizations in adjacent layers (**a**) parallel and (**b**) opposite: *black arrows* magnetizations, *white arrows* channels

in which the second expression applies at temperatures that are not too high. It shows that the additional temperature dependence is particularly high when ϱ^+ and ϱ^- are very different.

The Giant Magnetoresistive Effect: A Simplified Picture. The GMR effect deviates from the anisotropic magnetoresistive effect in that it does not depend on the angle between the current and $\boldsymbol{M}$. The spin-orbit coupling and the resulting anisotropy in the scattering probabilities play a minor role [123]. It is likely that many of the scattering centers in the multilayer stacks are located near the interfaces between the magnetic and nonmagnetic films. Furthermore, it should be expected that the large difference in the ϱ of minorities and majorities mentioned above might play a role.

A plausible picture of the origin of the GMR effect can readily be given. It is based on the decomposition of the current as given above. In Fig. 6.20, the cross-sections of a superlattice with the magnetizations in adjacent layers parallel (Fig. 6.20 a) and opposite (Fig. 6.20 b) are shown. In Fig. 6.20 b, there is the problem that a specific spin direction is a minority in one layer and a majority in its neighbor. Two current channels are therefore defined, one with spin orientation to the right and the other one to the left. These are marked by white arrows.

The situation in Fig. 6.20 a is simple because the majority electrons in all magnetic layers correspond to the channel with polarization to the right and all minorities belong to the opposite one. The resistivity ϱ_p in this situation follows from the shunt of the majority and minority channels with ϱ^+ and ϱ^-, respectively. Figure 6.20 b represents the situation with alternating $\boldsymbol{M}$'s. The

Table 6.6. Experimental data collected from the literature on $\alpha = \varrho^-/\varrho^+$ for diluted alloys with magnetic host Fe, Co and Ni marked by the *dark shadeded squares*. Each number in the *light shaded boxes* of a specific impurity indicates one information source from the literature

Ti	**V**	**Cr**	**Mn**	**Fe**	**Co**	**Ni**	**Cu**
0.25	0.12	0.17	0.09		1	3	
0.66	0.13	0.37	0.17		3.7	7	
Zr	**Nb**	**Mo**	**Tc**	**Ru**	**Rh**	**Pd**	**Ag**
		0.21		0.38	5.8		
Hf	**Ta**	**W**	**Re**	**Os**	**Ir**	**Pt**	**Au**
		0.24	0.31	0.33	9	8	

Ti	**V**	**Cr**	**Mn**	**Fe**	**Co**	**Ni**	**Cu**
1.4	1	0.3	0.8	12			
Zr	**Nb**	**Mo**	**Tc**	**Ru**	**Rh**	**Pd**	**Ag**
3.3	1	0.7		0.22	1		
Hf	**Ta**	**W**	**Re**	**Os**	**Ir**	**Pt**	**Au**
2.5	1.23	0.84	0.43	0.29	0.33		

Ti	**V**	**Cr**	**Mn**	**Fe**	**Co**	**Ni**	**Cu**
0.9	0.55	0.2	10	7	13		2.9
4.6	0.65	0.45	17	23	35		10
Zr	**Nb**	**Mo**	**Tc**	**Ru**	**Rh**	**Pd**	**Ag**
7.5	0.44	0.28		0.08	0.17	1	
	0.5	0.35		0.14	0.3	3	
Hf	**Ta**	**W**	**Re**	**Os**	**Ir**	**Pt**	**Au**
8.1	0.46	0.4	0.18	0.12	0.13	0.17	5.9
8.6	0.53	0.5	0.30	0.13	0.24	0.24	7

electrons in each channel are in half of the layers minorities and majorities in the layers with opposite $\boldsymbol{M}$. Consequently, the resistivity of each channel is $(\varrho^+ + \varrho^-)/2$. The relative change $\Delta\varrho/\varrho_\mathrm{p}$ in ϱ is then

$$\frac{\Delta\varrho}{\varrho_\mathrm{p}} = \frac{\varrho_0 - \varrho_\mathrm{p}}{\varrho_\mathrm{p}} = \frac{(\varrho^+ + \varrho^-)/4 - \varrho^+\varrho^-/(\varrho^+ + \varrho^-)}{\varrho^+\varrho^-/(\varrho^+ + \varrho^-)} = \frac{(1-\alpha)^2}{4\alpha} > 0\,, \quad (6.28)$$

where ϱ_0 is the resistivity in Fig. 6.14 b. Equation (6.20) suggests that ϱ_p is always smaller than ϱ_0 in the configuration with alternating $\boldsymbol{M}$'s. This need not always be true, as will be discussed later. Further, note that there is no question of interlayer coupling in the foregoing discussion and that the antiferromagnetic interlayer coupling only serves as a mechanism to get the alternating $\boldsymbol{M}$ configuration in the ground state at $H_0 = 0$, which can simply be transformed into a parallel arrangement by means of a uniform external field. On the other hand, there is a correlation between the GMR effect and the interlayer coupling because both originate from the difference in the scattering potential between the two spin directions of ferromagnetic impurities in a nonmagnetic metallic host. However, there is no one-to-one correspondence between these. No GMR effect arises from the interfaces when there are no scattering centers, i.e., no lattice imperfections, present here, while the interlayer coupling becomes frustrated when the atoms at the interfaces are randomly distributed in this region. Furthermore, the scattering leading to the GMR effect need not necessarily occur at the interfaces and may, in principle, also stem from bulk scattering only. It is obvious that the above simple approach provides only a very rough picture and that it is necessary to delve deeper into the scattering phenomena.

Element-Specific Scattering Properties. The question of whether the scattering leading to GMR occurs predominantly at the interfaces or in the bulk was hotly disputed for some years [124–129]. At present, the general opinion is that interface scattering dominates in the case of unalloyed magnetic layers and that bulk scattering has an impact in magnetic alloys. The meaning of α has been continuously emphasized. It is obvious from (6.28) that the values of α, measured in the past for the AMR effect, have a direct significance for the bulk contribution of GMR. In Table 6.6, the α's as collected mainly by Fert [130,131] and Dorleijn [120,121] during the seventies, have been reproduced in a structured form. Each number in the boxes refers to one information source from the literature.The correlation between α and the position in the periodic table, as discussed in Sect. 6.3.1, becomes evident. In general, the nonmagnetic elements to the left from the host material have α less than one. On the other hand, elements with more 3d electrons, i.e., in the columns to the right of the host, have $\alpha > 1$. Alloys of the magnetic elements always have $\alpha > 1$. Elements in the same column tend to have $\alpha < 1$, particularly those of the 6th row of the table.

It is obvious that the bulk α's have only limited meaning for the scattering events near the interfaces because of the strongly deviating electronic environment. The scattering processes within intermixed interfaces of magnetic and nonmagnetic elements have been extensively investigated by Inoue and coworkers on the basis of a tight-binding d-band model [132–134]. First, they evaluated the course of the potential of the 3d electrons across a smooth and perfect TM/FM interface (FM = Fe, Co and TM = nonmagnetic transition metal) for both the minorities and majorities. Next, they calculated the change in this potential when TM impurities were introduced in the monolayer adjacent to the interface in the FM host and also when FM atoms are located in the monolayer bounding the interface at the TM side. The case that the impurities are distributed over more than two monolayers was dis-

Table 6.7. The scattering potentials after Inoue et al. for the minority $^-$ and majority $^+$ d electrons of TM and FM impurities in FM and TM hosts.$\beta_{\mathrm{E,H}} = (V_{\mathrm{E,H}}{}^-/V_{\mathrm{E,H}}{}^+)^2$, where E refers to the impurities and H to the host. Scattering potentials: $V^{\sigma}_{\mathrm{E,H}}(\times 10^{-2}\,\mathrm{Ryd})$

Ti	**V**	**Cr**	**Mn**	**Fe**	**Co**	**Ni**	**Cu**	
18	14	9	3.5					$V_{\mathrm{TM,Fe}}{}^+$
12	7	0	−3.2					$V_{\mathrm{TM,Fe}}{}^-$
0.45	0.25	0	0.83					$\beta_{\mathrm{TM,Fe}}$
−15	−12	−9.2	−7.3					$V_{\mathrm{Fe,TM}}{}^+$
−15	−10	0	3.1					$V_{\mathrm{Fe,TM}}{}^-$
1	0.69	0	0.18					$\beta_{\mathrm{Fe,TM}}$
0.05	0.39	1	0.54				0.8	$\Delta_\varrho/\varrho_{\mathrm{AF}}$
Zr	**Nb**	**Mo**	**Tc**	Ru	Rh	**Pd**	**Ag**	
				5.9	5.4	2.5		$V_{\mathrm{TM,Fe}}{}^+$
				5	6.2	9.6		$V_{\mathrm{TM,Fe}}{}^-$
				0.71	1.3	15		$\beta_{\mathrm{TM,Fe}}$
				−5.7	−5.0	−1.9		$V_{\mathrm{Fe,TM}}{}^+$
				5.0	6.2	9.6		$V_{\mathrm{Fe,TM}}{}^-$
				0.77	1.5	26		$\beta_{\mathrm{Fe,TM}}$
				0.06	0.11	0.87		$\Delta_\varrho/\varrho_{\mathrm{AF}}$
Ti	**V**	**Cr**	**Mn**	**Fe**	**Co**	**Ni**	**Cu**	
22	19	17	15					$V_{\mathrm{TM,Co}}{}^+$
14	10	6.0	2.3					$V_{\mathrm{TM,Co}}{}^-$
0.40	0.28	0.12	0.02					$\beta_{\mathrm{TM,Co}}$
−18	−15	−13	−11					$V_{\mathrm{Co,TM}}{}^+$
−16	−13	−8.5	−4.2					$V_{\mathrm{Co,TM}}{}^-$
0.79	0.75	0.43	0.15					$\beta_{\mathrm{Co,TM}}$
0	0.18	0.52	0.88				0.86	$\Delta_\varrho/\varrho_{\mathrm{AF}}$
Zr	**Nb**	**Mo**	**Tc**	**Ru**	**Rh**	**Pd**	**Ag**	
				13	11	9.4		$V_{\mathrm{TM,Co}}{}^+$
				0	−4.6	−5.2		$V_{\mathrm{TM,Co}}{}^-$
				0	0.17	0.31		$\beta_{\mathrm{TM,Co}}$
				−10	−10	−9.6		$V_{\mathrm{Co,TM}}{}^+$
				−0.1	2.4	5.0		$V_{\mathrm{Co,TM}}{}^-$
				0	0.06	0.27		$\beta_{\mathrm{Co,TM}}$
				0.99	0.66	0		$\Delta_\varrho/\varrho_{\mathrm{AF}}$

cussed by Nakanishi et al. [136]. The calculated scattering potentials $V^{\sigma}_{\mathrm{E,H}}$ and the reciprocal of the relaxation time for each spin type are summarized in Table 6.7. The reciprocal of the life time $\tau^{\sigma^{-1}}$ follows from the relation

$$\tau^{\sigma^{-1}} = \frac{2\pi}{\hbar} \sum\nolimits_{\mathrm{E,H}} (V^{\sigma}_{\mathrm{E,H}})^2 \nu_{\mathrm{E,H}}(E_{\mathrm{F}}) C_{\mathrm{E}}\,, \tag{6.29}$$

where σ is + and − for majorities and minorities, respectively. E and H refer to the impurity and host material, respectively. C_{E} and $\nu_{\mathrm{E,H}}$ are, respectively, the defect density and the density of states of the impurity at the Fermi level E_{F}. The GMR effect follows from the above relation and from (6.24) and (6.28).

It is interesting to compare Table 6.7 with the bulk data in Table 6.6. The quantity $\beta_{\mathrm{E,H}} = (V^{-}_{\mathrm{E,H}}/V^{+}_{\mathrm{E,H}})^2$ represents the ratio of the scattering potentials of minority and majority spins and can be compared to α in Table 6.6. It should be noted that the general trends are equal. The majority spins scatter more strongly when the TM element is to the left of its magnetic counterpart and a gradual transition to a stronger minority contribution occurs upon moving to the right in the periodic table. Of course, the scattering potentials $V^{\sigma}_{\mathrm{TM,FM,H}}$ and $V^{\sigma}_{\mathrm{FM,TM}}$ have opposite signs. It is obvious from the data on Co/Ti that the state density of the impurities $\nu_{\mathrm{E,H}}$ cannot be ignored. The strong maxima in $\Delta\varrho/\varrho_{\mathrm{AF}}$ for Fe/Cr and Co/Ru are not as emphatically present in Table 6.6.

Coehoorn [137] considered theoretically the coverage of Ni/Cu interfaces with Co. He showed that the magnetic Ni moments near the interface are strongly reduced in the pure Ni/Cu system, i.e., the exchange splitting, being a prerequisite for the GMR effect, is very small. This effect is further enhanced when the interface is intermixed. A tiny Co coating of 0.5 monolayer suffices to increase the Ni moment to about its bulk value and thus effectively enhances the spin dependence of the interface scattering. An improvement in $\Delta\varrho/\varrho$ from a few percent to 20% at room temperature was experimentally demonstrated by Parkin [126], who inserted about two monolayers Co at each interface in the permalloy/Cu system. Takanashi et al. [138] alloyed the Cr in a Fe/Cr superlattice with Ti, V, Nb, Mn, Mo and Ru. They were not able to extract a correlation with the α of these elements from their data. This is mainly due to the larger imperfectness of the coupling due to alloying. On the other hand, Gurney et al. [139] inserted up to two ML of, first, Au, Ag, Al, Ge and Ir and of, second, V and Mn at the interfaces of Fe/Cr multilayers. The GMR reduces strongly when elements of the first group are added and changed hardly due to the V and Mn. The α of V and Mn is, like Cr, smaller than 1, so that this tendency agrees well with the above table. Coehoorn [137] also considered V and compared it with Cr at the interface with Fe from a theoretical point of view. Both V and Cr couple antiferromagnetically to the Fe, however, the atomic Fe moment is strongly reduced in the intermixing zone with V, and a drop in the scattering potential should lead to a smaller GMR contribution. No such atomic Fe-moment reduction occurs in the Fe/Cr intermixing zone due the frustration of the intrinsically antiferromagnetically coupled Cr at the stepped interface.

Up to now, we have considered the element-specific scattering properties of individual scattering centers. The consequences of their typical arrangement in multilayers have not been discussed so far.

Phenomenological GMR Theory of Multilayers. All phenomenological approaches hark back to the Fuchs–Sondheimer theory [140–142]. They considered a metallic plane-parallel slab with thickness D in the free-electron approach, at which electron scattering takes place at the outer boundaries and within the bulk. The spherical Fermi surface shifts into the direction of the electric field $\boldsymbol{E}$ (the x axis) over a disttance $\boldsymbol{\delta k}$, at which $\boldsymbol{\delta k}$ is proportional to the relaxation time τ, being generally different for majority and minority spins. In an unbounded homogeneous magnetic medium, the change in the occupation probability $g_{\boldsymbol{k}}^{\sigma}$ of state $\boldsymbol{k}$ with spin σ for a stationary current is given by

$$g_{\boldsymbol{k}}^{\sigma} = -\frac{\partial f_0}{\partial E_{\boldsymbol{k}}} e\tau^{\sigma} v_{\boldsymbol{k}} E\,, \tag{6.30}$$

where f_0 is the Fermi–Dirac distribution in the undisturbed state and $E_{\boldsymbol{k}}$ and $v_{\boldsymbol{k}}$ are the energy and drift velocity of state $\boldsymbol{k}$, respectively. By summing $g_{\boldsymbol{k}}$ over all states, one gets the conductivity in the form of (6.24). In the slab with perpendicular z axis, $g_{\boldsymbol{k}}^{\sigma}$ is a function of z due to reflections and diffuse scattering at the outer surfaces. For this situation, the following general relation applies

$$g_{\boldsymbol{k}}^{\sigma\pm}(z) = \frac{\partial f_0}{\partial v_x}\frac{e\tau^{\sigma}E}{m}\left[1 - F_{\boldsymbol{k}}^{\sigma\pm}\exp\left[-\frac{\pm z}{\tau^{\sigma}\mid v_z\mid}\right]\right], \tag{6.31}$$

in which the $+$ and $-$ signs apply to positive and negative v_z, respectively. The factor $F_{\boldsymbol{k}}^{\sigma\pm}$ follows from the boundary conditions at the slab outer surfaces. The velocity $\mid v_z \mid$ is smaller than the Fermi velocity v_{F} so that the interface scattering plays no role when the slab thickness $D \gg v_{\mathrm{F}}\tau^{\sigma} = \lambda^{\sigma}$. At diffuse interface scattering, the excursion of the Fermi surface from the ground is cancelled and $g_{\boldsymbol{k}}^{\sigma^{\pm}}$ is zero, implying that $g_{\boldsymbol{k}}^{\sigma^{+}}(-D/2)$ and $g_{\boldsymbol{k}}^{\sigma^{-}}(D/2)$ are zero for completely diffuse scattering at the outer surfaces. At specular reflection here, the relations $g_{\boldsymbol{k}}^{\sigma^{+}}(D/2) = g_{\boldsymbol{k}}^{\sigma^{-}}(D/2)$ and $g_{\boldsymbol{k}}^{\sigma^{+}}(-D/2) = g_{\boldsymbol{k}}^{\sigma^{-}}(-D/2)$ apply so that $F_{\boldsymbol{k}}^{\sigma^{\pm}} = 0$. In other words, (6.32) becomes equivalent to (6.30) and the conductivity is not affected by the limited thickness of the slab in the absence of diffuse surface scattering. In general, the reflection is incomplete and the ratio of the reflected and incoming wave – the specularity factor – is smaller than one.

The conductivity σ^{σ} follows from integration of (6.31) over all $\boldsymbol{k}$'s, i.e., $\boldsymbol{v}$'s, and z

$$\sigma^{\sigma} = \frac{e^2\tau^{\sigma}}{mD}\int \mathrm{d}z \int v_x \frac{\partial f_0}{\partial v_x}\left[1 - F_{\boldsymbol{k}}^{\sigma^{\pm}}\exp\left[-\frac{\pm z}{\tau^{\sigma}\mid v_z\mid}\right]\right]\mathrm{d}^3 v\,. \tag{6.32}$$

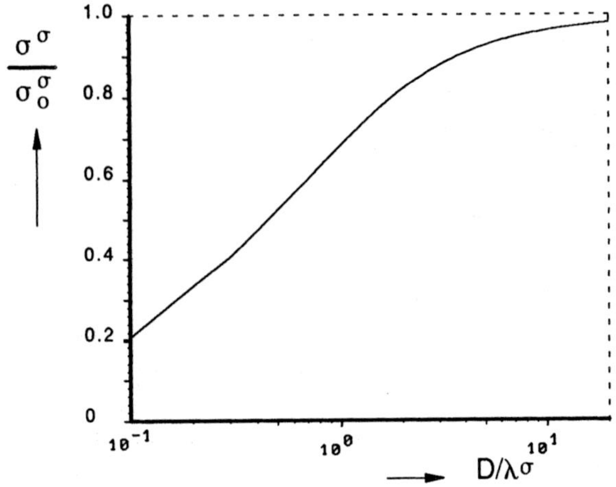

Fig. 6.21. The decrease in the conductivity with respect to the bulk value due to the diffuse scattering at the outer surfaces of a single film according to Fuchs–Sondheimer. σ^σ, mean film conductivity; σ_0^σ, bulk conductivity

The integration is over the Fermi sphere and can easily be performed. In the case of perfect diffuse surface scattering, Sondheimer obtained:

$$\sigma^\sigma = \sigma_0^\sigma \Bigg[1 - \frac{3}{4}\kappa^\sigma(1 - \frac{1}{12}\kappa^{\sigma^2})Ei(-\kappa^\sigma) - \frac{3}{8\kappa^\sigma}(1 - \exp(-\kappa^\sigma))$$

$$- \left(\frac{5}{8} + \frac{1}{16}\kappa^\sigma - \frac{1}{16}\kappa^{\sigma^2}\right)\exp(-\kappa^\sigma)\Bigg], \tag{6.33a}$$

where σ_0^σ is the bulk conductivity and Ei is

$$Ei(-u) = \int_{\mathrm{u}}^{\infty} -\frac{\Theta^{-t}}{t}\mathrm{d}t, \tag{6.33b}$$

and $\kappa = D/\lambda^\sigma$.

The relative change in the conductivity as a function of the ratio $\kappa = D/\lambda\sigma$ is given in Fig. 6.21 for perfect diffuse scattering at the outer boundaries. Note that σ^σ is infinite in slabs with finite thickness and no bulk scattering, i.e., $\sigma_0^\sigma = \infty$, notwithstanding the diffuse scattering at the outer surfaces. This is a deficiency of this semi-classical approach.

Scattering in multilayers will now be looked at with a brief sketch of the main steps of the method developed by Barnas et al. [143,144]. They considered a magnetic bilayer with one single interlayer (see Fig. 6.22).

They decomposed the stack in four plane parallel slabs, marked A, B, C and D, with one artificial interface at $z = 0$, where continuity conditions apply. Each film is uniformly magnetized with each medium characterized by its own relaxation time constants for both spin types. At the outer surfaces,

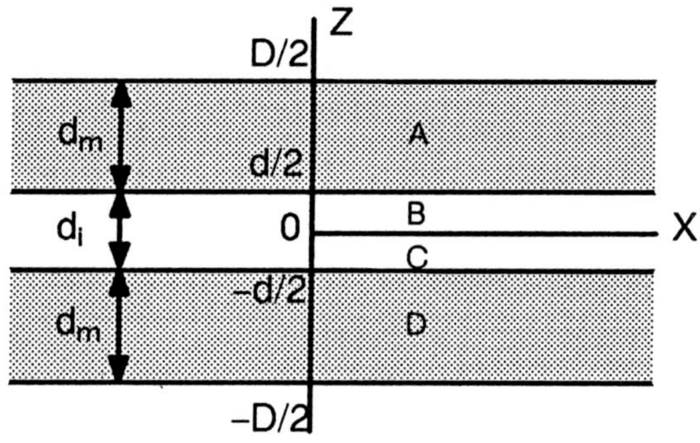

Fig. 6.22. Cross-section of a magnetic bilayer with interlayer with thickness d_i and the definition of the interfaces at which boundary conditions have to be satisfied. The spin channels of both magnetic layers are matched at the fictitious interface $X = 0$ by appropriate boundary conditions

the Fuchs boundary conditions apply, i.e.,

$$g_{\mathrm{kD}}^{\sigma^+}(-D/2) = p_{\mathrm{kD}}^{\sigma^-} g_{\mathrm{kD}}^{\sigma^-}(-D/2) \quad \text{and} \tag{6.34a}$$

$$g_{\mathrm{kA}}^{\sigma^-}(+D/2) = p_{\mathrm{kA}}^{\sigma^+} g_{\mathrm{kA}}^{\sigma^+}(D/2)\,, \tag{6.34b}$$

where $p_{\mathrm{kA}}^{\sigma^+}$ and $p_{\mathrm{kD}}^{\sigma^-}$ are the Fuchs specularity factors. At the interfaces $z = d/2$ and $z = -d/2$, Barnas et al. introduced the nondiffusive transmission T^σ and reflection R^σ coefficients, which are considered to be $\boldsymbol{k}$ and propagation-direction independent. The boundary conditions are then

$$g_{k\mathrm{C}}^{\sigma^+}(-d/2) = T^\sigma(-d/2) g_{k\mathrm{D}}^{\sigma^+}(-d/2) + R^\sigma(-d/2) g_{k\mathrm{C}}^{\sigma^-}(-d/2)\,, \tag{6.34c}$$

$$g_{k\mathrm{D}}^{\sigma^-}(-d/2) = T^\sigma(-d/2) g_{k\mathrm{C}}^{\sigma^-}(-d/2) + R^\sigma(-d/2) g_{k\mathrm{D}}^{\sigma^+}(-d/2)\,, \tag{6.34d}$$

$$g_{k\mathrm{A}}^{\sigma^+}(d/2) = T^\sigma(d/2) g_{k\mathrm{B}}^{\sigma^+}(d/2) + R^\sigma(d/2) g_{k\mathrm{A}}^{\sigma^-}(d/2)\,, \tag{6.34e}$$

$$g_{k\mathrm{B}}^{\sigma^-}(d/2) = T^\sigma(d/2) g_{k\mathrm{A}}^{\sigma^-}(d/2) + R^\sigma(d/2) g_{k\mathrm{B}}^{\sigma^+}(d/2)\,. \tag{6.34f}$$

Note that diffuse scattering occurs when $T^\sigma(z) + R^\sigma(z)$ is smaller than one. At the fictitious interface $z = 0$, the transformation of the majority and minority electrons of layer D into majorities and minorities of layer A is carried through. Defining the angle Θ between both $\boldsymbol{M}$'s, Barnas et al. posited the following relationships:

$$g_{k\mathrm{B}}^{+,+}(0) = \cos^2(\Theta/2) g_{k\mathrm{C}}^{+,+}(0) + \sin^2(\Theta/2) g_{k\mathrm{C}}^{-,+}(0)\,, \tag{6.35a}$$

$$g_{k\mathrm{C}}^{+,-}(0) = \cos^2(\Theta/2) g_{k\mathrm{B}}^{+,-}(0) + \sin^2(\Theta/2) g_{k\mathrm{B}}^{-,-}(0)\,, \tag{6.35b}$$

$$g_{k\mathrm{B}}^{-,+}(0) = \cos^2(\Theta/2) g_{k\mathrm{C}}^{-,+}(0) + \sin^2(\Theta/2) g_{k\mathrm{C}}^{+,+}(0)\,, \tag{6.35c}$$

$$g_{k\mathrm{C}}^{-,-}(0) = \cos^2(\Theta/2) g_{k\mathrm{B}}^{-,-}(0) + \sin^2(\Theta/2) g_{k\mathrm{B}}^{+,-}(0)\,, \tag{6.35d}$$

One should bear in mind that the $+/-$ sign of the g-parameter superscript at the left refers to the spin direction and the one at the right to the direction of v_z. Relationships (6.35) require modification when $\Delta\varrho/\varrho$ is high. By using (6.31) and the boundary conditions (6.34) and (6.37), we arrive at a coupled system of 16 linear equations with 16 unknown $F_{\boldsymbol{k}}^{\sigma\pm}$ for each $\boldsymbol{k}$. Fortunately, $F_{\boldsymbol{k}}^{\sigma\pm}$ depends only on the angle β between $\boldsymbol{k}$ and the z axis so that the linear system has to be numerically solved only for a limited number of values of β. The conductance of the stack is found by numerical integration of (6.32), at which the contribution of the different regions are simply added.

As already mentioned, the Fuchs–Sondheimer theory exhibits a deficiency at layer thicknesses D that are much smaller than the mean free path λ of the electrons, which betrays itself by a $D/\ln(\lambda/D)$ dependence of ϱ. The resistivity consequently vanishes at small D, notwithstanding the diffuse scattering at the outer surfaces. For these cases, a quantum-mechanical approach is required [145–147].

Dieny [148,149] argues in favour of the use of the conductance and conductivity because of the fundamental meaning of (6.31). However, it is still common practice to use $\Delta\varrho/\varrho_{\mathrm{P}}$, which can be evaluated by solving the problem for both the parallel and antiparallel magnetizations. This is a time-consuming task, which is complicated by the large number of unknown parameters. These parameters are not easily accessible for measurement. This becomes even more relevant when systems with many different layers are studied. In principle, the same procedure can be employed. However, the number of parameters and unknowns increase proportionally and the equation system becomes ill-conditioned because of the exponential function in (6.31) [150]. There is, therefore, a great need for analytic expressions which offer more physical insight into the pivotal factors for the GMR effect.

Analytic Expressions for the GMR. Barthélémy et al. [151] studied the Fe/Cr superlattices and derived analytic expressions for the cases with magnetic layer thicknesses that were either much larger or much smaller than the mean free path of the carriers.They considered only interface scattering as contributing to the signal and took the same mean free path for the bulk Fe and Cr, i.e., $\lambda = \lambda_{\mathrm{Cr}} = \lambda_{\mathrm{Fe}}$. Further, they confined themselves to thick Fe layers ($d_{\mathrm{Fe}} \gg \lambda$). Their pivotal equation is

$$\frac{\Delta\varrho}{\varrho_{\mathrm{AF}}} = \frac{3\lambda}{4d_{\mathrm{Fe}}}(T^+ - T^-)^2 \left\{E_3(d_{\mathrm{Cr}}/\lambda) - E_5(d_{\mathrm{Cr}}/\lambda)\right\} , \tag{6.36a}$$

where d_{Fe}, D_{Cr}, T^+ and T^- are the thicknesses of Fe and Cr and the transmissions of the majorities and minorities at the Fe/Cr interfaces, respectively. The functions E_3 and E_5 are exponential integral functions. They derived even simpler expressions for the two following situations:

$$\frac{\Delta\varrho}{\varrho_{\mathrm{AF}}} = \frac{3\lambda}{16d_{\mathrm{Fe}}}(T^+ - T^-)^2 \left\{1 - 8d_{\mathrm{Cr}}/(3\lambda)\right\} , \tag{6.36b}$$

for $d_{\rm Cr} \ll \lambda$, i.e., small Cr thickness, and

$$\frac{\Delta\varrho}{\varrho_{\rm AF}} = \frac{3\lambda}{2d_{\rm Fe}}(T^+ - T^-)^2 \exp(-d_{\rm Cr}/\lambda)/(d_{\rm Cr}/\lambda)^2 \,, \tag{6.36c}$$

for $d_{\rm Cr} \gg \lambda$, i.e., for large Cr thickness.

The strong dependence of the GMR effect on $d_{\rm Cr}$, represented by the factor $\exp(-d_{\rm Cr}/\lambda)/(d_{\rm Cr}/\lambda)^2$ in (6.36c) can be understood as follows. Electrons that leave the Fe medium I and that have a spin-dependent scattering probability at its Fe/Cr interface have a large nonspin discriminating scattering change inside the Cr. The scattering events at the opposite Fe/Cr interface contribute only relatively little and the correlation between the scattering events at both Fe/Cr interfaces vanishes. The $\lambda/d_{\rm Fe}$ dependence reflects the fact that only electrons that leave the Fe region with thickness λ adjacent to the interface have a sufficiently high relative scattering probability at its Fe/Cr interface to make a noticeable difference between both spin directions. The rest of the Fe is ineffective and is only a shunt, which reduces the signal.

Edwards et al. introduced a resistor model, which cannot easily cover situations in which the mean free paths are comparable to the layer thicknesses and in which many different layers are present [152,153].

Liu et al. [154] adopted an approach that does not suffer from Barthélémy's requirement of ferromagnetic layers that are thick in comparison to λ. They confined themselves to magnetic bilayers with equally thick magnetic layers. At the outer stack surfaces, they introduced the specularity factor R, which is zero at perfect diffuse and one in the absence of diffuse scattering. In other words, R replaces $p^{\sigma\pm}_{\rm kG}$ in (6.34b,6.34b). They reconsidered the Barnas theory, of which the main steps have been recapitulated above, and introduced the following simplifications, which are shared with Barthélémy.

- They assumed that only the scattering at the inner interfaces is spin-dependent.
- They put all reflection coefficients of the inner interfaces R^σ equal to zero and consequently, their factor $D^\sigma = 1 - T^\sigma$ represents the diffuse scattering at the interfaces.
- They defined the mean free paths of the magnetic layers λ_0 by

  $$\lambda_0 = (\lambda^+ + \lambda^-)/2 \,, \quad \text{at} \quad \text{which} \tag{6.37a}$$

  $$\lambda^\sigma = v_{\rm F}\tau^\sigma \,, \tag{6.37b}$$

  and required these to be equal for all three layers.

They confined themselves to either parallel or opposite magnetization in both layers, to which we shall refer by the symbols FM and AF, respectively. This simplifies (6.35). Notwithstanding all these simplifications, the ultimate expressions are very complicated as can be seen from their pivotal equations, which are duplicated below:

$$\sigma_{\rm FM} = \sigma^{+,+} + \sigma^{-,-} \,, \tag{6.38a}$$

$$\sigma_{\rm AM} = \sigma^{+,-} + \sigma^{-,+} \,, \tag{6.38b}$$

where the indices of the superscript of $\sigma^{\sigma 1,\sigma 2}$ indicate from left to right the spin type of the bottom and top layer, respectively. The $\sigma^{\sigma 1,\sigma 2}$'s follow from

$$\frac{\sigma^{\sigma_1,\sigma_2}}{\sigma_0} = 1 - \frac{3\lambda_0}{4D}\left(U_{\mathrm{mag}}^{\sigma_1,\sigma_2} - U_{\mathrm{int}}^{\sigma_1,\sigma_2}\right), \tag{6.39a}$$

with σ_0 given by (6.24) and (6.37), and D is again the total stack thickness. Finally, the U functions are

$$U_{\mathrm{mat}}^{\sigma_1,\sigma_2} = \int_0^1 du(1-u^2)u(1-E_{\mathrm{mat}})\, C_{\mathrm{mat}}^{\sigma_1,\sigma_2}/N^{\sigma_1,\sigma_2}, \tag{6.39b}$$

with the subscript mat is mag or int for a magnetic or the interlayer layer, respectively. The denominator is given by

$$N^{\sigma_1,\sigma_2} = 1 - T^{\sigma_1}T^{\sigma_2}R\,E_{\mathrm{mag}}^2 E_{\mathrm{int}}, \quad \text{with} \tag{6.39c}$$

$$E_{\mathrm{mag}} = \exp(-d_{\mathrm{m}}u/\lambda_0) \quad \text{and} \quad E_{\mathrm{int}} = \exp(-d_{\mathrm{i}}u/\lambda_0). \tag{6.39d}$$

The C parameters obey

$$C_{\mathrm{mag}}^{\sigma_1,\sigma_2} = (1-T^{\sigma_1})(1+T^{\sigma_2}E_{\mathrm{int}})(1+RE_{\mathrm{mag}}) + (1-R)(1+T^{\sigma_1}T^{\sigma_2}E_{\mathrm{mag}}E_{\mathrm{int}}), \tag{6.39e}$$

$$C_{\mathrm{int}}^{\sigma_1,\sigma_2} = (1-T^{\sigma_1})(1+T^{\sigma_2}RE_{\mathrm{mag}}^2)\eta^{\sigma_1,\sigma_2} + (1-R)T^{\sigma_1}E_{\mathrm{mag}}, \tag{6.39f}$$

$$\eta^{\sigma_1,\sigma_2} = \frac{(1-T^{\sigma_2^2}RE_{\mathrm{mag}}^2E_{\mathrm{int}})}{(1+T_1^{\sigma}T_2^{\sigma}RE_{\mathrm{mag}}^2E_{\mathrm{int}})}. \tag{6.39g}$$

Equations (6.38) and (6.39) contain all the information. However, they still do not betray the key parameters of the problem. Therefore, Liu et al. considered the case $R = 0$, i.e., perfect diffuse scattering at the exterior surfaces. Defining $\Delta\sigma = \sigma_{\mathrm{FM}} - \sigma_{\mathrm{AF}}$, they derived

$$\frac{\Delta\sigma}{\sigma_0} = \frac{3\lambda_0}{4D}(T^+ - T^-)^2 \int_0^1 du(1-u^2)u(1-E_{\mathrm{mag}})^2 E_{\mathrm{int}}. \tag{6.40}$$

This formula allows a number of conclusions. First, the GMR effect is very sensitive to the asymmetry in the interface scattering as follows from the $(T^+ - T^-)^2$ term. Note that the same dependence is also present in the Barthélémy's equation (6.36). The course of the relation as a function of $\kappa_{\mathrm{m}} = d_{\mathrm{m}}/\lambda_0$ and $\kappa_{\mathrm{i}} = d_{\mathrm{i}}/\lambda_0$ is given in Fig. 6.23. The GMR effect vanishes at very small d_{m} due to the $(1 - E_{\mathrm{mag}})^2$ term. The reason is the growing impact of the interlayer conductance and the diffuse scattering at the exterior surfaces. At large thickness of the ferromagnetic layer the $\Delta\sigma$ reduces again because $\Delta\sigma$ originates from the interface scattering and the bulk scattering in the magnetic layers does not contribute to the signal. The magnetic bulk provides an inactive shortcut only, as has previously been concluded from (6.36). $\Delta\sigma$ is maximum when d_{i} is vanishing (see E_{int}), which simply reflects

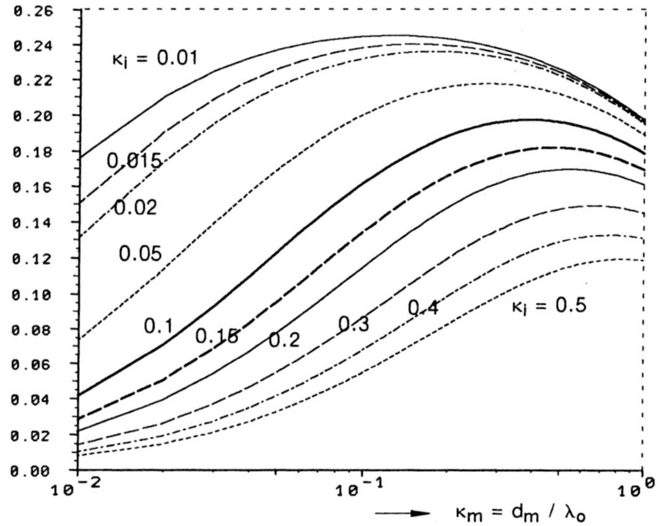

Fig. 6.23. The function of (6.40) for the magnetic bilayer with perfect diffuse scattering at the outer boundaries, i.e., $\lambda_0/(2d_\mathrm{m}+d_\mathrm{i})\int_0^1 \mathrm{d}u(1-u^2)u(1-E_\mathrm{mag})^2E_\mathrm{int}$, with $\kappa_\mathrm{i}=d_\mathrm{i}/\lambda_0$ and $\kappa_\mathrm{m}=d_\mathrm{m}/\lambda_0$

the minimization of the relative contribution of the bulk scattering of the interlayer and the maximum correlation between the scattering events at both interfaces.

It has been recognized that a multilayer with an infinite number of periods can be simulated by taking R equal to one in the magnetic bilayer [144], at which the magnetic layer thickness is taken as half the actual one. Dieny has extensively studied the dependence of the conductance of multilayers on the number of multilayer periods and on the physical parameters of the individual layers and interfaces [148]. In general, the GMR effect increases with the number of repetitions and approaches asymptotically a maximum. Partly, this behavior is caused by a vanishing impact of the diffuse scattering at the outer surfaces. Figure 6.21 shows that this occurs when the total stack thickness is about 5 to 10 times the longest mean free path (very often λ^+). In the case of a dominant interface contribution, it is advantageous to have the density of the interfaces as high as possible. Of course, there are practical limitations, such as the demands on the decoupling and the integrity of the magnetic layers.

In the so-called exchange-biased systems, the stack configuration is limited to two ferromagnetic layers, in which the magnetization in one layer is fixed by an antiferromagnetic layer with high resistivity. In this case, the GMR signal is limited and the AMR contribution cannot be neglected. Rijks et al. [155] studied this in the framework of the Fuchs–Sondheimer theory, in which they introduced a mean free path that depends on the angle between

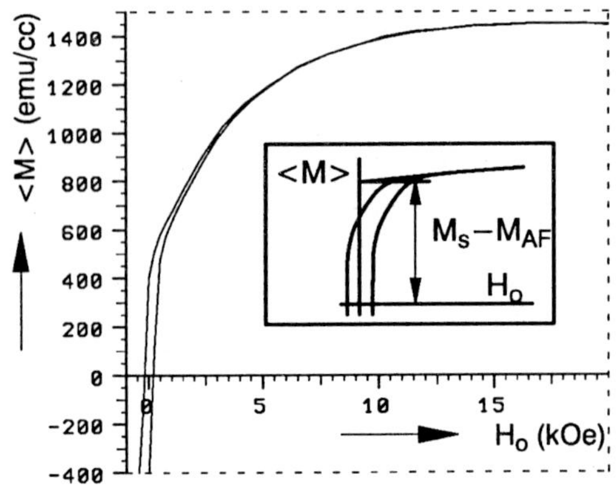

Fig. 6.24. The $\langle M \rangle$–H curve of an antiferromagnetically coupled multilayer system with defects manifesting themselves by an increase in the remanence. Stack geometry: glass/Fe60 Å/[Co10 Å/Cu10 Å] × 100

the electron velocity and the magnetization. They concluded that the AMR and GMR effects can simply be superimposed in most situations.

It should be emphasized that we have confined ourselves to a discussion of in-plane currents, frequently referred to as CIP (current in plane). To date, there is also interest in systems with transport perpendicular to the stack (CPP, i.e., current perpendicular) [156–160], because the characteristic length is the spin-flop mean path. This length is about an order of magnitude larger than the mean free path for diffuse scattering. This has the advantage of allowing larger film thicknesses without loss of signal.

6.3.2 Micromagnetics of Multilayer Systems

Defects in Multilayer Systems. We will now discuss another aspect of multilayers which has a major impact on the magnetoresistive response. In Sect. 6.3.1 on the phenomenological GMR theory, it was emphasized that, in the case of a dominant interface scat tering, the highest $\Delta\varrho/\varrho$ should be expected when the spatial density of interfaces is as high as possibly. In this regime of thin layers of typical 10 Å thickness, coupling defects often occur in the form of ferromagnetic exchange shortcuts between neighboring magnetic layers. These ferromagnetic bridges often cause a dramatic reduction in the signal. Figure 6.24 shows a typical manifestation of defects in the M–H curve of antiferromagnetically coupled stacks.

The curve exhibits a significant remanence. In this case, it is partly due to the Fe/Co buffer. However, the surplus in the remanence originates in the Co of the Co/Cu multilayer. The procedure to estimate the remanence of the periodic multilayer system as indicated in the inset of Fig. 6.24. We assume that the magnetic buffers are decoupled and that they can rotate without interaction with the superlattices. By subtracting the remanence

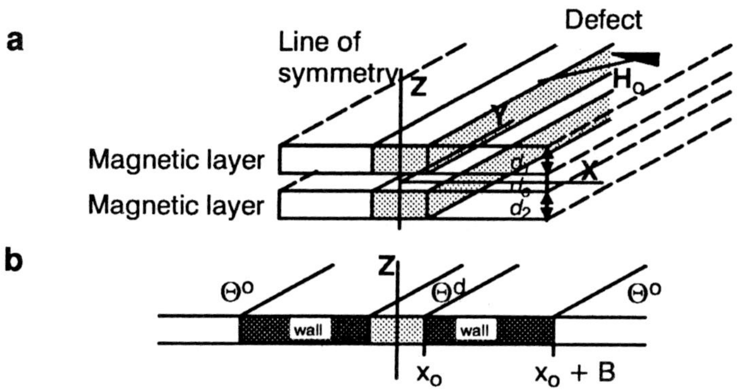

Fig. 6.25. (**a**) A perspective view of the magnetic bilayer with the line defect marked by the hatched bars. (**b**) The domain walls in one of the films matching $\boldsymbol{M}$ of the defect and of the exterior regions

contribution of the buffer from the total remanence $\langle M^r \rangle$, it was seen that, generally, the layers of the stack also contribute to $\langle M^r \rangle$.

The portion of the stack that is perfectly AF coupled is represented by the factor F_{AF}. F_{AF} follows from

$$F_{\mathrm{AF}} = \frac{(\langle M^{\mathrm{s}} \rangle - \langle M^{\mathrm{r}} \rangle) d_{\mathrm{m}} - (M_{\mathrm{Bu}}^{\mathrm{s}} - M_{\mathrm{Bu}}^{\mathrm{r}}) d_{\mathrm{Bu}}}{\langle M^{\mathrm{s}} \rangle d_{\mathrm{m}} - M_{\mathrm{Bu}} d_{\mathrm{Bu}} - (1 - (-1)^{N_{\mathrm{St}}}) M_{\mathrm{St}} d_{\mathrm{St}} / 2} \tag{6.41}$$

in which the superscripts ^{r} and ^{s} refer to the remanence and the saturated state (measured at 20 kOe in one present experiments), respectively. The total magnetic thickness d_{m}, including the buffer, is taken to determine the mean stack magnetization $\langle M \rangle$. M_{Bu} and M_{St} are the magnetizations of the (Fe/Co) buffer and the magnetic stack layer (Co), respectively. d_{St} and d_{Bu} are, respectively, the thicknesses of the individual Co stack layers and of the buffer. The term with $M_{\mathrm{St}} d_{\mathrm{St}}$ in the denominator refers to the fact that the magnetic moment of one stack layer, also in defect-free stacks, is always expected to contribute to $\langle M^{\mathrm{r}} \rangle$ when the number N_{St} of magnetic layers in the superlattice is odd. It is obvious that the latter modifications are particularly relevant for stacks with low N_{St}, i.e., with low total magnetic moment.

Determination of F_{AF} is an attempt to estimate the fraction of the multilayer that contributes to the GMR signal. Thus, it is tried to improve the quantitative evaluation of the scattering characteristics by taking advantage of the information provided by the M–H curves. The underlying assumption is that there is a one-to-one correspondence between the remanence and the loss in the GMR signal. In the next section, this underlying hypothesis is verified by considering defects from the micromagnetic point of view.

The Micromagnetics of Defects in Multilayers. In the foregoing discussion, we misleadingly suggested that the coupling defects are signified by parallel magnetizations in adjacent layers. It will be seen that this is generally not true. To illustrate this, one should consider a magnetic bilayer that contains one single infinitely long line defect. The magnetization far away from this defect is governed by the physical parameters of the outside region and the external field. In the defect, these parameters are different and the magnetization deviates from its external pendant. These different magnetizations imposed in regions of the same layer have to be matched by domain walls [161–163]. These domain walls extend from the defect into the defect-free region, i.e., the effective defect dimension is larger than the real one. Furthermore, the magnetizations in the effective defect region are certainly not parallel and also contribute to the GMR effect. In other words, one has to reconsider the correlation between remanence and the loss in GMR signal again.

We shall now consider the theoretical aspects of single line defects in magnetic bilayers by using a Ritz-type of model, i.e., a model in which some characteristics of the wall shape are predefined. However, the defect energy can be minimized by adjusting a set of free wall-shape parameters. The stack consists of an upper and a lower magnetic layer, defined by subscripts 1 and 2, respectively, and of a nonferromagnetic intermediate layer. The thicknesses of the magnetic films are given by d_i and the spacer thickness by d_{c}. The stack contains a line defect, indicated by two hatched bars in Fig. 6.25, along the y axis of a Cartesian coordinate system. The magnetic layers are characterized by the in-plane magnetization $\boldsymbol{M}_i(x)$ with magnitude $M_{\mathrm{s}i}$ and direction Θ_i with respect to the x axis ($i = 1, 2$). These layers have in-plane uniaxial anisotropy with constants $K_{\mathrm{u}i}$ and direction $\Phi_{\mathrm{u}i}$ of the easy axis and also cubic anisotropy with constants $K_{\mathrm{c}i}$, whereby $\Phi_{\mathrm{c}i}$ defines the directions of preferential $\boldsymbol{M}$. The final bulk-film parameter is the exchange constant A_i. The exchange coupling is given by J_{bil} and J_{biq} for the linear and biquadratic contribution, respectively. In general, the parameters are different for both layers and for the defect and the outer regions. Those applying to the defects are marked with superscript ^{d} and those applying to the outer regions with $^{\circ}$. The external field $\boldsymbol{H}_0$ is in-plane with direction Θ.

The surface-energy density e_{s} of the stack is given by [163]

$$\begin{aligned} e_{\mathrm{s}} = \sum\nolimits_{i=1}^{2} d_i \Bigg\{ & A_i \left(\frac{\mathrm{d}\theta_i}{\mathrm{d}x}\right)^2 (1 + \sin^2\theta_i) + K_{\mathrm{u}i}\sin^2(\Phi_{\mathrm{u}i} - \theta_i) \\ & + \frac{K_{\mathrm{c}i}}{4}\sin^2(2(\Phi_{\mathrm{c}i} - \theta_i)) - M_{\mathrm{s}i}H_0\cos(\Theta - \theta_i) \Bigg\} \\ & + 2\pi d_{\mathrm{c}} \left(d_1 M_1 \frac{\mathrm{d}\theta_1}{\mathrm{d}x}\sin\theta_1 + d_2 M_2 \frac{\mathrm{d}\theta_2}{\mathrm{d}x}\sin\theta_2 \right)^2 \\ & - J_{\mathrm{lin}}\cos(\theta_1 - \theta_2) - J_{\mathrm{biq}}\cos^2(\theta_1 - \theta_2) \, . \end{aligned} \tag{6.42}$$

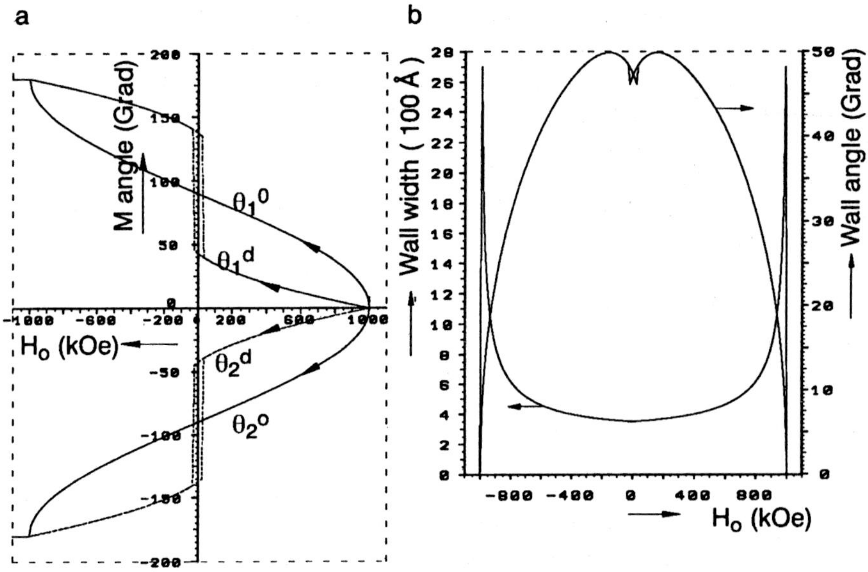

Fig. 6.26. The magnetization angles (**a**) and the wall width and wall angle as a function of H_0 (**b**) for an isotropic magnetic bilayer with a single defect. M angles: θ_1^{o}, θ_2^{o} in exterior region, θ_1^{d}, θ_2^{d} in the middle of the defect. Thickness of magnetic layers is 10 Å, $J_{\mathrm{bil}} = -0.05\,\mathrm{erg/cm^2}$, $A = 1.10^{-6}\,\mathrm{erg/cm}$, $M_{\mathrm{s}} = 1000\,\mathrm{emu/cm^3}$, Defect: $J_{\mathrm{bil}} = -6\,\mathrm{erg/cm^2}$ with width 5 Å

The term preceded by the thickness of the intermediate layer d_{c} represents the energy contribution of the asymmetrical part of the surface-charge distributions in the thinfilm approach, which arise in the domain-wall regions. The material parameters are assumed to be constant in both the exterior and defect regions and thus can only change in jumps at the defect edges at $x = \pm x_0$, where $2x_0$ is the defect width.

A one-dimensional wall model is used in which θ_i varies along the x axis. The Ritz method is applied and the wall shape is predefined by

$$\begin{aligned}
&\theta_i(x) = \theta_i^{\mathrm{d}} + (\theta_i^{\mathrm{o}} - \theta_i^{\mathrm{d}})(x\,\mathrm{sign}(x/\mid x\mid) - x_0)/B_i \\
&\qquad\quad \text{when} \quad 0 < x\,\mathrm{sign}(x/\mid x\mid) - x_0 < B_i\,, \\
&\theta_i(x) = \theta_i^{\mathrm{d}} \quad \text{when} \mid x \mid \le x_0\,, \\
&\theta_i(x) = \theta_i^{\mathrm{o}} \quad \text{when} \mid x \mid \ge x_0 + B_i \quad \text{with } i = 1, 2\,,
\end{aligned} \tag{6.43}$$

where B_i is the wall width.

The wall energies have been derived by analytically integrating (6.42) along the wall width, while the surface-energy density in the remaining region follows directly from (6.42), at which the bulk-exchange and the magnetic charge terms can be omitted.

The Θ's are found by numerically minimizing the total energy. Hysteresis curves are traced by gradually changing H_0, at which the starting vector at a specific H_0 is defined by the outcomes at the previous H_0 step(s). (For further details, see [163].)

Here, we shall confine ourselves to polycrystalline materials in which there are no preferential orientations in the plane due to the small size of the crystal lites and bulk-exchange coupling. Therefore, the K_{ci} and K_{ui} are set at zero. $\boldsymbol{H}_0$ is along the x axis. Examples of anisotropical cases can be found else where [163,164]. Figure 6.26 a shows the M directions in the outer regions θ_1^o and θ_2^o and in the middle of the defect θ_1^d and θ_2^d in a stack where $J_{bil} = -0.05\,erg/cm^2$ in the outer and $6\,erg/cm^2$ in the defect region. The width of the defect is 5 Å. Of course, θ_1^o and θ_2^o exhibit typical AF behavior. The magnetizations in the defect partly behave ferromagnetically as can be deduced from the jumps in θ_1^d and θ_2^d on reversal of the field. However, it can be also seen that the defect magnetizations are pulled into the direction of the outer magnetizations by the bulk exchange and also exhibit an antiferromagnetic component. At zero field, the defects magnetizations are at about 45° to the remanence direction (the x-axis). In other words, there is no question of pure ferromagnetic defect behavior in the present example. The remanence at $H_0 = 0$ vanishes completely when the J_{bil} of the defect is reduced to $4\,erg/cm^2$. Next, we will look at the lateral extension of the walls.

Figure 6.26 a shows both the wall width and the wall angle as a function of the field.The wall angle is significantly smaller than $|\ \theta_1^o\ |$, which demonstrates the dominance of the AF coupling of the outer region due to the small defect width. The total width of the defect walls at $H_0 = 0$ is about 800 Å, i.e., a factor of about 150 larger than the defect width. This demonstrates the disastrous impact of defects.

It was already indicated that the defect width and its J_{bil} have an impact on the deflection from the pure ferromagnetic defect behavior. Figure 6.27 a illustrates this for a bilayer with a defect width of 40 Å and J_{bil} of $10\,erg/cm^2$. The curve is only traced in one direction from positive to negative H_0. Indeed, the deflection of the defect M is much smaller now (10° at $H_0 = 0$). However, the total defect-wall width in this situation is still 700 Å, i.e., the effective defect width is still very large (see Fig. 6.27 b). Figures 6.27 a (and 6.26 a) betray a number of mode branches: (1) The saturated one above $|\ H_0\ |=$ 1 kOe, (2) the AF one for $-690 < H_0 < 1000$ Oe in Fig. 6.27 a, (3) the asymmetrical one between −690 and 760 Oe and, finally, the counterpart of branch 2. The occurrence of branch 3 demonstrates that, even in the simple isotropic case, unexpected mode branches are possible that are typical of multilayer systems. On this branch, both defect $\boldsymbol{M}$'s are still rougly aligned but rotated by 50° with respect to H_0.

We will now return to the impact of the defects and the validity of the 1-F_{AF} factor for correlating the remanence and the loss in the GMR signal.

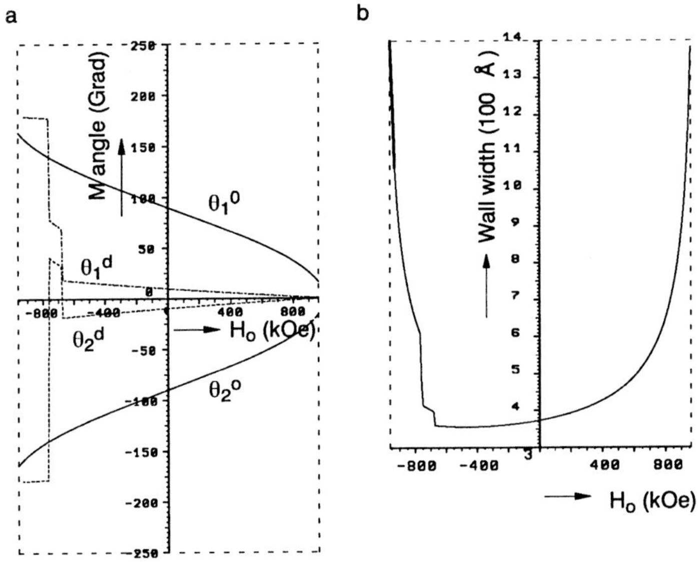

Fig. 6.27. The magnetization angles (**a**) and the wall width as a function of H_0 (**b**) for an isotropic magnetic bilayer with a single defect. M angles: θ_1^{o}, θ_2^{o} in exterior region, θ_1^{d}, θ_2^{d} in the middle of the defect. Thickness of magnetic layers is 10 Å, $J_{\mathrm{bil}} = -0.05\,\mathrm{erg/cm^2}$, $A = 1.10^{-6}\,\mathrm{erg/cm}$, $M_{\mathrm{s}} = 1000\,\mathrm{emu/cm^3}$. Defect: $J_{\mathrm{bil}} = 10\,\mathrm{erg/cm^2}$ with width 40 Å

The normalized GMR signal is given by (see (6.35))

$$\Delta\varrho_{\mathrm{n}}(\theta_1,\theta_2) = (1-\cos(\theta_1-\theta_2))/2\,. \tag{6.44a}$$

The additional change $\Delta\varrho_{\mathrm{n}}^{\mathrm{d}}$ in $\Delta\varrho_{\mathrm{n}}$ due to a line defect and its bounding walls follows from

$$\begin{aligned}\Delta\varrho_{\mathrm{n}}^{\mathrm{d}} = 2\int_{\mathrm{wall}}\Delta\varrho_{\mathrm{n}}(\theta_1,\theta_2)\mathrm{d}x + \Delta\varrho_{\mathrm{n}}(\theta_1^{\mathrm{d}},\theta_2^{\mathrm{d}})2x_0\\ -\Delta\varrho_{\mathrm{n}}(\theta_1^{0},\theta_2^{0})2(x_0+B)\,,\end{aligned} \tag{6.44b}$$

where $2x_0$ is the defect width.

A similar expression applies to the normalized remanence $m_{\mathrm{n}}^{\mathrm{d}}$ of the defect:

$$m_{\mathrm{n}}(\theta_1,\theta_2) = \frac{(\cos(\theta_1)d_1M_1 + \cos(\theta_2)d_2M_2)}{(d_1M_1+d_2M_2)}\,, \tag{6.45a}$$

where d_1 and d_2 are the magnetic layer thicknesses. The impact of the defect follows from

$$\begin{aligned}m_{\mathrm{n}}^{\mathrm{d}} = 2\int_{\mathrm{wall}} m_{\mathrm{n}}(\theta_1,\theta_2)\mathrm{d}x + m_{\mathrm{n}}(\theta_1^{\mathrm{d}},\theta_2^{\mathrm{d}})2x_0\\ -m_{\mathrm{n}}(\theta_1^{\mathrm{o}},\theta_2^{\mathrm{o}})2(x_0+B)\,.\end{aligned} \tag{6.45b}$$

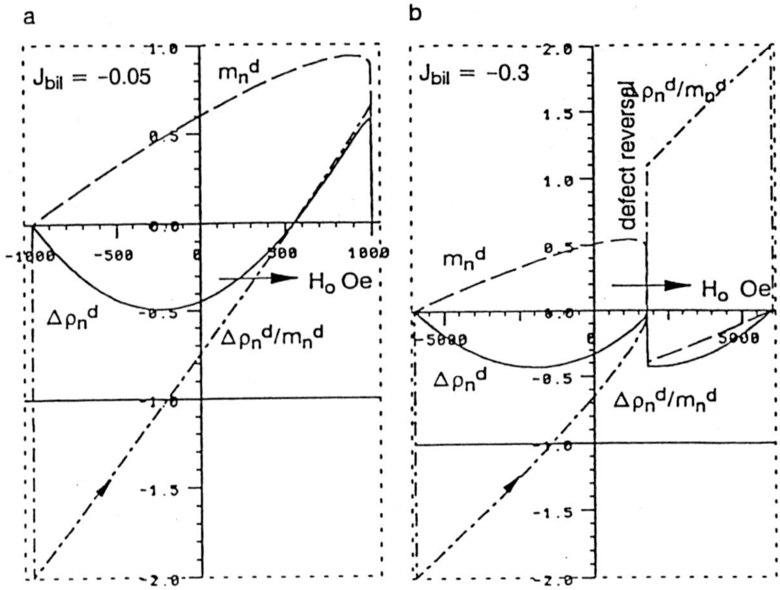

Fig. 6.28. **a, b** The ratio $\Delta\varrho_{\rm n}^{\rm d}/m_{\rm n}^{\rm d}$ of the change in the normalized GMR and in the normalized magnetization along H_0 due to a single ferromagnetic coupling defect with a width of 10 Å and $J_{\rm bil} = 10\,{\rm erg/cm^2}$. The bilinear coupling constants $J_{\rm bil}$ of the outer regions are $-0.05\,{\rm erg/cm^2}$ (**a**), and $-0.3\,{\rm erg/cm^2}$ (**b**).
$d_1 = d_2 = 10$ Å, $A_1 = A_2 = 10^{-6}$ erg/cm, $M_{\rm s1} = M_{\rm s2} = 1000\,{\rm emu/cm^3}$.
$m_n^{\rm d}$ is the surplus normalized M component along the initial H_0 direction.
$\Delta\rho_n^{\rm d}$ is the surplus normalized GMR resistivity

Equations (6.44a) and (6.45a) have a different functional dependence on the angles. The ratio $\Delta\varrho_{\rm n}^{\rm d}/m_{\rm n}^{\rm d}$ will, therefore, generally not be equal to one, as assumed when the simple $F_{\rm AF}$ approach is employed. Figure 6.28 shows the course of $\Delta\varrho_{\rm n}^{\rm d}/m_{\rm n}^{\rm d}$ along the hysteresis curve of a sandwich with a defect width of 10 Å and an $J_{\rm bil}$ of $10\,{\rm erg/cm^2}$ in the defect, while the bilinear coupling coefficient for the outer regions is -0.05 and $-0.3\,{\rm erg/cm^2}$ for Figs. 6.28 a and 6.28 b, respectively. The curves are traced from minus to plus saturation. It can be seen that $\Delta\varrho_{\rm n}^{\rm d}/m_{\rm n}^{\rm d}$ varies between -2 and $+2$. Let us now look at Fig. 6.28 a in more detail. Starting from minus saturation, it is obvious from (6.44a) and (6.45a) that, for small $\mid \theta_i^{\rm d} - \theta_i^{\rm o} \mid$, the impact on the normalized GMR is twice as large as on the normalized M-component along H_0. The shape of the $\Delta\varrho_{\rm n}^{\rm d}$ and $m_{\rm n}^{\rm d}$ curves is quite different. The $m_{\rm n}^{\rm d}$ curve increases monotonically up to high H_0, while $\Delta\varrho_{\rm n}^{\rm d}$ exhibits a zero crossing at about $H_0 = 550$ Oe. This zero crossing indicates that the weight of the regions with a higher $\Delta\varrho$ compared to the defectless situation is compensated by regions with lower $\Delta\varrho$. This occurs when the magnetizations in the outer regions start to return to a parallel situation and the defect is strong enough to keep the

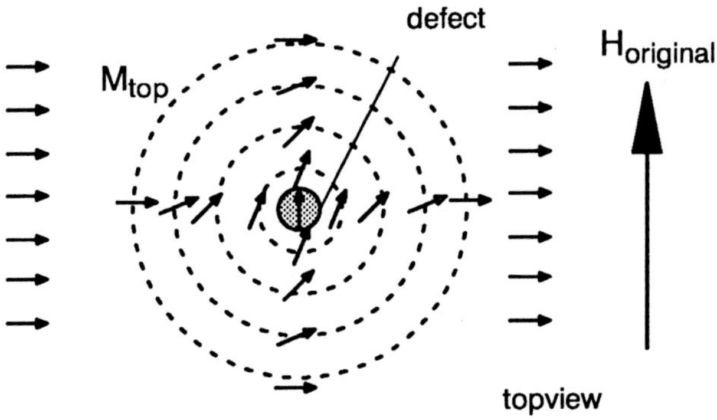

Fig. 6.29. Schematic plot of the magnetization distribution in the top layer of a sand wich around a circular ferromagnetic defect in an AF coupled stack with equally thick magnetic layers at $H_0 = 0$

defect $\boldsymbol{M}$ along the original saturation direction. Important for the present method is the value at remanence, i.e., at $H_0 = 0$. At $H_0 = 0$, zero, $\Delta\varrho_n^d/m_n^d$ is 0.75 and thus not equal to -1. This implies that the loss in the GMR signal due to the defect is overestimated by 25% when the remanence in the B–H curves are used as reference. What is the origin of this difference of 25%? Let us assume that the $\Theta_i^d = 0$ and with $\Theta_i^o = \pm\pi/2$ at $H_0 = 0$. One can derive the following simple expression for $\Delta\varrho_n^d/m_n^d$ when $d_1M_1 = d_2M_2$ by using (6.43–6.45):

$$\Delta\varrho_n^d/m_n^d = -\frac{1+2x_0/B}{4/\pi+2x_0/B}\,, \tag{6.46a}$$

in which the defect-wall width can be approximated by

$$B = \pi/2(-3dA/J_{\rm bil})^{1/2}\,, \tag{6.46b}$$

by assuming $d = d_1 = d_2$ and $A = A_1 = A_2$. Taking the parameters of Fig. 6.28 a, (6.46) yields that B is 380 Å and $\Delta\varrho_n^d/m_n^d$ is -0.79. The latter value is rather close to its counterpart in Fig. 6.28 a, so that it is concluded that the difference in the functional dependence of (6.44a) and (6.45a) is the main reason for the deviation. Equation (6.46a) suggests that $\Delta\varrho_n^d/m_n^d$ approaches -1 when B is smaller, i.e., when e.g. $-J_{\rm bil}$ is larger. This is not confirmed by Fig. 6.28 b, in which $\Delta\varrho_n^d/m_n^d$ is -0.63, while (6.46) predicts -0.80. This is caused by the larger deflection of defect magnetization, which is subjected to a much stronger torque by the thinner (160 Å) and stiffer defect wall. The larger forces exerted by the outer region on the defect are reflected also by the reversal of the defect magnetization at $H_0 = 1700$ Oe in Fig. 6.28 b. In other words, we should expect $\Delta\varrho_n^d/m_n^d$ to be closer to -1 when the defect width is larger and the ferromagnetic interlayer coupling is strong, so that Θ_i^d will be smaller and $2x_0$ larger.

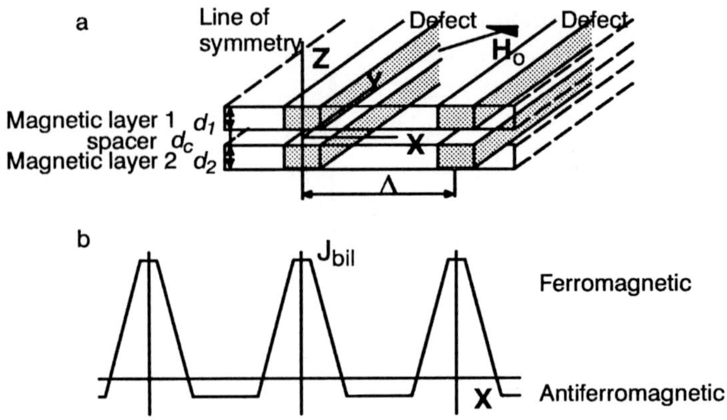

Fig. 6.30. (**a**)Arrangement of the periodic array of line defects in a sandwich of two magnetic layers. (**b**) A possible course of the the interlayer coupling constant

The requirement to the ferromagnetic coupling is satisfied in the case of ferromagnetic bridges between the magnetic layers, generally considered to be the main origin of the remanence. It is hard to judge whether the dimensions of the defect are sufficiently large in a real system. The above arguments show that the F_{AF}, as derived from magnetization measurements, cannot generally directly used for correcting the magnetoresistive signal. A higher accuracy of the assessment of the scattering properties is achieved by defining F'_{AF} as

$$F'_{\mathrm{AF}} = 1 + \Delta\varrho_{\mathrm{n}}^{\mathrm{d}}/m_{\mathrm{n}}^{\mathrm{d}} - \Delta\varrho_{\mathrm{n}}^{\mathrm{d}}/m_{\mathrm{n}}^{\mathrm{d}} F_{\mathrm{AF}} \tag{6.47a}$$

$$\text{with} \quad \Delta\varrho = \Delta\varrho_{\mathrm{meas}}/F'_{\mathrm{AF}}\,, \tag{6.47b}$$

where $\Delta\varrho_{\mathrm{meas}}$ is the measured resistivity change. Equation (6.47) states that the relationship between $\Delta\varrho$, $\Delta\varrho_{\mathrm{meas}}$ and F_{AF} is nonlinear. Note that the remanence of the defect according to Fig. 6.28 is 0.6 the saturation. The defects start to overlap when the total remanence exceeds about 0.5, implying that the above approach has to be modified. It should be emphasized that quantitative deviations from the above estimations are to be expected since real defects are generally not of line type. Figure 6.29 gives a schematic impression of the M distribution of the top layer around a circular defect.

Assuming the rotational symmetry as indicated in Fig. 6.29, one obtains in the same order of approximation as in (6.46a) the following estimation for $\Delta\varrho_{\mathrm{n}}^{\mathrm{d}}/m_{\mathrm{n}}^{\mathrm{d}}$:

$$\Delta\varrho_{\mathrm{n}}^{\mathrm{d}}/m_{\mathrm{n}}^{\mathrm{d}} = -\frac{\pi(x_0+B)^2 - \pi B(2x_0+B)/2 - 2B^2/\pi}{4\{B^2(1-2/\pi) + x_0 B + \pi x_0^2/4\}}\,, \tag{6.48}$$

so that $-1 < \Delta\varrho_{\mathrm{n}}^{\mathrm{d}}/m_{\mathrm{n}}^{\mathrm{d}} < -0.65$. The remanence varies between 0.45 for relatively wide edge walls and 1 for the pure defect. In general, the defect magnetization will deflect stronger due to the larger torque exerted by the surrounding edge wall, as compared to the line defects. As a consequence,

(6.48) will be less accurate than its counterpart (6.46a). The remanence varies between 0.45 for relatively wide edge walls and 1, i.e., the defects start to overlap at even smaller remanence than for line defects. Its should be expected that $\Delta\varrho_{\mathrm{n}}^{\mathrm{d}}/m_{\mathrm{n}}^{\mathrm{d}}$ approaches -2 in the case of a high density of strong defects (see Fig. 6.30). In this case of overlap, an alternative approach is needed [165]. Further, modification of the definition of F_{AF} is required in the case of anisotropic media.

The method presented above has the advantage of being fast so that a large number of situation can easily be scanned. It is a Ritz method with a restricted number of free parameters, as has been frequently used for estimating domain-wall configurations and for evaluating the changes in domain structures with predefined geometrical characteristics. Though helpful, it is no real micromagnetic method and its accuracy is limited. The micromagnetic approach provides a formalism that gives the equilibrium distributions of the magnetization in the macroscopical limit. The torque $\boldsymbol{\tau}$ satisfies

$$\boldsymbol{\tau} = \boldsymbol{i} \times \boldsymbol{H}_{\mathrm{eff}} \,, \tag{6.49a}$$

both within the object volume V and at its surface S. Equation (6.49a) states that the unit vector $\boldsymbol{i}$ along $\boldsymbol{M}$ has to be parallel to the effective micromagnetic field $\boldsymbol{H}_{\mathrm{eff}}$. In the two-dimensional situation with the y axis being the direction of invariance, $\boldsymbol{H}_{\mathrm{eff}}$ is given by (see [166,167])

$$\begin{aligned} \boldsymbol{H}_{\mathrm{eff}} &= \frac{2A_i}{M_{\mathrm{s}i}} \left\{ \frac{\partial^2 i_x}{\partial x^2} + \frac{\partial^2 i_x}{\partial z^2}, \frac{\partial^2 i_y}{\partial x^2} + \frac{\partial^2 i_y}{\partial z^2}, \frac{\partial^2 i_z}{\partial x^2} + \frac{\partial^2 i_z}{\partial z^2} \right\} \\ &\quad - \frac{1}{M_{\mathrm{s}i}} \left\{ \frac{\partial e_{\mathrm{a}i}}{\partial i_x}, \frac{\partial e_{\mathrm{a}i}}{\partial i_y}, \frac{\partial e_{\mathrm{a}i}}{\partial i_z} \right\} + \boldsymbol{H}_0 + \boldsymbol{H}_{\mathrm{d}} \,, \end{aligned} \tag{6.49b}$$

$$\boldsymbol{H}_{\mathrm{d}} = -\left\{ \frac{\partial \Phi}{\partial x}, 0, \frac{\partial \Phi}{\partial z} \right\} , \quad \text{in} \quad V \,, \tag{6.49c}$$

where $e_{\mathrm{a}i}$, $\boldsymbol{H}_{\mathrm{d}}$ and Φ are the volume anisotropy-energy density, the demagnetizing field and the magnetic potential, respectively. i_x, i_y and i_z are the direction cosine of $\boldsymbol{M}$, and the subscript i refers to the magnetic medium. At the surface S, the following boundary conditions apply:

$$\begin{aligned} \boldsymbol{\tau}_S &= \boldsymbol{i}_i \times \boldsymbol{H}_s = 0 \quad \text{on} \quad \mathrm{S} \\ \boldsymbol{H}_S &= -\frac{2A_i}{M_{\mathrm{s}i}} \frac{\partial \boldsymbol{i}_i}{\partial n} + \frac{J_{\mathrm{bil}} + 2J_{\mathrm{biq}}(\boldsymbol{i}_i \cdot \boldsymbol{i}_j)}{M_{\mathrm{s}i}} \boldsymbol{i}_j - \frac{2K_{\mathrm{s}i}(\boldsymbol{i}_i \cdot \boldsymbol{n}_i)}{M_{\mathrm{s}i}} \boldsymbol{n}_i \,, \end{aligned} \tag{6.49d}$$

where subscript $j = i \pm 1$, i.e., j refers to a magnetic medium adjacent to i, and where K_{s} represents the so-called surface anisotropy, a consequence of the symmetry breaking at the interface. Finally, $\boldsymbol{n}_i$ is the outward pointing unit vector of magnetic medium i. Of course, the boundary conditions for Φ have to be simultaneously satisfied. Only in a few very rare cases can this set of equations be analytically solved. The main difficulty in solving this boundary value problem is caused by the nonlocal magnetostatic fields.

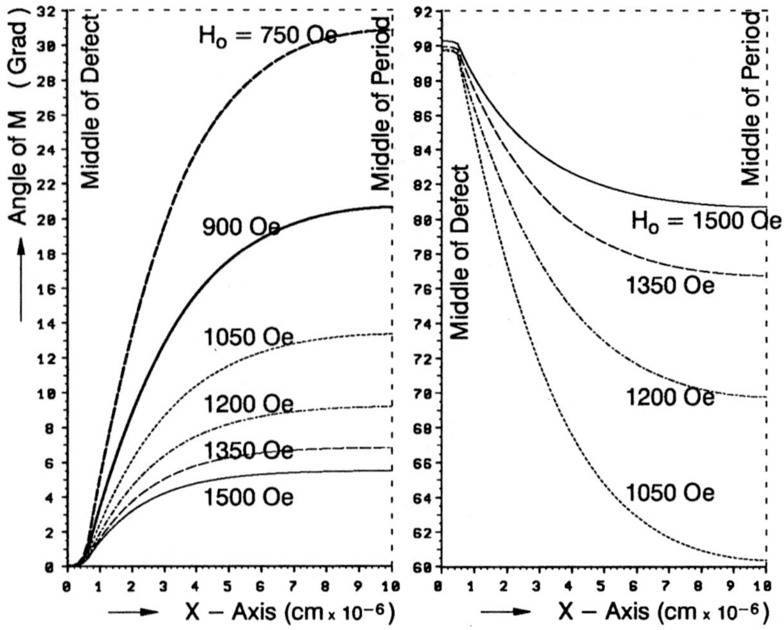

Fig. 6.31. H_0 normal to (**a**) and parallel (**b**) to the line defects. $d_1 = d_2 = 10\,\text{Å}$, $d_c = 20\,\text{Å}$. Defect: width 100 Å, mutual distance $\Lambda = 2000\,\text{Å}$, $J_{\text{def}} = 10\,\text{erg/cm}^2$, $M_{s1} = M_{s2} = 1000\,\text{erg/cm}^3$. Anisotropy: $K_{u1} = K_{u2} = 10^5\text{erg/cm}^3$, $\Phi_{Ku1} = -\Phi_{Ku2} = 20°$ in plane with respect to the x axis, $A_1 = A_2 = 10^{-6}\,\text{erg/cm}$. Interlayer exchange constant of the outer region: $J_0 = -0.05\,\text{erg/cm}^2$

In periodic geometries, fast Fourier transform (FFT) is most frequently employed for evaluating $\boldsymbol{H}_d$ [168]. The zero padding technique is used to extend the applicability to nonperiodic structures.

To evaluate the impact of defects on the GMR signal, we cannot confine ourselves to a single defect, but should assume a certain density of these. Therefore, we shall solve the above boundary-value problem for a periodic arrangement of line defects in a plane-parallel magnetic bilayer. The geometry is presented by Fig. 6.30. In principle, the defects might be characterized by deviations in all the parameters of (6.49) and the transition from the defect to the defect-free regions may be linear as indicated in Fig. 6.30 b. In this case, however, we shall confine ourselves to ferromagnetic bridges without the linear transition zones. The coupling strength of the defects is 10 erg/cm^2, being typical for coupling bridges with heights of 20 Å (see (6.23) with $A = 10^{-6}\,\text{erg/cm}$). The layers are 20 Å apart, a typical decoupling thickness of Cu, and a small residual AF-coupling is assumed in the defect-free regions.

Figure 6.31 shows the spatial dependence of the $\boldsymbol{M}$ angle in layer I for the in-plane $\boldsymbol{H}_0$ both perpendicular and parallel to the line defects. For both

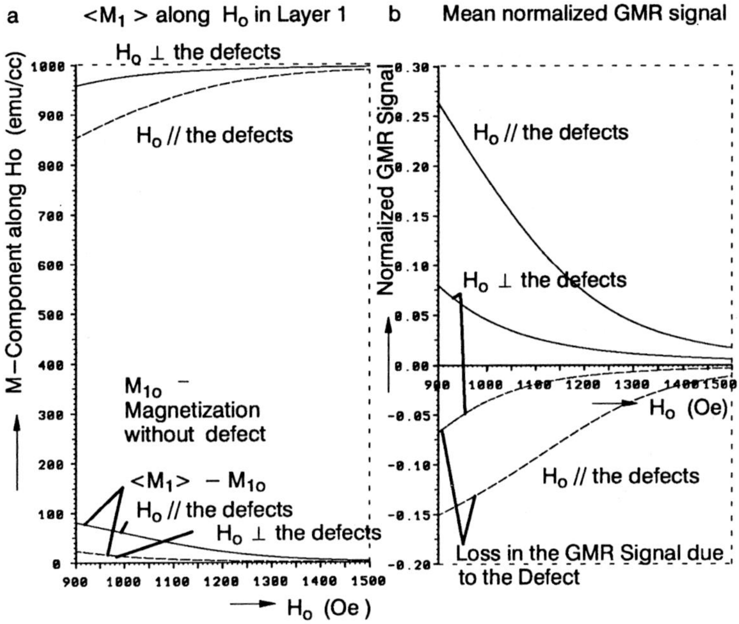

Fig. 6.32. a, b The impact of the defects on $\boldsymbol{M}_1$ along $\boldsymbol{H}_0$ (**a**), the mean normalized GMR effect of the sandwich (**b**). $d_1 = d_2 = 10\,\text{Å}$, $d_c = 20\,\text{Å}$. Defects: width $100\,\text{Å}$, mutual distance $\Lambda = 2000\,\text{Å}$, $J_{\text{def}} = 10\,\text{erg/cm}^2$, $M_{s1} = M_{s2} = 1000\,\text{erg/cm}^3$. Anisotropy: $K_{u1} = K_{u2} = 10^5\,\text{erg/cm}^3$, $\Phi_{\text{Ku1}} = -\Phi_{\text{Ku2}} = 20°$ in plane with respect to the x axis, $A_1 = A_2 = 10^{-6}\,\text{erg/cm}$. Interlayer exchange constant of the outer region: $J_0 = -0.05\,\text{erg/cm}^2$

directions, $\boldsymbol{M}$ remains parallel to $\boldsymbol{H}_0$ within the defects since the torques exerted by the wall configuration are too weak. The course of the $\boldsymbol{M}$ angles is smooth and a linear region is only found directly adjacent to the defect, and not in the total wall region, as assumed in (6.43) of the Ritz approach presented previously. The width of the region is however predicted rather well by the Ritz method. The larger deflection of $\boldsymbol{M}$ from saturation that occurs with $\boldsymbol{H}_0$ parallel to the line defects originates in the uniaxial anisotropies present in both layers.

Figure 6.32 a shows the $\boldsymbol{M}$ component along $\boldsymbol{H}_0$, again with $\boldsymbol{H}_0$ parallel and normal to the defects, as a function of H_0. The stronger reduction in M with $\boldsymbol{H}_0$ parallel to the defects is caused by the anisotropies. Further, the mean deviation $\langle M_1 \rangle - M_{10}$ of $\boldsymbol{M}$ along $\boldsymbol{H}_0$ due to the defects has been plotted in Fig. 6.32 a. Here, M_{10} refers to the defect-free situation. Note that the defect impact is significant. Figure 6.32 b shows the mean normalized GMR signal as a function of H_0 for both $\boldsymbol{H}_0$ directions mentioned above. A dramatic reduction in the signal by 40% of the ideal signal occurs, though 5% of the area is occupied by defects only. Bear in mind that, according to

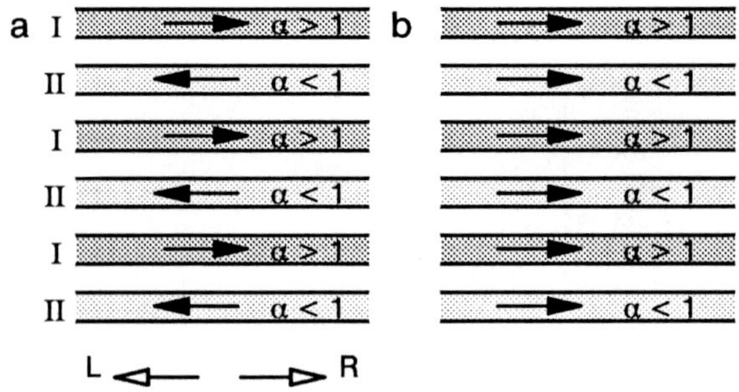

Fig. 6.33. **a, b** The inverse GMR effect originating in a difference in the bulk α's of two types of magnetic layers I and II: *black arrows* magnetization, *white arrows* channels

the previous study with the Ritz method, small defect widths already suffice to fix M within the defects. This demonstrates the low level of the defect density which can be tolerated.

6.3.3 Experimental Data on the GMR Effect

In this section, the experimental data and the theoretical insights developed above will be compared. First, the relationship $\Delta\varrho/\varrho = (\alpha - 1)^2/(4\alpha)$, as given by (6.28), is considered in detail.

The Reversed Magnetoresistive Effect. Equation (6.28) suggests that the resistivity in the parallel situation is always smaller than in the state with the magnetizations of adjacent layers opposite. This statement is generally true. However, there are exceptions. This can easily be seen by considering Fig. 6.33 with an unusual choice of the α's.

It is assumed that, in this example, only bulk scattering plays a role. The values of α of consecutive layers in the stack alternate from smaller to larger than one or vice versa. This can easily be achieved by alloying the magnetic layers. Taking Fe as the basic material, it is obvious from Table 6.6 that Fe diluted with Cr has $\alpha < 1$ and with Pt $\alpha > 1$. To begin with, we consider the resistivity of the situation with alternating magnetization directions (Fig. 6.33 a). First, we focus on the electrons in channel R with spins to the right. In the layers with magnetization to the right, marked I, these electrons are majorities and experience the lowest resistivity ϱ_1^{min}. The same electrons arriving in layers II become minorities. However, they are only weakly scattered with resistivity ϱ_2^{min} since α is smaller than one here. Roughly speaking, the electrons of channel I experience a mean resistivity $\varrho_{\text{R}}^{\text{min}}$ equal to $(\varrho_1^{\text{min}} + \varrho_2^{\text{min}})/2$. The opposite applies to channel L in Fig. 6.33 a. The resistivity ϱ_{L} is the maximum $(\varrho_1^{\text{max}} + \varrho_2^{\text{max}})/2$. Let us now turn to the parallel situation in Fig. 6.33 b. Channels R and L have resistivities $(\varrho_1^{\text{min}} +$

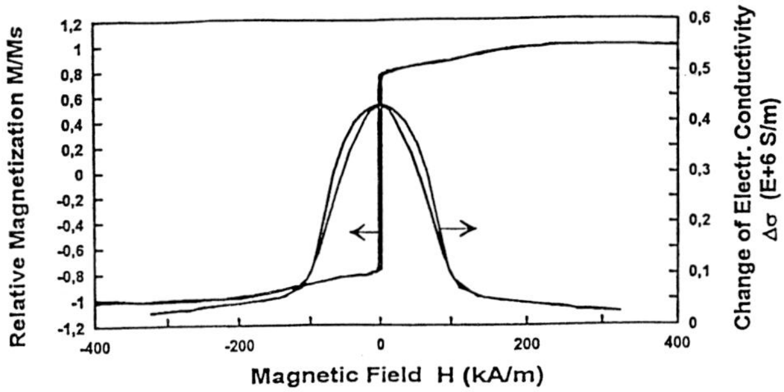

Fig. 6.34. $\Delta\varrho/\varrho_{\text{P}}$ of a stack made by DC magnetron sputtering for a glass/Fe60 Å/[Co10 Å/Cu10 Å] × 3

$\varrho_2^{\max})/2$ and $(\varrho_1^{\max} + \varrho_2^{\min})/2$, respectively. It can easily be verified that

$$\frac{\Delta\varrho}{\varrho_{\text{P}}} = \frac{\varrho_0 - \varrho_{\text{P}}}{\varrho_{\text{P}}} = -\frac{(\varrho_1^{\max} - \varrho_1^{\min})(\varrho_2^{\max} - \varrho_2^{\min})}{(\varrho_1^{\max} + \varrho_2^{\min})(\varrho_2^{\max} + \varrho_1^{\min})} . \tag{6.50}$$

Contrary to (6.28), $\Delta\varrho$ is negative and is called the inverse magnetoresistive effect.

It should be pointed out here that Levy [145,146] in his quantum-mechanical theory uses the spin asymmetry parameter p rather than α, at which both are related by

$$p = -(1 - \sqrt{\alpha})/(1 + \sqrt{\alpha}) , \quad \text{where} - 1 \leq p \leq 1 , \tag{6.51a}$$

so that (6.50) becomes

$$\frac{\Delta\varrho}{\varrho_{\text{P}}} = \frac{16 p_1 p_2}{[(1+p_1)^2(1-p_2)^2/\varrho_2^+ + (1+p_2)^2(1-p_1)^2/\varrho_1^+](\varrho_1^+ + \varrho_2^+)} , \tag{6.51b}$$

where the subscript refers again to the different magnetic layers. From this equation, which applies to both effects, it is immediately clear that the inversed GMR effect only occurs when the p's have opposite sign.

Up to now, we have confined ourselves to bulk scattering. Of course, similar effects can be achieved with surface scattering by coating the ferromagnetic interfaces with layers with appropriate α. Such layers need not be thicker than one to two monolayer. As Table 6.7 shows, Fe/Cr/Pd/Fe might be a candidate.

An experimental confirmation of this idea was given by Barthélémy et al. [169] for a stack with different interlayer materials (Fe12 Å/Cr4Å/Fe12 Å/Cu16 Å/Fe15 Å/Cu)× 5.

Experimental GMR Data for Various Material Combinations. Figure 6.34 shows the typical curve of the GMR effect versus the external field of

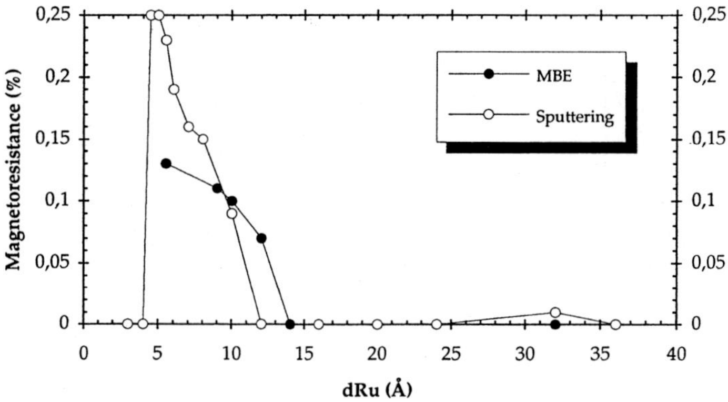

Fig. 6.35. a, b $\Delta\varrho/\varrho_p$ as a function of the Ru thickness of sandwiches made by (**a**) RF magnetron sputtering (glass/Ru50 Å/Co30 Å/Ru d_{Ru}/Co30 Å/Ru50 Å) and (**b**) MBE (mica/Ru150 Å/Co30 Å/Ru d_{Ru}/Co30 Å/Ru100 Å)

a glass/Fe60 Å/[Co10 Å/Cu10 Å] × 3 multilayer with perfectly antiferromagnetic coupling. $\Delta\varrho/\varrho_p$ of similar Co/Cu stacks with 50 instead of 3 repetitions exceeds 70% at room temperature, to date belonging to the highest values achieved [170]. The angle Φ between the M's of adjacent layers in a stack with $J_{biq} = 0$ follows from (6.10), yielding $\Phi = 2\cos^{-1}\{D_m M_s H_0/(4J_{bil})\}$. Using (6.44a), we get the parabolic dependence of $\Delta\varrho$ on H_0, as given in

$$\frac{\varrho(H_0) - \varrho_p}{\varrho_p} = \Delta\varrho\left(1 - \left(\frac{d_m H_0 d_m}{4J_{bil}}\right)^2\right), \quad \text{where} \tag{6.52a}$$

$$\Delta\varrho = \varrho_0 - \varrho_p, \quad \text{and} \tag{6.52b}$$

$$| H_0 | < H_s = \frac{4J_{bil}}{d_m m_s}, \tag{6.52c}$$

and d_m is the magnetic layer thickness. Note that the shape is indeed parabolic in the middle, but with tails at high $| H_0 |$. This deviation is caused by the non uniform interlayer-coupling strength, as discussed in the context of coupling (see Fig. 6.11).

Though the large $\Delta\varrho/\varrho_p$ is attractive, the large saturation field H_s limits a wide spread application for such material systems. The sensitivity S, defined by $\Delta\varrho/\varrho_p$ per Oe, is much too small. S is typically 1%/Oe in the case of the AMR effect in permalloy, whereas it is about 0.01 at the flanks of the curve in Fig. 6.34.

A large number of multilayer systems have been explored during recent years. Table 6.8 provides a selection from the literature with representative data. It is striking that the Co/Cu and Fe/Cr are the supreme systems, with effects of more than 100% at low temperatures. These values agree with the data in Table 6.7, which applies to the residual resistance. At room temperature, $\Delta\varrho/\varrho$ of Fe/Cr is a factor of 20 smaller, presumably being caused

Table 6.8. The magnitude of the GMR effect and the sensitivity for a number of multilayers

System	$\Delta\varrho/\varrho$%	$\Delta\varrho/\varrho$% /Oe	Thickness (Å) magn. layer	interlayer	Rep.	T	Refs.
Co/Ag	22	8.10^{-3}	6	18	70	300	[81][d]
Co/Cu	50	2.10^{-2}	8	20	30	300	[171][b]
	80	$2.10^{-}2$	40	9.3	16	4.5	[82][a]
	70	10^{-2}	10	10	50	300	[17][a]
	120	2.10^{-2}	10	10	50	77	O[a]
	51	6.10^{-3}	10	10	16	300	O[e]
Co/Re	0.1	$\ll$	30	5	bi	300	[85][b]
Co/Rh	0.3	2.10^{-5}	30	9	bi	4	O[a]
	> 0.7	$\ll$	30	10	bi	4	O[d]
Co/Ru	6	10^{-3}	20	4	20	4.5	[86][a]
	0.3	$\ll$	30	6	bi	300	O[b]
Fe/Au	2	2.10^{-2}	9.6	33	19	300	[172][d]
Fe/Cr	100	5.10^{-3}	30	9	60	4.2	[29][d]
	35	3.10^{-3}	20	8	30	4.5	[86][a]
	5	10^{-2}	30	12	40	293	[173][d]
Fe/Cu	13	2.10^{-2}	15	15	60	4.2	[92][b]
Fe/Mo	1.5	5.10^{-3}	30	9	60	4.2	[93][d]
Ni/Ag	28	2.10^{-2}	8	11	2.10[c]	4.2	[174][b]
Per/Ag	17	0.15	12.5	10	2.10[c]	300	[95][b]
Per/Au	12	0.2	30	21	11	300	[175][d]
Per/Cu	18	10^{-2}	10	10	20	300	[176][e]
	14	6.10^{-3}	11	10	10	300	O[e]
	16	3.10^{-2}	15	9	14	300	[97][a]

[a] DC magnetron sputtering; O own data
[b] sputtering; *bi* magnetic bilayer
[c] e-beam deposition; Per permalloy
[d] *MBE*; Rep. repetition number
[e] *IBS*; T temperature (K)

by the fact that the majority 3d band of Fe intersects the Fermi level. The sensitivity S of these systems is poor, which indicates that the expected correlation between coupling strength and GMR really exists. However, there are also strong deviations from Inoue's table. For example, Co/Ru falls short of Inoue's expectations. This applies equally to MBE-grown and sputtered samples, as can be seen in Fig. 6.35. This system is particularly conspicuous because the interlayer-coupling strength for the same stacks belongs to the highest reported to date (see Table 6.2 and Fig. 6.9). However, one should pay attention to the so-called phase of the coupling in Co/Ru, which is such that J starts at negative values, i.e., antiferromagnetically, at the first monolayer of Ru [177]. The coupling strength is usually measured in the first AF maximum. For Ru, this maximum occurs at small thicknesses, where (6.5) has the highest value. In other words, $|J|$ is not the right measure for the potentials v_m in (6.6). On the other hand, Inoue predicts high scattering potentials, and stacks with 10 Å Ru layers also still have a J of about $-1\,\mathrm{erg/cm^2}$. In other words, the dilemma of the simultaneous occurrence of weak GMR activity and strong interlayer coupling is still there. Of course, the GMR effect requires scattering centers and it might be that the interfaces are of a quality that is too high. This is not a plausible explanation, since, judging from the lower J, the sputtered samples exhibit a clear coupling frustration and no improvement of the GMR. Moreover, NMR and HTEM analyses of these stacks demonstrated a high degree of interface intermixing as already discussed in Sect. 6.2.2. In short, the contradiction between the coupling strength and the magnetoresistivity of Co/Ru is still an unsolved problem.

With regard to the sensitivity S in Table 6.8, it can be seen that, in general, the systems with the noble metals Au and Ag combined with the alloy permalloy are distinguished by reasonable GMR effects (about 20%) and low interlayer coupling. This combination makes them very attractive for applications. However, the growth of these stacks constitutes a real problem, e.g., the permalloy/Ag system has to be deposited at very low temperatures (77 K) in order to attain the required integrity of the layers [64,95]. This is partly due to the noble nature of these metals, implying that they are immiscible, and that it is difficult to get a Frank–van de Merwe–type of growth in the superlattices. In Table 6.8, it is remarkable that the systems with high GMR are all immiscible, such as Fe/Cr, Co/Cu, Co/Ag, permalloy/Au, so that we are inclined to believe that large intermixing of the interfaces is not desirable. Perhaps, the reduction in the atomic magnetic moment in such intermixed regions, resulting in smaller scattering potentials, is the reason for this [137].

Scattering Properties of Co/Cu Stacks with a Fixed Number of Periods. In this section, the magnetoresistivity of Co/Cu multilayers with Cu thicknesses close to the first AF maximum will be considered. The analysis will be based on the conductivity change originating in the portion that contributes to the GMR signal generation, i.e., the AF one. Thus, we shall consider $\Delta\sigma/F_{\mathrm{AF}}$ instead of $\Delta\sigma$ and shall rely on (6.41) for evaluating F_{AF}. The Cu layers are 10 Å thick, so that the interlayer-coupling constant J is about $-0.3\,\mathrm{erg/cm^2}$. In Sect. 6.3.2, it has been shown that the F_{AF} correction works well, though the impact of the defects is overestimated by approximately 25%. We will simply ignore this 25%, since we do not know much about the actual dimensions and the overlaps of the real defects. The buffer is weakly coupled to the Co/Cu stack, as has been shown in Sect. 6.2.2 on the J dependence on the Cu index in the Co/Cu system (see Fig. 6.11). It is assumed that remanence $M_{\mathrm{Bu}}^{\mathrm{r}}$ equals $M_{\mathrm{Bu}}^{\mathrm{s}}$ of the Fe/Co buffer. Experimental evidence for this can be found elsewhere [170]. The remanence of the Co/Cu stack does not originate in biquadratic coupling since the Cu thickness of

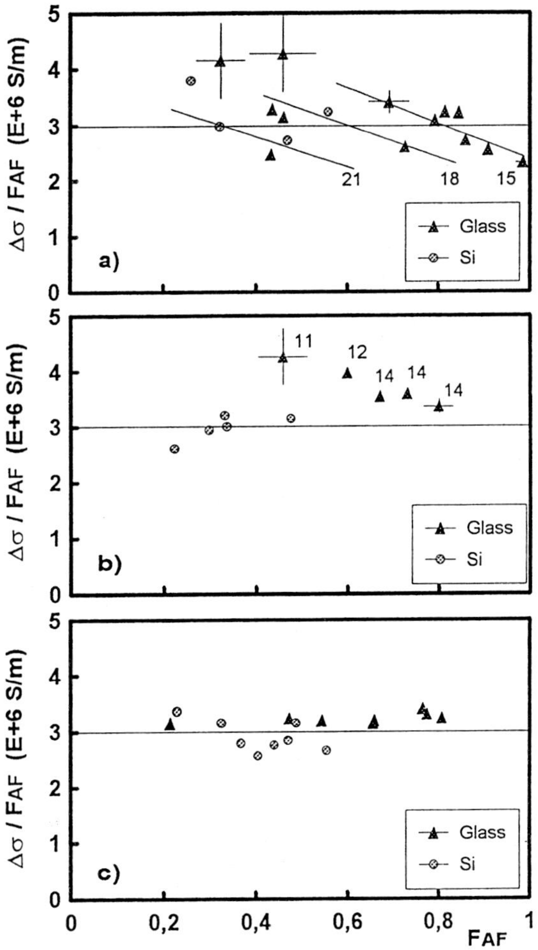

Fig. 6.36. **a–c** $\Delta\sigma/F_{\mathrm{AF}}$ versus F_{AF} near the first AF maximum of the interlayer thickness of Co/Cu systems made by *DC* magnetron sputtering. Stack geometry: substrate/Fe60 Å/[Co d_{Co} Å/Cu d_{Cu} Å] × 50. Parameters varied: d_{Cu}, d_{Co} and the *Ar* pressure during deposition in (**a**), (**b**) and (**c**), respectively

10 Å is just in the middle of the AF range (see also Sect. 6.2.2), where the $J_{\mathrm{biq}}/J_{\mathrm{bil}}$ ratio is small. Also, the contribution of the rotational hysteresis of the Co/Cu multilayer is of minor importance since the coercivity of the pure Co fcc (100), being typically 200 Oe, is negligible compared to the saturation fields H_{s}, which are of the order of 15 kOe at10 Å Co thickness.

In Fig. 6.36, Co/Cu multilayers with a fixed number of repetitions ($N_{\mathrm{Co}} = 50$) that are deposited by DC magnetron sputtering on both thermal oxidized silicon and glass substrates are considered. The Cu thickness is always close to 10 Å, i.e., to the first AF maximum. Figure 6.36 c shows the impact of the

Ar pressure P_{Ar} at a d_{co} and D_{cu} equal to 10 Å. F_{AF} varies between 0.2 and 0.8 by varying P_{Ar} between 1.4 and 15 mTorr [178]. F_{AF} tends to decrease gradually when P_{Ar} becomes larger than 10 mTorr. This is probably caused by the thermalization of the sputtered atoms by collisions with Ar, leading to a reduction in the mobility in the sputtered atoms at the substrate. Columnar growth, undulating interfaces and voids are the consequences which give rise to coupling defects. Notwithstanding the large F_{AF} span, $\Delta\sigma/F_{\mathrm{AF}}$ exhibits an astonishingly constant level for both the Si (●) and the glass substrate (▲). This suggests that the kinetic energy of the particles that hit the substrate is not high enough to cause intermixing. The low impact of the substrate bombardment might also be a consequence of the immiscibility of the binary Co-Cu system. On the other hand, the layer growth on a more macroscopic scale, e.g., crystal lite size and void generation, is strongly affected by the Ar pressure. Their density need not be high in order to have significant impact on F_{AF}, as has been explained in the theoretical discussion in Sect. 6.3.2.

Therefore, large changes in F_{AF} can occur without dramatic consequences for ϱ_{p} and the X-ray spectrum. Suzuki et al. [179] artificially influenced the intermixing by codeposition of Co and Cu near the interface zones. They found a drop in H_{s}, i.e., in J, when the thickness d_{i} of the intermixing zone is increased up to 1 Å. This is plausible as previously discussed for the coupling frustration in the sputtered Co/Ru sandwiches. Further, it could be seen from their B-H curves that F_{AF} decreases as d_{i} increases (see their Fig. 1). Therefore, one has to conclude from their d_{i} thickness range with constant $\Delta\varrho$ below a certain critical value of d_{i} that $\Delta\sigma/F_{\mathrm{AF}}$ increases as the interface roughness grows. At larger d_{i} (> 1 Å), $\Delta\sigma$ drops strongly. The reduction in $\Delta\varrho/\varrho_{\mathrm{p}}$ as d_{i} grows is caused by both the increase in ϱ_{p} and the decrease in F_{AF}. This example demonstrates the importance of the coupling volume corrections. It should be mentioned that Le Dang et al. [180] also come to the same conclusion. They concluded from NMR spectra that $\Delta\varrho/\varrho_{\mathrm{p}}$ is lowest at very sharp interfaces with perfect AF coupling and $F_{\mathrm{AF}} = 1$. They emphasized that there is no one-to-one correspondence between F_{AF} and $\Delta\varrho/\varrho_{\mathrm{p}}$, and that the scattering characteristics are just as important.

The impact of the Cu thickness variations is demonstrated in Fig. 6.36 a where P_{Ar} is 7 mTorr and D_{Co} is 10 Å. F_{AF} varies over an even larger span up to 1. The scattering in $\Delta\sigma/F_{\mathrm{AF}}$ is much larger. A structure only becomes visible when resistivity ϱ_{p} is indicated by numbers (in μΩ cm) and by the contours (solid lines) of approximate constant ϱ_{p} in the plot. Samples with equal F_{AF} exhibit the highest $\Delta\sigma/F_{\mathrm{AF}}$ at the lowest ϱ_{p}. Obviously, the scattering centers that are responsible for the difference in ϱ_{p} contribute relatively weakly to $\Delta\sigma$. Given a certain line of constant ϱ_{p}, d_{Cu} tends to increase when going from left to right. At small d_{Cu}, $\Delta\sigma/F_{\mathrm{AF}}$ is higher, which might stem from the higher density of scattering interfaces relative to the inactive scattering sides inside the bulk of the Cu.

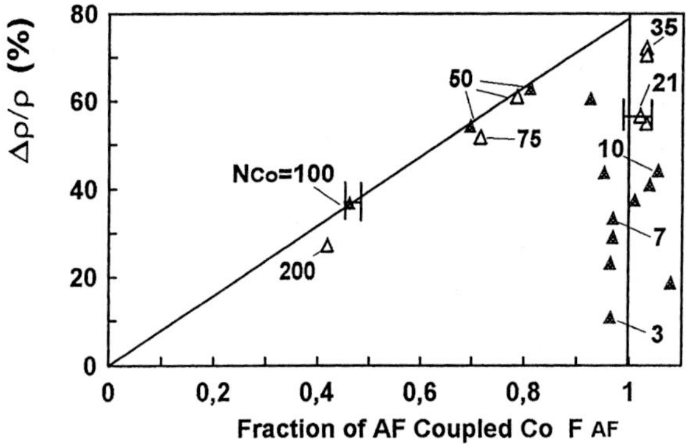

Fig. 6.37. $\Delta\varrho/\varrho$ versus F_{AF} of Co/Cu with varying number of repetitions. Stack geometry: glass/Fe60 Å/[Co10 Å/Cu10 Å] $\times N_{\mathrm{Co}}$. ▲: $P_{\mathrm{Ar}} = 6.5$ mTorr. △: $P_{\mathrm{Ar}} =$ 13 mTorr

In Fig. 6.36 b where P_{Ar} is 7 mTorr and d_{Cu} is 10 Å, d_{Co} varies with ϱ_{p} being indicated again for the glass samples (▲). For these, $\Delta\sigma/F_{\mathrm{AF}}$ becomes larger at large d_{Co}. This cannot be explained in terms of a higher density of scattering interfaces. A possible explanation might be the portion of uncoupled regions at large d_{Co}. Uncoupled regions combine high remanence and relatively high $\Delta\sigma$. At large d_{Co}, the intrinsic forces of the Co come into prominence, because the interlayer-coupling depends hardly on d_{Co} and, consequently, H_{s} becomes smaller. The Co layers tend to behave more like uncoupled ones and the coherence in the magnetization configurations of neighboring Co layers vanishes. In other words, adjacent regions in neighboring layers have different $\boldsymbol{M}$ directions during the switching phase.

Scattering Properties of the Co/Cu Stacks with Varying N_{Co}. In Sect. 6.2.2, we discussed the dependence of J on the interlayer position in the stack and demonstrated J to be small near the buffer and to increase monotonically with growing distance to the substrate. We shall now consider the scattering properties of the same series of stacks with d_{Co} and d_{Cu} equal to 10 Å. Figure 6.37 shows the very striking relationship between $\Delta\varrho/\varrho_{\mathrm{p}}$ and F_{AF}.

The antiferromagnetic coupling is almost perfect and F_{AF} is unity for stacks with less than about 35 Co/Cu pairs. Up to this upper boundary, $\Delta\varrho/\varrho_{\mathrm{p}}$ increases monotonically. Above this repetition number, strong degeneration of the stack occurs and the additional layers hardly contribute to the GMR effect. This degeneration is probably due to the development of a columnar structure and surface roughness, for which deviating coupling properties occur at the edges of the columns. In other words, parallel channels exist with strongly different magnetoresistive properties. This provides an argument, in addition to Dieny's ones, for considering σ rather than ϱ.

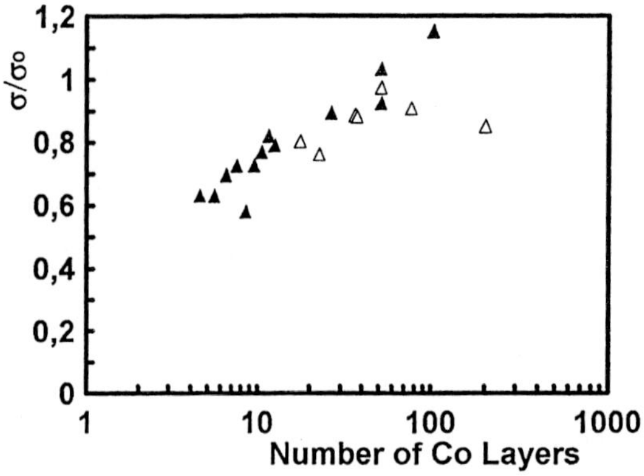

Fig. 6.38. σ_p versus the total thickness including buffer of Co/Cu stacks with varying number of repetitions. Stack geometry: glass/Fe60 Å/[Co10 Å/Cu10 Å] $\times N_\mathrm{Co}$. ▲: $P_\mathrm{Ar} = 6.5\,\mathrm{mTorr}$. △: $P_\mathrm{Ar} = 13\,\mathrm{mTorr}$

Figure 6.38 shows σ_p in the saturated state as a function of N_Co, i.e., of the total thickness. It is tempting to compare this course with the Fuchs–Sondheimer theory of the single slab because σ_p increases with stack thickness, indicating diffuse scattering at the outer stack surfaces. Thus, we can estimate the longest mean free path of the two spin types of current carriers. According to Inoue (see Table 6.7), these are the majority electrons. By comparing Figs. 6.38 and 6.21, we can discern the same trend and can assess λ^+ at 150 Å. Such a value is more frequently cited in the literature [148], though there are also measurements with values three times smaller than the present mean free path [181]. At small N_Co, the σ–versus–N_Co curve is also affected by the difference in λ of the Fe buffer, being rather small, and the Co/Cu mean free path.

Figure 6.39 demonstrates the rise in $\Delta\sigma/F_\mathrm{AF}$ as the number of repetitions increases. The rise can be simply understood by considering the following approximate relationship:

$$\Delta\sigma = \sigma_\mathrm{p}\,\alpha\,\frac{1}{p+p^\mathrm{P}} - \frac{1}{p+p^0} = \frac{p^0 - p^\mathrm{P}}{p^2 + p(p^\mathrm{P}+p^0) + p^\mathrm{P}p^0}\,, \tag{6.53}$$

where p is the mean scattering probability at sides with no magnetoresistive activity, p^0 and p^P are the mean scattering probabilities of active sides in the parallel and antiparallel magnetization configurations, respectively. At small N_Co, p is high because of the diffuse scattering at the outer boundaries and the large thickness of the Fe buffer, which hardly shows any magnetoresistive bulk effect. Furthermore, p^P and p^0 are much smaller than their counterparts for Co/Cu multilayer with $N_\mathrm{Co} = \infty$. From these arguments and (6.53), the strong rise in $\Delta\sigma/F_\mathrm{AF}$ with N_Co is evident.

There are a few strongly deviating points in both Figs. 6.38 and 6.39, for instance point $N_\mathrm{Co} = 7$, which has a much too low σ_p. This stack is also

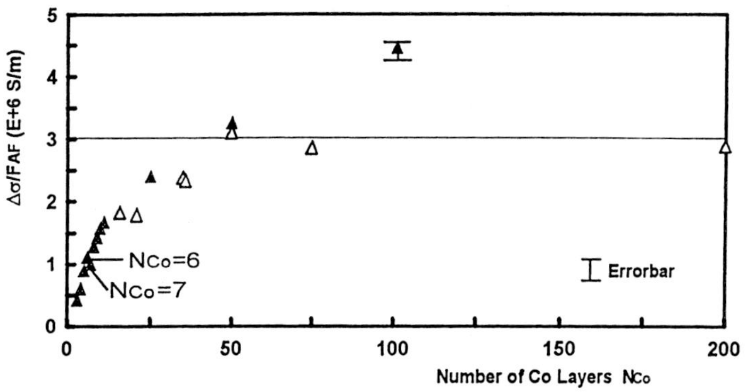

Fig. 6.39. $\Delta\sigma/F_{\mathrm{AF}}$ versus N_{Co} of Co/Cu stacks with varying number of repetitions. Stack geometry: glass/Fe60 Å/[Co10 Å/Cu10 Å] $\times N_{\mathrm{Co}}$

signified by a relatively small $\Delta\sigma/F_{\mathrm{AF}}$. $\Delta\sigma/F_{\mathrm{AF}}$ is even smaller than for the $N_{\mathrm{Co}} = 6$ stack. As previously discussed, stacks with relatively high ϱ have a small GMR signal, presumably due to a reduction in the mean free path λ^+ of the majorities by magnetoresistive weakly active scattering centers. $N_{\mathrm{Co}} = 21$ provides a similar but less striking example. Another strongly deviating stack with $N_{\mathrm{Co}} = 101$ has high $\Delta\sigma/F_{\mathrm{AF}}$. Figure 6.38 shows that this stack features a combination of the highest σ and $\Delta\sigma/F_{\mathrm{AF}}$ and confirms the impact of σ quoted above.

GMR Effect in Stacks with Biquadratic Coupling. In Sect. 6.2.2, two options for explaining the shapes of the M–H curves as a function of the Ru interlayer thickness in Co/Ru sandwiches were studied. We argued that the small width of the transition zone from the purely ferromagnetic behavior of the curves without remanence excludes the possibility of superposition of independent regions. Such a sharp transition does not always occur, so that it is desirable to have additional criteria for distinguishing between both aforementioned alternatives. The magnetization curves give us the projection of $\boldsymbol{M}$ on the field $\boldsymbol{H}_0$. In the case of superposition of regions in isotropic stacks, the magnetization curve can be reconstructed by a sum of the form

$$\langle M_{\|}\rangle/M_{\mathrm{s}} = \sum\nolimits_{i=1}\{f_0 b_0 + F_i f(H_0/H_{\mathrm{s}i})\}\,,$$

$$f(H_0/H_{\mathrm{s}i}) = \begin{cases} H_0/H_{\mathrm{s}i}\,, & \text{for } \mid H_0 \mid < H_{\mathrm{s}i}\,,\\ H_0/\mid H_0\mid\,, & \text{for } \mid H_0 \mid \geq H_{\mathrm{s}i}\,, \end{cases} \tag{6.54}$$

$b_0 = \pm 1$, depending on the branch, where the F_i's are uniquely determined. The factor F_0 gives the sum of the areas with ferromagnetic coupling and b_0 signifies the sign of the remanence. Implicitly, it is assumed that decoupled regions have no significant impact. The GMR effect provides an additional

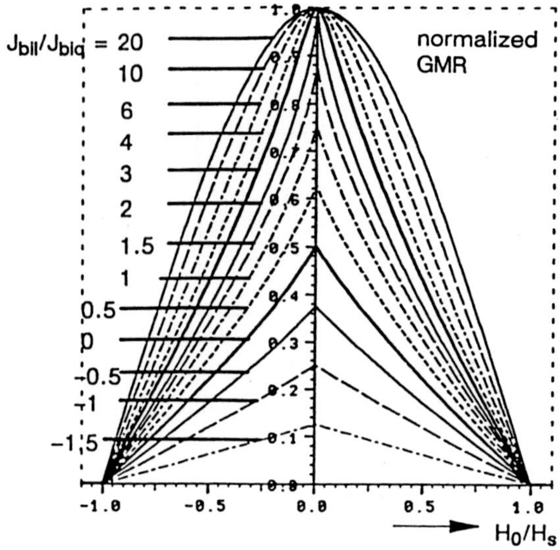

Fig. 6.40. The normalized GMR curves as a function of H_0/H_s for various J_{bil}/J_{biq}

tool for analyzing the coupling, since it represents the angle enclosed by the $\boldsymbol{M}$'s of adjacent layers rather than the projections of $\boldsymbol{M}$ along $\boldsymbol{H}_0$. Given the F_i's, the GMR response is uniquely determined. In the case of biquadratic coupling, there are only two parameters, namely J_{bil} and J_{biq}, which modulate both measurement curves in a different fashion, thus providing an ideal consistency test. It can easily be verified that the biquadratic and the superposition model give rise to different correlations between M–H and the GMR curve. Let us consider the situation with a normalized remanence C. In the superposition model, the normalized GMR signal is $1 - C$, while the biquadratic model yields $1 - C^2$. Both results are equal only when $C = 0$ or 1.

The typical shape of the GMR curves can be derived from (6.19–6.21) and (6.44a) and has been plotted as a function of J_{bil}/J_{biq} in Fig. 6.40. The curves are parabolic when $\mid J_{biq} \mid$ is small and become more acute and triangle like when $\mid J_{biq} \mid$ increases. The signal gradually drops when J_{bil}/J_{biq} becomes smaller than 2.

Experimental curves of Co/Ru sandwiches with varying Ru interlayer thicknesses have been collected in Fig. 6.41. The ratio J_{bil}/J_{biq}, as determined from the magnetization curves, has been added to the plots. A comparison of Figs. 6.40 and 6.41 shows a striking agreement between the shape of the curves with comparable J_{bil}/J_{biq}. It is obvious that one should not compare the heights of the signals, since the thickness of the intermixed Ru layer is not constant. The Ru layer is strongly intermixed and contains a large portion of Co atoms that are only weakly coupled and that behave paramagnetically. These scattering centers provide equal scattering probabilities for both spin types and cause the longest mean free path to be strongly reduced. As

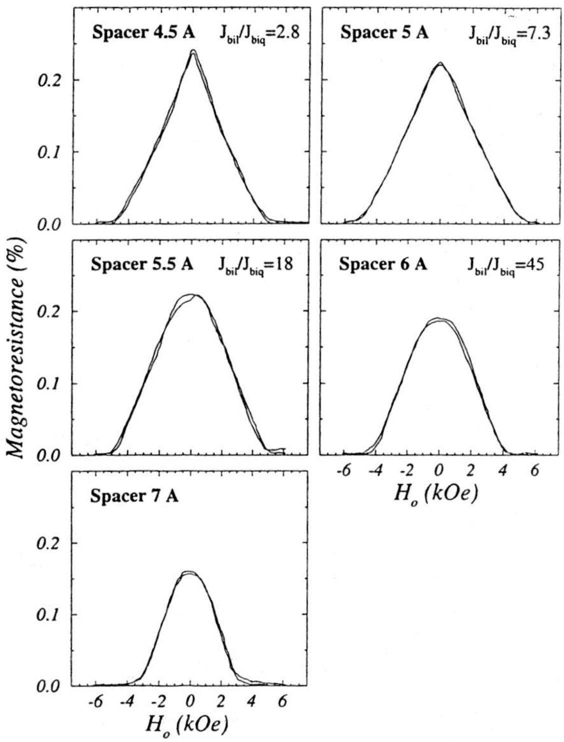

Fig. 6.41. The measured magnetoresistive curves as a function of the Ru thickness for Co/Ru sputtered sandwiches: glass/Ru50 Å/Co30 Å/Ru d_{Ru} Å/Co30 Å/Ru50 Å. $J_{\mathrm{bil}}/J_{\mathrm{biq}}$ as derived from magnetization measurements have been indicated

a consequence, the GMR signal for these stacks is very small, so that the anisotropic magnetoresistive effect might contribute to the shape. One can extract the pure GMR contribution by summing the magnetoresistive signals for perpendicular directions of the current. The AMR signal is very small in these systems with relatively high resistivity due to the Ru.

Summary. The reversed GMR effect reveals itself only when the spin-scattering asymmetry factors $-(1-\sqrt{\alpha})/(1+\sqrt{\alpha})$ of adjacent magnetic layers have opposite signs. Therefore, at least three different materials are required to get this effect. In simple stacks with two materials, the normal GMR effect will be met. In general, this effect tends to be largest when the two materials are immiscible, the moment at the magnetic/nonmagnetic layer interface is high, the stack resistivity is small and no ferromagnetic coupling bridges are present.

The correction factor F_{AF} is useful for extracting the scattering data from resistivity measurements in stacks with sufficiently high AF interlayer coupling and with not too thick ferromagnetic layers. GMR curves can be used as an additional experimental tool, giving access to the nature of the interlayer coupling.

6.3.4 Prospects for Sensors Based on the GMR Effect

In the previous sections, we discussed the physics of magnetic superlattices and reviewed a number of typical aspects of measuring physical parameters, such as interlayer coupling, surface anisotropy and magnetoresistivity. Apart from scientific inquisitiveness, the boom in the interest in these phenomena is motivated by the potentials of the GMR effect for sensor applications. Here, a number of sensor principles based on the GMR effect is briefly screened and their potentials and limitations in the light of the phenomena which have come up for discussion in the previous sections will be indicated.

In Sect. 6.3.1, we emphasized that the GMR effect originates in the mutual orientations of the magnetization in regions that are separated by electrically conductive spacers with typical thicknesses well below the longest mean free path of the current carriers. For sensor applications, one requires the magnetization orientations to be controlled by externals fields. The alignment of the moments of all the regions can easily be accomplished by applying an external field of sufficient magnitude. In principle, this is indeed very simple. However, a well controlled adjustment of the moments by magnetic fields is less evident and a number of schemes have been developed for this purpose. Actually, one method has already been mentioned, namely the antiferromagnetic coupling of the magnetic layers in stacks. However, before assessing this principle, one should briefly discuss a scheme, which is not covered by the previous sections on multilayers and which yet exhibits the same GMR effect, namely the heterogeneous alloys.

The so-called granular materials are also systems of current interest. In these materials, magnetic particles with typical diameters of tens of Å are imbedded in nonmagnetic hosts. These systems can be produced far more simply than the layered systems, in particular in the case of systems with immiscible elements. After deposition, usually by sputtering, the samples are annealed and a segregation of the elements takes place, i.e., the magnetic atoms that were originally almost randomly distributed congregate and constitute magnetic particles separated by the host material [182–184]. It has recently been shown [185,186] that these types of systems can also be prepared by rapid solidification from a melt and by subsequent heat treatment.

The sensitivity of these systems is very poor due to the small dimensions of the magnetic precipitates. The magnetic grains are superparamagnetic at room temperature. At zero field, the magnetic moments of these are randomly oriented due to thermal agitation and, generally, ϱ is highest in that state. A large field is required to align the moments. There are two strategies for improving the sensitivity. In the first place, the dimensions of the magnetic particles can be increased. As a consequence, the mutual distance between them will inevitably increase and the volume density of surface scatterers will be low. As a result, the signal drops. Furthermore, the grain's shape anisotropy and the dispersion in both the easy axis directions and the anisotropy strengths become more dominant, leading to inherent rotational hysteresis. When the magnetic grains become very large, the magnetic particles will even become multidomain. In the second place, an attempt is made to increase the blocking temperature by producing samples with small particles that have very uniform dimensions and that are close enough to exhibit a significant RKKY interaction.

In most granular systems used so far [187,188], the grains are still superparamagnetic and exhibit low sensitivity. It has still to be clarified whether the heterogeneous systems have prospects for the future.

The antiferromagnetically coupled multilayer systems, in which the interlayer coupling causes the alternating alignment of magnetizations in the ground state, constitute the second category that has to be discussed. In general, these systems suffer from a low sensitivity due to the strength of the interlayer coupling. Presently, only systems based on the noble metals Ag and Au and permalloy have sensitivities that can compete with the classical AMR sensors. Unfortunately, these materials are not simple to deposit and the systems are generally hampered by ferromagnetic bridges between the magnetic layers. Their sensitivity S is directly proportional to the interlayer-coupling constant. This gives another more fundamental argument against the antiferromagnetically coupled systems. Small deviations in the interlayer thickness and/or changes in the interface intermixing already cause the coupling constant to vary significantly. The specifications of the various applications are different, however, for most applications, the tolerance windows of the production process are very narrow and it is questionable whether these can be satisfied.

Two alternatives were proposed that are both based on multilayers of which the integrity is suppressed either by heat treatment or by selection of appropriate material combinations and film thicknesses. The Hylton's system [189,190] has gained most interest. It is based on a Ag/permalloy superlattice. As prepared, these systems exhibit hardly any GMR signal due to ferromagnetic coupling bridges. Upon annealing at about 350°C, the closed permalloy films break up into flat disks separated by Ag. The advantage of the system is in the confinement of the lateral extent of coupling bridges, i.e., the defect walls discussed previously, to the two ferromagnetically coupled disks. The remaining decoupled disks contribute fully to the signal, which increases to about 15% at room temperature. Due to the disk shape of the permalloy islands, the sensitivity of the system is high and comparable to AMR permalloy sensors. Surprisingly, the magnetization in adjacent "layers" is opposite when H_0 is zero. As explained by Slonczewski [191], this is due to magnetostatic coupling, the strength of which sensitively depends on the lateral distance between the disks. It will be difficult to control the coupling strength and, along with it the sensor sensitivity.

It was already emphasized above, that the sensor response should not directly depend on any coupling constant between layers or disks. The so-called hard-soft system and its alternative versions, e.g., the exchange-biased systems, offer a scheme for fulfilling this requirement. The hard-soft system in the basic form consists of two types of magnetic layers: First, a hard magnetic one with, in principle, rigid magnetization and, second, a detection layer, being soft magnetic with eventually an imposed easy axis for getting the desired sensor characteristics. These layers are separated by interlayers with relatively high thicknesses for which use is made of the rapid decline in the RKKY coupling with increasing separation between the magnetic layers. Furthermore, the interlayer thickness is selected near the crossover thickness between the ferromagnetic and antiferromagnetic coupling regimes. The base period of a stack consists of consecutively a hard, a decoupling, a soft and another decoupling layer. In order to get higher signals, this base unit is repeated a number of times in the sensor pile.

The scheme discussed above works in principle, but, it is difficult to get rid of the hysteresis in the sensor signal. Of course, there are the hysteresic phenomena that are typical of domain structures in soft magnetic elements [192–195,167]. In a number of applications, one is basically measuring in the saturated state of the detection layer, so that domain effects do not harm that much. In this context, angle and threshold detection of the magnetic field should be mentioned. In most circumstances, however, the magnitude of the field has to be measured with maximum sensitivity, which is incompatible with a sample saturated by the detection or bias field. Therefore, the usual precautions for domain suppression in the AMR sensors have also to be implemented in the GMR sensors. Still, it will be hard to obtain a really uniform magnetization state in the detection layer. The reasons for this are of a rather fundamental nature. The sensors of interest are polycrystalline in order to get detection layers with soft magnetic properties. The idea is to cancel the crystallographic preferential axis by averaging over a large number of crystallites by adequately enhancing the stiffness of the magnetization due to bulk-exchange and lateral magnetostatic couplings. Nanocrystallites are required to provide the desired softness of the detection layer. Unfortunately, these crystallites extend themselves over the entire stack thickness or a large portion of it. In other words, the crystallites in the hard and soft layers have the same size, and a reduction of this size effectively demolishes the hard characteristics, i.e., the magnetization in the hard layer will be no longer fixed. The conflicting requirements upon the softness and hardness of the layers is a real challenge with respect to applications. The second problem with these systems originates in the thickness of the hard layers. The magnetostatic coherence length scales with the layer thickness. In the thin film systems of the past, the film thicknesses were such that the magnetostatic coherence length is much larger than its exchange counterpart. This does not apply any longer for the present films so that the stiffness of the magnetization becomes much smaller and the well-known magnetization ripple is no longer a small disturbance of the magnetization direction [109,196]. Consequently, the magnetization in the hard layers exhibits a strong stochastic component correlated to the randomness of the crystallites. The accompanying stray fields affect the magnetization of the nearby detection layer and cause a significant increase in its coercivity.

Much more progress has been made with the so-called exchange-biased systems, in which the hard layer has been modified and consists of an antiferromagnetic layer, usually FeMn, with a ferromagnetic layer directly attached to it [197–199]. Both layers are ferromagnetically coupled. The magnetization is adjusted by applying a field and increasing the temperature to a level at which the moments in the antiferromagnetic layer can orientate themselves according to the magnetization in the saturated ferromagnetic layer. The magnetization configuration is frozen in the antiferromagnetic layer upon cooling to room temperature. Alternatively, the FeMn layer is deposited in a magnetic field on top of the ferromagnetic layer. The spin configuration in the antiferromagnetic layer is very stiff and can hardly be affected by external fields. In principle, it might be rotated by the torque exerted by the field via the magnetic top layer when the antiferromagnetic layer thickness is too small. The uniform state of the ferromagnetic layer during the adjustment phase is well main tained in this layer due to the direct exchange coupling between antiferromagnet and ferromagnet. Since, the hardness of this layer pair does not originate from the ferromagnetic layer itself, it is possible to use (111) crystal textures. Thus, it is possible to strongly reduce the impact of the random anisotropy of the crystallites in the case of in-plane orientation of the magnetizations, which is a prerequisite. To get the (111) texture, the stacks are grown on Ta buffers with the detection layer directly on top of it. The GMR signals are limited to about 8% because the use of repetitions makes no sense [198]. This is caused by the very high resistivity ($90\,\mu\Omega\,\mathrm{cm}$) of FeMn so that it insolates very effectively by reducing the difference in the mean-free paths of both spin-type electrons that pass the FeMn layer.

The discussion so far was focussed on the FeMn which exhibits disadvantages with respect to the temperature stability of the γ crystallographic phase and its sensitivity to oxidation. The latter hampers the application as read heads in magnetic drives in which the front edge of the read sensor including the FeMn layer is unprotected and directly subjected to the atmosphere. Furthermore, thick FeMn layers are required to get the blocking temperatures in the acceptable range. Therefore alternative antiferromagnetic layers like NiMn [200,201], IrMn [202,203], PtPdMn [204] and PtMn [205] have been explored. Some of these (NiMn, PtPdMn and PtMn) require heat treatments at elevated temperatures (far into the 200°C range) in order to achieve the desired γ phase and/or to induce an anisotropy for pinning the spins of the antiferromagnet. NiMn and PtPdMn are characterized by a high exchange coupling to the pinned layer, the bias-coupling strength. NiMn has the advantage of having blocking temperatures higher than the 300°C, being far sufficient for most applications. Presently, in particular, IrMn is gaining a lot of interest because it does not require any annealing and is signified by a high bias-coupling strength and blocking temperatures of about 250°C. Also antiferromagnetic oxides like NiO [206,207] and α–Fe_2O_3 [208–210] are currently extensively studied. These have the advantage of being strong electron reflectors so that GMR signals of up to 27% have been achieved in the dual-sensor scheme [210], i.e., a sensor with two pinned layers enclosing one single detection layer in the middle [217]. NiO is not very attractive because of its poor rigidity and bias-coupling strength and because of its low blocking temperature. The α–Fe_2O_3 requires a heat procedure at elevated temperatures, which affects both the signal height and the magnetic response of the various layers.

Another interesting alternative is to replace the antiferromagnet by a ferrimagnet like TbCo [211]. TbCo has a composition close to its compensation point so that its moment is very small, resulting in a high stiffness. It has the advantage of coupling much stronger to its permalloy/Co cover layer than Mn-based antiferromagnets, so that the cover layer is able to retain its orientation up to high fields (700 Oe). This has very beneficial consequences to the accuracy of angle detecting systems and to sensor robustness against faults in general. Of course, the high rigidity is only given close to the compensation temperature. However, it should be emphasized that the high oxidation rate of Tb limits its use to the so-called yoke type of heads.

For head applications, the most desired sensor characteristic is higher sensitivity, which is a prerequisite for future generations of storage media. These will have extremely high storage densities and the sensors must have very small dimensions. The very thin detection layers, with thicknesses of typically 50 Å, can be realized with very small sensor widths without reducing the sensitivity due to an unaffordable high in-plane demagnetization factor. This gives the required improvement in the sensitivity compared to the standard AMR per malloy components of a factor of 10 to 20, as has already been demonstrated [199]. One of the implementation of GMR elements is in reading heads based on exchange-biased systems. However, a problem arises at higher data density which require sensor stripes smaller than 1 µm in width. The pinned layer causes high magnetostatic fields that give rise to a strong biasing of the response curve of the detection layer. Therefore, it is becoming common practice to combine antiferromagnets with the AAF structures (see Sect. 6.1) which also provide superior robustness at, in particular, higher temperatures [205,212–214]. Applied in the dual-sensor scheme, there is the additional advantage of reducing the bias field in the sensor layer caused by the read current, the density of which is continuously increasing from generation to generation of read heads.

A potential advantage of GMR over AMR is the two times larger span of the field angle that can be measured uniquely. It is fairly simple to design sensors that cover the entire arc of 360° at which the detection layers always remain saturated [215,216]. For angle detection, the AAF sensor is superior to the alternative schemes in that it is very insensitive to variations in the strength of the external field. For many other applications however, the advantage of GMR is in the larger amplitude of the signal, compared with other established solutions. This implies that systems with a number of effective repetitions have to be developed. The exchange-biased schemes contain the antiferromagnetic layers, which must have sufficiently large thicknesses to withstand the torques exerted by the attached ferromagnetic layer. The resistivity of the metallic antiferromagnets (and also of TbCo, which is amorphous) is very high for both types of spin carriers so that multiperiodicity makes no sense. Therefore, dielectric antiferromagnets like NiO and a–Fe_2O_3 with reflecting surfaces for the electrons at the interface with the pinned layer(s) provide the required signal enhancement by effectively increasing the number of periods via the boundary conditions (see Sect. 6.3.1). Optimal signal height in sensors with AAF hard subsystems are obtained by using the dual sensor configuration [217].

Acknowledgments

We would like to thank our partners who have contributed to the present work. We would like to mention W. Andrä and R. Mattheis from the IPHT in Jena, who developed the atomic layer model, and A. Dinia, K. Ounadjela, D. Stoeffler, and S. Zoll from the IPCMS in Strasbourg for their contribution on the Co/Ru systems and Co/Rh. Finally, we would like to express our gratitude to our colleagues G. Gieres, G. Rupp, and M. Vieth. Further, we would like to thank S. Schmeußer for allowing us to use his IBS data. We greatfully acknowledge D. Berkov, R. Mattheis and S. Zoll for carefully reading the manuscript.

References

1. W. Thompson: Proc. R. Soc. London A **8**, 546 (1857)
2. D.A. Thompson, L.T. Romankiw, A.F. Mayadas: IEEE Trans. Magn. MAG-**11**, 1039 (1975)
3. R. Hunt: IEEE Trans. Magn. MAG-**7**, 1039 (1971)
4. F.J. Jeffers, H. Karch H: IEEE Trans. Magn. MAG-**20**, 703 (1984)
5. N. Smith, D. Wachenschwanz: IEEE Trans. Magn. MAG-**23**, 2494 (1987)
6. F.J. Jeffers, D. Wachenschwanz, D. Phelps, J. Freeman: IEEE Trans. Magn. MAG-**23**, 2088 (1987)
7. W.D. Kehr, J.L. Nix, D.P. Camps: IEEE Trans. Magn. MAG-**24**, 2615 (1988)
8. G. Kelley, J. Freeman, H. Copenhaver, R. Kefcham, E. Valstyn: IEEE Trans. Magn. MAG-**17**, 2615 (1981)

9. W.F. Druyvesteyn, J. van Ooyen, L. Postma, E. Raemaekers, J.J.M. Ruigrok, J. de Wilde: IEEE Trans. Magn. MAG-**17**, 2884 (1981)
10. W.F. Druyvesteyn, L. Postma, G.H. Somers, J. de Wilde: IEEE Trans. Magn. MAG-**19**, 1748 (1981)
11. Y. Nagata, T. Fukazawa, K. Wada, K. Tosaki, T. Aoi: IEEE Trans. Magn. MAG-**23**, 2500 (1987)
12. S. Tumanski, M. Stabrowski: IEEE Trans. Magn. MAG-**20**, 963 (1984)
13. U. Dibbern: Sensors & Actuators **4**, 221 (1983)
14. U. Dibbern: Sensors & Actuators **10**, 127 (1986)
15. K.J. Eijkel, J.H. Fluitman: IEEE Trans. Magn. MAG-**22**, 955 (1986)
16. C.H. Bajorek, H. Krongelb, L.T. Romankiw, D.A.Thompson: IEEE Trans. Magn. MAG-**12**, 813 (1976)
17. J.P.J. Groenland: IEEE Trans. Magn. MAG-**20**, 969 (1984)
18. W. Kwiatkowski, S. Tumanski: J. Phys. E **19**: Sci. Instrum., 502 (1986)
19. J.S. Gau: J. Mater. Sci. Eng. B **3**, 337 (1989)
20. K. Miyashita, T. Takahashi, M. Yamanaka: IEEE Trans. Magn. MAG-**23**, 2182 (1987)
21. M. Sei, M. Sakakihara, H. Suzuki, IEEE Trans. Magn. MAG-**23**, 2185 (1987)
22. R. Campbell: IEEE Trans. Magn. **26**, 2029 (1990)
23. S. Yamashita, H. Kadu, M. Ikeda: IEEE Trans. Magn. Jpn. **5**, 711 (1990)
24. P. Grünberg: European Patent Application 0 346 817 (1988)
25. G. Binasch, P. Grünberg, F. Saurenbach, W. Zinn: Phys. Rev. B **39**, 4828 (1989)
26. F. Saurenbach, J. Barnas, G. Binasch, M. Vohl, P. Grünberg, W. Zinn: Thin Solid Films **175**, 317 (1989)
27. M. Vohl, J. Barnas, P. Grünberg: Phys. Rev. B **39**, 12003 (1989)
28. F. Saurenbach, U. Walz, L. Hinchey, P. Grünberg, W. Zinn: Appl. Phys. Lett. **63**, 3473 (1988)
29. M.N. Baibich, J.M. Broto, A. Fert, F. Nguyen, F.F. Petroff, P. Etitenne, G. Creuzet, A. Friederich, J. Chazelas: Phys. Rev. Lett. **61**, 2472 (1988)
30. A. Barthélémy, A. Fert, M.N. Baibich, S. Hadjoudj, F. Petroff, P. Etitenne, R. Cabanel, S. Lequien, F. Nguyen Van Dau, J. Creuzet: J. Appl. Phys. **67**, 5908 (1990)
31. R.F.C. Farrow, C.H. Lee, S.S.P. Parkin: IBM J. Res. & Develop. **34**, 903 (1990)
32. Y.Y. Huang, G.P. Felcher, S.S.P. Parkin: J. Magn. Magn. Mater. **99**, L31 (1991)
33. S.S.P. Parkin: Phys. Rev. Lett. **66**, 3598 (1991)
34. S.S.P. Parkin, R. Bhadra, K.P. Roche: Phys. Rev. Lett. **66**, 2152 (1991)
35. G. Rupp, K. Schuster: J. Magn. Magn. Mater. **121**, 416 (1993)
36. M.A. Ruderman, C. Kittel: Phys. Rev. **96**, 99 (1956), Progr. Theor. Phys. Jpn. **16**, 45 (1954)
37. K. Yosida: Phys. Rev. **106**, 893 (1957)
38. P. Bruno, C. Chappert: Phys. Rev. Lett. **76**, 1602 (1991)
39. P. Bruno, C. Chappert: Phys. Rev. B **46**, 261 (1992)
40. R. Coehoorn: Phys. Rev. B **44**, 9331 (1991)
41. C. Chappert, J.P. Renard: Europhys. Lett. **15**, 553 (1991)
42. D.M. Deaven, D.S. Rokhsar, M. Johnson: Phys. Rev. B **44**, 5977 (1991)
43. L.M. Roth, H.J. Zeiger, T.A. Kaplan: Phys. Rev. **149**, 519 (1966)

44. M.T. Johnson, R. Coehoorn, J.J. de Vries, N.W.E. McGee, J. aan de Stegge, P.J.H. Bloemen: Phys. Rev. Lett. **69**, 969 (1992)
45. M.T. Johnson, S.T. Purcell, N.W.E. McGee, N.W.E., R. Coehoorn, J. aan de Stegge, W. Hoving: Phys. Rev. Lett. **68**, 2688 (1992)
46. J. Unguris, R.J. Celotta, D.T. Pierce: Phys. Rev. Lett. **67**, 140 (1991)
47. A. Fuß, S. Demokritov, P. Grünberg, W. Zinn: J. Magn. Magn. Mater. **103**, L221 (1992)
48. W.R. Bennett, W. Schwarzacher, W.F. Egelhoff: Phys. Rev. Lett. **65**, 3169 (1990)
49. P. Bruno: Europhys. Lett. **23**, 615 (1993)
50. J. Barnas: J. Magn. Magn. Mater. **111**, L215 (1992)
51. J. Barnas: J. Magn. Magn. Mater. **128**, 171 (1994)
52. P.H.J. Bloemen, M.T. Johnson, M.T.H. van de Vorst, R. Coehoorn, J.J. de Vries, R. Jungblut, J. aan de Stegge, A. Reinders, W.J.M. de Jonge: Phys. Rev. Lett. **72**, 764 (1994)
53. S.N. Okuno, K. Inomata: Phys. Rev. Lett. **72**, 1553 (1994)
54. P.J.H. Bloemen, M.T. Johnson, M.T.H. van de Vorst, R. Coehoorn, A. Reinders, J. aan de Stegge, R. Jungblut, W.J.M. de Jonge: J. Magn. Magn. Mater. **148**, 193 (1995)
55. D. Stoeffler, F. Gautier: Progr. Theor. Phys. Suppl. **101**, 139 (1990)
56. D. Stoeffler, F. Gautier: Phys. Rev. B **44**, 10389 (1991)
57. F. Herman, J. Sticht, M. van Schiffgaarde: J. Appl. Phys. **69**, 4783 (1991)
58. F. Herman, J. Sticht, M. van Schiffgaarde: Mat. Res. Soc. Symp. Proc. **231**, 195 (1991)
59. C.L. Fu, A.J. Freeman: J. Magn. Magn. Mater. **54–57**, 777 (1986)
60. D. Stoeffler, F. Gautier: J. Magn. Magn. Mat. **121**, 259 (1993)
61. D. Stoeffler, F. Gautier: J. Magn. Magn. Mat. **140–44**, 529 (1995)
62. M. Rührig, A. Hubert: Phys. Stat. Solidi A **125**, 635 (1991)
63. C.J. Gutierrez, J.J. Krebs, M.E. Filipkowski, G.A. Prinz: J. Magn. Magn. Mater. **116**, L305 (1992)
64. B. Rodmacq, K. Dumesnil, Ph. Mangin, M. Hennion: Phys. Rev. B **48**, 3356 (1993)
65. P. Bruno: J. Magn. Magn. Mater. **121**, 248 (1993)
66. J. Barnas, P. Grünberg: J. Magn. Magn. Mater. **121**, 326 (1993)
67. D.Stoeffler, F. Gautier: "Magnetic Properties of Ferro/Antiferro-Magnetic-based Sandwiches" in: Magnetism and Structure in S ystems with Reduced Dimensions, ed. by R.F.C. Farrow (Plenum Press, New York,1993)
68. M. Freyss and D.Stoeffler: private communication
69. J.C. Slonczewsk: Phys. Rev. Lett. **67**, 3172 (1991)
70. J.C. Slonczewski: J. Appl. Phys. **73**, 5957 (1993)
71. S. Zoll, H.A.M. van den Berg, J.P. Jay, H.J. Elmers, C. Meny, P. Panissod, D. Stoeffler, A. Dinia, K. Ounadjela: J. Magn. Magn. Mater. **156**, 231 (1996).
72. P. Grünberg, R. Schreiber, Y. Pang, M.B. Brodsky, H. Sowers: Phys. Rev. Lett. **57**, 2442 (1986)
73. B. Hillenbrands: Phys. Rev. B **41**, 530 (1991)
74. J.F. Cochran, J.R. Dutcher: J. Appl. Phys. **64**, 6092 (1988)
75. J.J. Krebs, P. Lubitz, A. Chaiken, G.A. Prinz: Phys. Rev. Lett. **63**, 1645 (1989)
76. B. Heinrich, Z. Celinski, J.F. Cochran, W.B. Muir, J. Rudd, Q.M. Zhong, A.S. Arrott, K. Myrtle, J. Kirschner: Phys. Rev. Lett. **64**, 673 (1990)

77. B. Heinrich, J.F. Cochran, M. Kowalewski, J. Kirschner, Z. Celinski, K. Myrtle: Phys. Rev. B **44**,9348 (1991)
78. S.S.P. Parkin, D. Mauri: Phys. Rev. B **44**, 7132 (1991)
79. B. Dieny, J.P. Gavigan: J. Phys.-Condens. Matter **2**, 187 (1990)
80. W. Folkerts: J. Magn. Magn. Mater. **94**, 302 (1991)
81. S. Araki, K. Yasul, Y. Narumiya: European Patent Appl. 0 504 473 A1 (1992)
82. S.S.P. Parkin, A. Modak, D.J. Smith: Phys. Rev. B **47**, 9136 (1993)
83. R. Mattheis, W. Andrä, L. Fritzsch, A. Hubert, M. Rührig, F. Thrum: J. Magn. Magn. Mater. **121**, 424 (1993)
84. P.J.H. Bloemen, W.J.M. de Jonge, R. Coehoorn: J. Magn. Magn. Mater. **121**, 306 (1993)
85. P.P. Freitas, L.V. Melo, I. Trindade, M. From: J. Magn. Magn. Mater. **116**, 92 (1992)
86. S.S.P. Parkin, N. More, K.P. Roche: Phys. Rev. Lett. **64**, 2304 (1990)
87. K. Ounadjela, A. Arbaoui, A. Herr, Poinsot, A. Dinia, D. Muller, P. Panissod: J. Magn. Magn. Mater. **104–7**, 1896 (1992)
88. Z. Celinski, B. Heinrich: J. Magn. Magn. Mater. **99**, L25 (1991)
89. P. Grünberg, S. Demokritov, A. Fuß, R. Schreiber, J.A. Wolf, S.T. Purcell: J. Magn. Magn. Mater. **104–7**, 1734 (1992)
90. S.S.P. Parkin, A. Mansour, G.P. Felcher: Appl. Phys. Lett. **58**, 1473 (1991)
91. J.E. Mattson, M.E. Brubaker, C.H. Sowers, M. Conover, Z. Qiu, S.D. Bader: Phys. Rev. B **44**, 9378 (1991)
92. F. Petroff, A. Barthélémy, D.H. Mosca, D.K. Lottis, A. Fert, P. A. Schroeder, W.P. Pratt, S. Loloee, S. Lequien: Phys. Rev. B **44**, 5355 (1991)
93. M.E. Brubaker, J.E. Mattson, C.H. Sowers, S.D. Bader: Appl. Phys. Lett. **58**, 2306 (1991)
94. B. Rodmacq, B. George, M. Vaezzadeh, Ph. Mangin: Phys. Rev. B **46**, 1206 (1992)
95. B. Rodmacq, G. Palumbo, Ph. Gerard: J. Magn. Magn. Mater. **118**, L11 (1993)
96. X. Bian, J.O. Ström-Olsen, Z. Altounian, Y. Huai, R.W. Cochrane: Appl. Phys. Lett. **62**, 3525 (1993)
97. S.S.P. Parkin: Appl. Phys. Lett. **60**, 512 (1992)
98. S. Zoll, A. Dinia, D. Stoeffler, H.A.M. van den Berg, A. Herr, R. Poinsot, J.C. Ousset: J. Magn. Magn. Mater. **165**, 442 (1996)
99. Y. Wang, P.M. Levy, Fry: Phys. Rev. Lett. **65**, 2732 (1990)
100. A. Dinia, K. Ounadjela, A. Arbaoui, G. Suran, D. Muller, P. Panissod: J. Magn. Magn. Mater. **104–7**, 1871 (1992)
101. W. Andrä, H.A.M. van den Berg, H. Danan, R. Mattheis, G. Rupp, S. Schmidt: J. Magn. Magn. Mater. **148**, 223 (1995)
102. W. Andrä, H. Danan: Phys. Stat. Sol. (a) **102**, 367 (1987)
103. W. Andrä: IEEE Trans. Magn. MAG-**2**, 560 (1966)
104. W. Appel, W. Andrä: Phys. Stat. Sol. (a) **105**, 589 (1988)
105. M.M.H. Willekens, H.J.M. Swagten, A.M. Duif, P.J.H. Bloemen, R.J.T. van Kempen, S.K.J. Lenczowski, W.J.M. de Jonge, Proc. Mat. Res. Symp. **313**, 129 (1993)
106. E.E. Fullerton, J.E. Mattson, S.R. Lee, C.H. Sowers, Y.Y. Huang, G. Felcher, S.D. Bader, F.T. Parker: J. Magn. Magn. Mater. **117**, L301 (1992)
107. L. Néel: Compt. Rend. Ac. Sci. **241**, 1445 (1962)

108. L. Néel: Compt. Rend. Ac. Sci. **241**, 1676 (1962)
109. R. Kergoat: Distribution et méchanismus de aimentation dans certaines multicouche douches (Ph. D. Thesis, Univ. Paris Sud, 1994)
110. S. Zhang, S.M. Zhou, K. Ounadjela, P. Wigen: Phys. Rev. B **50**, 6094 (1994)
111. D. Muller, K. Ounadjela, P. Vennegues, V. Pierron-Bohnes, J.P. Arbaoui, A. Dinia, P. Panissod: J. Magn. Magn. Mater. **104–7**, 1873 (1992)
112. K. Ounadjela, D. Muller, A. Dinia, J.P. Arbaoui, P. Panissod: Phys. Rev. B **45**, 7768 (1992)
113. A. Dinia, S. Zoll, K. Ounadjela: J. Appl. Phys. **75**, 6548 (1994)
114. S. Zoll, H.A.M. van den Berg, K. Ounadjela, D. Stoeffler, A. Dinia: J. Magn. Magn. Mater. **148**, 145 (1995).
115. S. Zoll, H.A.M. van den Berg, K. Ounadjela, D. Stoeffler, A. Dinia: J. Appl. Phys. **79**, 2601 (1996)
116. R. Mattheis, W. Andrä, L. Fritzsch, J. Langer, S. Schmidt: J. Appl. Phys. **76**, 6510 (1994)
117. N.H. Mott: Proc. Roy. Soc. A **153**, 699 (1936)
118. J.M. Ziman: Electrons and Phonons (Claradon Press, Oxford 1960) p. 275
119. M.B. Stearns: J. Appl. Phys. **72**, 5354 (1992)
120. J.W.F. Dorleijn: Philips Res. Rep. **31**, 287 (1976)
121. J.W.F. Dorleijn. A.R. Miedema: J. Phys. F **5**, 487 (1975)
122. I.A. Campbell, A. Fert: in: Ferromagnetic Materials, ed. by E.P. Wohlfarth (North-Holland Publ. Comp., Amsterdam, 1982) p. 782
123. J. Smit: Physica **16**, 612 (1951)
124. B.A. Gurney, P. Baumgart, D.R. Wilhoit, B. Dieny, V.S. Speriosu: J. Appl. Phys. **70**, 5867 (1991)
125. P. Baumgart, B.A. Gurney, D.R. Wilhoit, T. Nguyen, B. Dieny, V.S. Speriosu: J. Appl. Phys. **69**, 4792 (1991)
126. S.S.P. Parkin: Appl. Phys. Lett. **61**, 1358 (1992)
127. V.S. Speriosu, J.P. Nozieres, B.A. Gurney, B. Dieny, H. Lefakis, T.C. Huang: J. Magn. Magn. Mater. **121**, 386 (1993)
128. V.S. Speriosu, J.P. Nozieres, B.A. Gurney, B. Dieny, T.C. Huang, H. Lefakis: Phys. Rev. B **47**, 11579 (1993)
129. B.Dieny, V.S. Speriosu, S. Metin, S.S.P. Parkin, B.A. Gurney, P. Baumgart: J. Appl. Phys. **69**, 4774 (1993)
130. A. Fert, I.A. Campell: J. Phys. F **6**, 849 (1976)
131. O. Jaoul, I.A. Campell, A. Fert: J. Magn. Magn. Mater. **5**, 23 (1977)
132. J. Inoue, H. Itoh, S. Maekawa: J. Magn. Magn. Mater. **121**, 344 (1993)
133. J. Inoue, H. Itoh, S. Maekawa: Phys. Rev. B **47**, 5809 (1993)
134. J. Inoue, A. Oguri, S. Maekawa: J. Magn. Magn. Mater. **104–7**, 1883 (1992)
135. J. Inoue, H. Itoh, S. Maekawa: J. Phys. Soc. Jpn. **61**, 1149 (1993)
136. H. Nakanishi, A. Okiji, H. Kasai: Jpn. J. Appl. Phys. **32**, L447 (1993)
137. R. Coehoorn: J. Magn. Magn. Mater. **121**, 432 (1993); R. Coehoorn, A. de Vierman, J.P.W.B. Duchateau: J. Magn. Magn. Mater. **121**, 266 (1993)
138. K.Takanashi, Y. Obi, N. Tsuda, H. Fujimori: J. Phys. Soc. Jpn. **61**, 4148 (1992)
139. B.A. Gurney, D.R. Wilhuit, V.S. Spuriosu, L.L.Sanders: IEEE Trans. Magn. MAG-**26**, 2747 (1990)
140. K. Fuchs: Proc. Roy. Soc. **34**, 100 (1938)
141. E.H. Sondheimer: Advan. Phys. **1**, 1 (1952)

142. T. van Peski-Tinbergen, A.J. Dekker: Physica **29**, 917 (1963)
143. R.E. Camley, J. Barnas: Phys. Rev. Lett. **63**, 664 (1989)
144. J. Barnas, A. Fuß, R.E. Camley, P. Grünberg, W. Zinn: Phys. Rev. B **42**, 8110 (1990)
145. P.M. Levy, S. Zhang, A. Fert: Phys. Rev. Lett. **65**, 1643 (1990)
146. S. Zhang, P.M. Levy: Mat. Res. Soc. Symp. Proc. Vol. **231**, 255 (1992)
147. A. Vedyayev, B. Dieny, N. Ryshanova: Europhys. Lett. **19**, 329 (1992)
148. B. Dieny: J. Phys.: Condens. Matter **4**, 8009 (1992)
149. B. Dieny, J.P. Nozieres, V.S. Speriosu, B.A. Gurney, D.R. Wilhoit: Appl. Phys. Lett. **61**, 2111 (1992)
150. D. Stoeffler: private communication
151. A. Barthélémy, A. Fert: Phys. Rev. B **43**, 13124 (1991)
152. D.M. Edwards, J. Mathon, R.B. Muniz: IEEEE Trans. Magn. MAG-**27**, 3548 (1991)
153. J. Mathon: Contemp. Phys. **32**, 143 (1991)
154. M. Liu, D.Y. Xing: Phys. Rev. B **47**, 12272 (1993)
155. T.G.S.M. Rijks, R. Coehoorn, M.J.M. de Jong, W.J.M. de Jonge: Phys. Rev. B **51**, 283 (1995)
156. S. Zhang, P. Levy: Phys. Rev. B **47**, 6776 (1993)
157. W.P. Pratt, S.F. Lee, Q. Yang, P. Holody, R. Loloee, P.A. Schroeder, J. Bass: J. Appl. Phys. **73**, 5326 (1993)
158. J. Barnas, A. Fert: Phys. Rev. B **49**, 12835 (1994)
159. A. Brataas, G.E.W. Bauer: Europhys. Lett. **26**, 117 (1994)
160. T. Valet, A. Fert: J. Magn. Magn. Mat. **121**, 378 (1993)
161. B. Dieny, J.P. Gavinan, J.P. Rebouillat: Mat. Res. Soc. Symp. Proc. **51**, 35 (1989)
162. R. Ribas, B. Dieny: Phys. Lett. A **167**, 103 (1992)
163. H.A.M. van den Berg, S. Schmeußer: IEEE Trans. Magn. MAG-**29**, 3099 (1993)
164. M. Labrune, J. Miltat: J. Magn. Magn. Mater. **151**, 231 (1995)
165. D.V. Berkov, K. Ramstock, A. Hubert: Phys. Stat. Sol. (a) **137**, 207 (1993)
166. W.F. Brown: Micromagnetics (Robert E. Krieger, Huntington 1978)
167. H.A.M. van den Berg: Micromagnetics and Domains in Soft-Ferromagnetic Media (Delft Univ. Press, Delft, 1984)
168. D.V. Berkov, K. Ramstöck, A. Hubert: Phys. Stat. Sol. (a) **137**, 207 (1993)
169. A. Barthélémy, L.G. Pereira, L. Steren, J.M. George, F. Petroff, A. Fert, R. Loloee, P. Holody, P.A. Schroeder: Proc. 14th Int. Coll. Magnetic Films and Surfaces, Düsseldorf, Germany (1994), 19
170. D.H. Mosca, F. Petroff, A. Fert, P.A. Schroeder, W.P. Pratt, R. Loloee: J. Magn. Magn. Mat. **94**, L1 (1991)
171. H.A.M. van den Berg, G. Rupp: IEEE Trans. Magn. MAG-**30**, 809 (1994)
172. K. Shintaku, Y. Daitoh, T. Shinjo: Phys. Rev. B **17**, 14584 (1993)
173. W. Folkerts, W. Hoving, W. Coene: J. Appl. Phys. **71**, 362 (1992)
174. B. Rodmacq, M. Vaezzadeh, B.George, Ph. Mangin: J. Magn. Magn. Mater. **121**, 213 (1993)
175. S.S.P. Parkin, R.F.C. Farrow, R.F. Marks, A. Cebollada, G.R. Harp, R.J. Savoy: Phys. Rev. Lett. **72**, 3718 (1994)
176. R.Takatani, T. Dei, Y. Sugita: J. Appl. Phys. **73**, 6375 (1993)
177. D. Stoeffler, F. Gautier, J. Magn. Magn. Mat. **140–4**, 529 (1995)

178. G. Rupp, H.A.M. van den Berg, IEEE Trans. Magn. **MAG–29**, 3102 (1993)
179. M. Suzuki, Y. Taga: J. Appl. Phys. **74**, 4660 (1993)
180. K. Le Dang, P. Veillet, E. Velu, S.S.P. Parkin, C. Chappert: Appl. Phys. Lett. **63**, 108 (1993)
181. B.A. Gurney, V.S. Speriosu, J.P. Nozieres, H. Lefakis, D.R.Wilhoit, O.M. Need: Phys. Rev. Lett. **71**, 4023 (1993)
182. J.Q. Xiao, J.S. Jiang, C.L Chien: Phys. Rev. B **43**, 8089 (1991)
183. J.Q. Xiao, J.S. Jiang, C.L Chien: Phys. Rev. Lett. **68**, 3749 (1992)
184. A.E. Berkowitz, J.R. Mitchell, M.J. Carey, A.P. Young, S. Zhang, F.E. Spada, F.T. Parker, A. Hutten, G. Thomas: Phys. Rev. Lett. **68**, 3745 (1992)
185. J. Wecker, R. von Helmolt, L. Schultz, K. Samwer: Appl. Phys. Lett. **62**, 1985 (1993)
186. J. Wecker, R. von Helmolt, L. Schultz, K. Samwer: IEEE Trans. Mag. MAG–**29**, 3087 (1993)
187. K. Takanashi, K. Sugawara, K. Hono, H. Fujimori: J. Appl. Phys. **76**, 6790 (1994)
188. A. Maeda, M. Kume, S. Oikawa, K. Kuroki: J. Appl. Phys. **76**, 6793 (1994)
189. T.L. Hylton, K.R. Coffey, M.A. Parker, J.K. Howard; Science **261**, 1021 (1993)
190. M.A. Parker, T.L. Hylton, K.R. Coffey, J.K. Howard: J. Appl. Phys. **76**, 6790 (1994)
191. J.C. Slonczewski: J. Magn. Magn. Mater. **129**, L123-8 (1994)
192. H.A.M. van den Berg: IBM J. Res. & Develop. **33**, 540 (1989)
193. H.A.M. van den Berg, F.A.N. van der Voort: IEEE Trans. Magn. MAG–**21**, 1936 (1985)
194. F.A.N. van der Voort, H.A.M. van den Berg: IEEE Trans. Magn. MAG–**23**, 250 (1987)
195. B.N. Filippov, G.A. Shmatov, A.B. Dichenko: J. Magn. Magn. Mater. **88**, 316 (1990)
196. R. Kergoat, J. Miltat, T. Valet, R. Jérome: J. Appl. Phys. **76**, 7087 (1995)
197. R. Jungblut, R. Coehoorn, M.T. Johnson, J. aan de Stegge, A. Reinders: J. Appl. Phys. **75**, 6659 (1994)
198. T. Anthony, J. Brug, S. Zhang: IEEE Trans. Magn. MAG–**30**, 3819 (1994)
199. W. Folkerts, J. Kools, Th. Rijks, R. Coehoorn, M. de Nooijer, G. Somers, J. Ruigrok, L. Postma: IEEE Trans. Magn. **30**, 3813 (1994)
200. T. I-Fei, K.J. Duxstad: J. Appl. Phys. **85**, 5858 (1999)
201. S. Mao, N. Amin, E. Murdock: J. Appl. Phys. **83**, 6807 (1998)
202. H. Iwasaki, A.T. Saito, O. Tsutai: IEEE Trans. Magn. MAG–**33**, 2875 (1997)
203. S. Mao, J. Giusti, N. Amin, E. Murdock: J. Appl. Phys. **85**, 6112 (1999)
204. G.W. Anderson, Y. Huai, L. Miloslavsky, C.X. Qian, Yi Feng: J. Appl. Phys. **85**, 6109 (1998)
205. M. Saito, N. Hasegawa, F. Koike, H. Seki, T. Kuriyama: J. Appl. Phys. **85**, 4928 (1999)
206. L. ChihHuang, H. Matsuyama, R.L. White, T.C. Anthony: IEEE Trans. Magn. MAG–**31**, 2609 (1995)
207. L. ChihHuang, T.J. Regan, R.L. White, T.C. Anthony: J. Appl. Phys. **81**, 3989 (1997)
208. Y. Kawaka, Y. Sugita, M. Satomi, H. Sakakima: J. Appl. Phys. **85**, 5024 (1999)
209. J. Fujikata, K. Hayashi, M. Saito, M. Nakada: IEEE Trans. Magn. MAG–**34**, 954 (1998)

210. Y. Sugita, Y. Kawaka, M. Satomi, H. Sakakima: Jap. J. Appl. Phys. **37**, 5984 (1998)
211. J.L. Leal, N.J. Oliveira, L.M. Rodrigues, A.T. Sousa, P.P. Freitas: IEEE Trans. Magn. MAG–**30**, 3831 (1994)
212. J.L. Leal, M.H. Kryder: J. Appl. Phys. **83**, 3720 (1998)
213. Y. Huai, J. Zhang, G.W. Anderson, P. Rana, S. Funada, C.Y. Zhao, S. Tran: J. Appl. Phys. **85**, 5528 (1999)
214. J.L. Leal, M.H. Kryder: IEEE Trans. Magn. MAG–**35**, 800 (1999)
215. H.A.M. van den Berg, W. Clemens, G. Gieres, G. Rupp, W. Schelter, M. Vieth: IEEE Trans. Magn. **32**, 4624 (1996)
216. H.A.M. van den Berg, G. Rupp, G. Gieres, M. Vieth, J. Wecker, S. Zoll: IEEE Trans. Magn. **34**, 1336 (1998)
217. N.N. Persat: La Magnétorésistance Géante dans les Multicouches Magnétiques et ses Applications dans les Capteurs (Ph. D. Thesis, Univ. Paris 7-Denis Diderot, 1998)

7 Nonlinear Magnetooptics for Magnetic Thin Films

Th. Rasing

Some of the most exciting recent discoveries in magnetic thinfilm research, like the Giant Magnetoresistance (GMR) and the oscillatory exchange coupling are related to the properties of multilayers of alternating ferromagnetic and paramagnetic layers. As the interfaces between these layers appear to play an essential role for these phenomena and consequently for the device properties based on them, a detailed study of the magnetic interface properties is needed.

There exist several techniques (e.g., spin-polarized photoemission spectroscopy [1], spin-polarized electron energy loss spectroscopy [2], and spin-polarized low-energy electron diffraction [3] to study the magnetic properties of clean surfaces. Unfortunately, (polarized) electrons are difficult to use for studying buried interfaces due to their short mean free path. Since interfaces between thin metallic films are accessible by light, an optical technique would have a significant advantage. The Magneto-Optical Kerr Effect (MOKE) is well known and frequently employed, both for fundamental studies as well as applications. This linear optical technique is based on changes in the linear susceptibility as a function of the applied magnetic field. As this results in a (small) rotation of the polarization of light travelling through a magnetic material, it yields a probe for its magnetization. The use of modulation techniques can yield a very high sensitivity and applications to surfac and thin film studies lead to the introduction of SMOKE: surface magnetooptical Kerr effect. However, though very sensitive and even applicable to monolayers, MOKE is not interface-specific but essentially a bulk probe.

Second Harmonic Generation (SHG) is a nonlinear optical technique that derives its interface sensitivity from the breaking of symmetry at boundaries between centrosymmetric media [4–7]. Though second harmonic generation was one of the first nonlinear optical effects discovered after the invention of the laser [8], it took another two decades before the development of SHG as a tool for surface and interface science started.

Though possible surface effects were already discussed in one of the early papers by Bloembergen [9], at that time they were considered as minor, small contributions. The development of surface science and the pioneering work by the Shen group in Berkeley has changed this picture drastically. Shen's book on nonlinear optics gives a good introduction to this field [4]. In the past decade, SHG has been proven to be an extremely versatile surface and interface probe. Applications range from the study of the structure and symmetry of semiconductor and metal surfaces and interfaces [6,10], molecular adsorbates, in situ studies of molecular adsorption, desorption and diffusion [11], electric field dynamics at metal-semiconductor interfaces [12] and the electrochemistry at electrolytic interfaces [13]. As an optical, interface-sensitive technique, SHG is particularly attractive for the study of buried interfaces [5]. It was not until 1989 that Pan et al. [15] and Hübner and Bennemann [16] showed that SHG would also be sensitive to the magnetic properties of surfaces and interfaces, i.e., SHG would yield a nonlinear, surface- and interface-sensitive technique equivalent of the MOKE.

The origin of MOKE lies in the spin-orbit coupling that acts like a magnetic field on the current induced by the electromagnetic field of the incident light [14]. This should also hold for the nonlinear contributions of the induced current, that are the origin of SHG, leading to a Nonlinear Magneto-Optical Kerr Effect (NOMOKE). Based on symmetry arguments, Ru-Pin Pan et al., indeed, showed that the presence of a magnetization would lead to new, nonzero *surface* contributions to the nonlinear optical response [15]. At almost the same time, Hübner and Bennemann independently calculated the nonlinear magnetooptical spectrum of Ni, based on spin-depentent band-structure calculations [16]. They demonstrated that this should, indeed, lead to observable effects, with magnetic contributions to the nonlinear tensor coefficients of more than 10%. The first experimental evidence for a magnetization-induced second-harmonic generation (MSHG) effect was given by Reif et al. for a clean Fe(110) surface [17], whereas Spierings et al. showed the first MSHG results from buried Co/Au interfaces [18].

A clear demonstration of the surface and interface sensitivity of the MSHG technique was given by Wierenga et al. [19] by in situ MSHG and MOKE studies of the Co/Cu(100) system. These studies also showed that the nonlinear magnetooptical effects are quite sizeable: magnetic contrasts of over 50% were observed. This indicates that the magnetization-induced tensor elements are of the same order of magnitude as the nonmagnetic ones. This is in strong contrast to the linear optical response, where the magnetization induces very small off-diagonal tensor elements. However, though the relative effects are very large (up to $10^3\times$ the linear equivalents), the absolute SHG signals originating from surface or interface layers are very small (the efficiency is of the order of 10^{-15}!). The development of mode-locked fs Ti:sapphire lasers has therefore strongly boosted the practical applicability of the MSHG technique. Pustogowa et al. also predicted the appearance of large enhancements of the nonlinear Kerr angle [20]. Experimentally, the first observation of a NOMOKE rotation was made by Aktsipetrov et al., who measured a nonlinear Kerr rotation between 1° and 4° for a $Y_{2.5}Bi_{0.5}Fe_5O_5$ garnet film [21] and by Reif et al. [22] who found a nonlinear rotation of 14° for the Heussler alloy. Even larger effects were subsequently observed in experiments on thin Fe/Cr films and on single-crystalline Fe whiskers, showing enhancements of the nonlinear versus the linear MOKE rotation of over three orders of magnitude [23].

In this chapter, the various aspects of nonlinear magnetooptics and its potential for magnetic thinfilm research and applications will be treated. After a brief dicussion of the linear MOKE technique, the effects of the presence of a magnetization on SHG will be considered, both from a phenomenological and a microscopic point of view. We will treat the various configurations of the nonlinear Kerr effect (longitudinal, transversal and polar), and compare them with their linear equivalents. The combination of a large probing depth and the interface specificity of this nonlinear optical technique offers interesting possibilities for the study of multilayers. However, this requires a detailed understanding and analysis of the nonlinear optical response of such metallic multilayers [24]. This will therefore be treated in some detail. After a short discussion of some experimental aspects, we will describe various experimental studies, addressing both the interface sensitivity as well as the enormous enhancements of the Kerr rotation and the apparent extreme sentitivity of (M)SHG to confined quantum-well states in very thin (magnetic) metallic films. A brief outline of future developments and possible applications will terminate this chapter.

7.1 The Magnetooptical Kerr Effect

The presence of a time-reversal symmetry-breaking magnetization $\boldsymbol{M}$ introduces nonreciprocal optical effects, like polarization rotation. For metallic, reflecting media, these effects are summarized under the title magnetooptical Kerr effects. (For transparent media, this leads to the so-called Faraday effects). One distinguishes three different configurations: the polar, longitudinal and transversal Kerr effect (Fig. 7.1), because the relative orientation of $\boldsymbol{M}$ with respect to the polarization as well as the optical plane of the light plays a role.

Though the origins of NOMOKE and MOKE are the same (the spin-orbit coupling), it will be shown that there are some very basic differences between the two effects, leading to enormous enhancements of the NOMOKE versus the MOKE rotation. For that reason, we will first briefly describe the well-known linear Kerr rotation, before introducing its nonlinear equivalent.

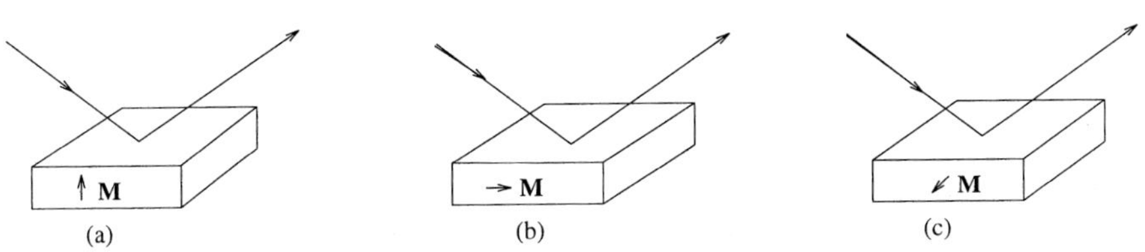

Fig. 7.1. a–c Kerr effect configurations, (**a**) polar, (**b**) longitudinal, (**c**) transversal

The Linear Magneto Optical Kerr Effect (MOKE)

In a medium the $\boldsymbol{E}$ field should satisfy the wave equation

$$\nabla \times \nabla \times \boldsymbol{E}(\omega) + \frac{\overleftrightarrow{\varepsilon}(\omega)}{c^2} \frac{\delta^2}{\delta t^2} \boldsymbol{E}(\omega) = 0 \,. \tag{7.1}$$

It can be shown [25] that with a magnetization along the y axis (the longitudinal geometry) the frequency-dependent linear dielectric tensor $\overleftrightarrow{\varepsilon}(\omega)$ is given by

$$\overleftrightarrow{\varepsilon}(\omega) = \begin{pmatrix} \varepsilon_0(\omega) & 0 & -\varepsilon_1(\omega) \\ 0 & \varepsilon_0(\omega) & 0 \\ \varepsilon_1(\omega) & 0 & \varepsilon_0(\omega) \end{pmatrix} .$$

The diagonal components $\varepsilon_0(\omega)$ have purely nonmagnetic character (are symmetric under magnetization reversal), whereas the elements $\varepsilon_1(\omega)$ are magnetic contributions (antisymmetric under magnetization reversal). With this form of the dielectric tensor the eigenmode solutions of the wave equation are left- and right-handed circularly polarized waves $\boldsymbol{E}^{\pm}(\omega) = E^{\pm} \exp[\mathrm{i}\omega(t - N_{\pm} z/c)]$, with as eigenvalues the complex refraction indices

$$N_{\pm}^2 = \varepsilon_0(\omega) \pm \mathrm{i}\varepsilon_1(\omega) \sin\theta_{\mathrm{t}} \,, \tag{7.2}$$

in which θ_{t} denotes the angle of refraction of the fundamental beam. Thus, the magnetism is included in the different indices of refraction for left- and right-polarized light. The complex Kerr angle is defined as

$$\tan\psi_{\mathrm{K}} = \Phi_{\mathrm{K}} + \mathrm{i}\varepsilon_{\mathrm{K}} = \mathrm{i}\frac{E_{\mathrm{r}}^{+} - E_{\mathrm{r}}^{-}}{E_{\mathrm{r}}^{+} + E_{\mathrm{r}}^{-}} \,, \tag{7.3}$$

where Φ_{K} is the Kerr angle and ε_{K} is the Kerr ellipticity. To express the reflected field amplitude $E_{\mathrm{r}}^{\pm}$ in terms of the incident field amplitude $E_{\mathrm{i}}^{\pm}$ the Fresnel formulae can be used:

$$\frac{E_{\mathrm{r}}^{\pm}}{E_{\mathrm{i}}^{\pm}} = \frac{N_{\pm}\cos\theta_{\mathrm{t}} - \cos\theta_{\mathrm{i}}}{N_{\pm}\cos\theta_{\mathrm{t}} + \cos\theta_{\mathrm{i}}} = \frac{N_{\pm}^2\cos\theta_{\mathrm{t}} - \sqrt{N_{\pm}^2 - \sin^2\theta_{\mathrm{i}}}}{N_{\pm}^2\cos\theta_{\mathrm{t}} + \sqrt{N_{\pm}^2 - \sin^2\theta_{\mathrm{i}}}} \,, \tag{7.4}$$

in which θ_{i} is the angle of incidence. Since for linearly polarized incident light $E_{\mathrm{i}}^{+} = E_{\mathrm{i}}^{-}$ an expression for the complex linear Kerr angle for the longitudinal geometry can be derived:

$$\tan\Phi_{\mathrm{K}}^{(1)}(\omega) = -\frac{1}{\sqrt{1+\chi_0}} \frac{\chi_1}{\chi_0} \frac{\sin\theta_{\mathrm{i}}\cos\theta_{\mathrm{i}}}{\sqrt{\cos^2\theta_{\mathrm{i}} + \chi_0(\omega)}} \frac{\cos(2\theta_{\mathrm{i}}) + \chi_0(\omega)}{\cos(2\theta_{\mathrm{i}}) + \chi_0(\omega)\cos^2\theta_{\mathrm{i}}} \,, \tag{7.5}$$

in which $\varepsilon_1 = \chi_1$ and $\varepsilon_0 = 1 + \chi_0$.

A similar expression can be deduced for the polar configuration. In the transversal geometry, it follows that there is no Kerr rotation, however, the *intensity* of the reflected light does depend on the magnetization [26].

7.2 Surface Second-Harmonic Generation

Second-Harmonic Generation (SHG) arises from the nonlinear polarization $P(2\omega)$ induced by an incident laser field $E(\omega)$. This polarization can be written as an expansion in $E(\omega)$:

$$\boldsymbol{P}(2\omega) = \boldsymbol{\chi}^{(2)}\boldsymbol{E}(\omega)\boldsymbol{E}(\omega) + \boldsymbol{\chi}^{(3)}\boldsymbol{E}(\omega)\boldsymbol{\nabla}\boldsymbol{E}(\omega) + \dots , \tag{7.6}$$

where $\boldsymbol{\chi}^{(n)}$ is the nth order nonlinear optical susceptibility tensor. The lowest-order term describes an electric dipole source. Symmetry considerations show that this contribution is zero in a centrosymmetric medium, thus limiting electric dipole radiation to the interfaces where inversion symmetry is broken (Fig. 7.2). The bulk second harmonic can now be described in terms of the much smaller electric quadrupole-like contributions [second term in (7.6)]. However because of the large volume difference between interface and bulk this does not necessarily mean that the total bulk second-harmonic signal is negligible. Interface sensitivity needs to be verified for any given system. In particular, for insulating materials, and for semiconductors using below bandgap excitation, these bulk effects can be large and sometimes even dominating. The very effective screening in metals usually limits the symmetry breaking to the first atomic layers [27] though sizeable bulk contributions have been observed as well [28,29]. Experimentally, surface sensitivity can often be demonstrated by in situ surface modification, like oxidation or CO adsorbtion. A complete theoretical treatment of the various contributions to surface SHG has been given by Guyot-Sionnest et al. and more recently by Mills in the 2nd edition of his text book [30].

As will be shown in Sect. 7.3, in the case of magnetic surfaces or interfaces, the presence of the magnetization will not affect the inversion symmetry but will introduce extra nonzero surface-tensor elements. Those will change sign when the direction of $\boldsymbol{M}$ will be reversed, which will lead to a nonlinear polarizability that will depend on $\boldsymbol{M}$:

$$\boldsymbol{P}(2\omega, \pm\boldsymbol{M}) = (\boldsymbol{\chi}^{(2),+} \pm \boldsymbol{\chi}^{(2),-})\boldsymbol{E}(\omega)\boldsymbol{E}(\omega) . \tag{7.7}$$

The presence of these new tensor components that are odd in the magnetization will lead to the phenomena of magnetization-induced second-harmonic generation.

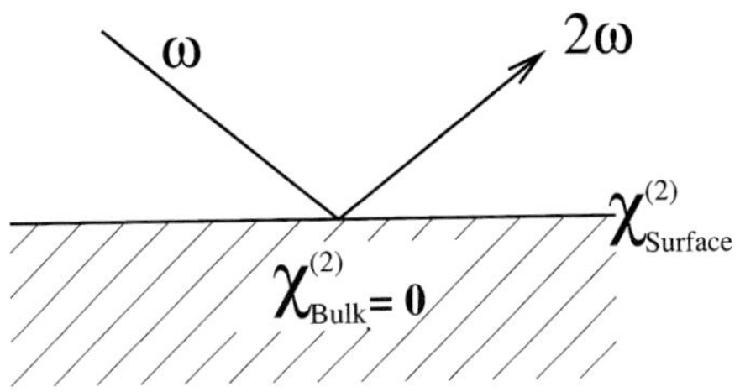

Fig. 7.2. A schematic view of surface second-harmonic generation

7.3 Magnetization-Induced Second-Harmonic Generation

Similar to the linear case, the presence of a magnetization $\boldsymbol{M}$ will affect the nonlinear optical susceptibility tensor $\boldsymbol{\chi}^{(2)}$. Some of the first theoretical treatments of the effect of a magnetization on the nonlinear optical properties already appeared more than two decades ago for systems without inversion symmetry [31]. In that case, the presence of a magnetization lowers the symmetry of the crystal and gives rise to new nonzero tensor elements. Here we are concerned with the response of metallic thin films, where the bulk possesses inversion symmetry. For these systems Ru-Pin Pan et al. demonstrated how the presence of a magnetization leads to new nonzero tensor elements. In the following we will first explain how these can be derived from simple symmetry arguments, and how they lead to nonlinear equivalents of the various MOKE effects.

7.3.1 The Nonlinear Tensor Components of Magnetized Materials

As already mentioned before, for materials that possess inversion symmetry, the nonlinear susceptibility tensor $\boldsymbol{\chi}^{(2)}$ vanishes for the bulk and only gives a nonzero surface contribution due to the symmetry breaking at the surface. The presence of a magnetization $\boldsymbol{M}$ does not break the inversion symmetry, since $\boldsymbol{M}$ is an axial vector that conserves its orientation under inversion. However, $\boldsymbol{M}$ does lower the symmetry of the surface, introducing extra nonzero tensor elements [15]. These elements can be derived in the following way: Consider the (100) surface of a cubic crystal, and the longitudinal configuration, i.e., the magnetization parallel to the surface and in the plane of incidence (Fig. 7.1).

In the absence of $\boldsymbol{M}$, the (100) surface has 4 mm symmetry. With $\boldsymbol{M} = (M,0,0)$, the m_y mirror symmetry becomes m_y^-, whereas the m_x mirror is unaffected by the magnetization. The symmetry operations then become

$$\left.\begin{aligned} m_x^+ o\chi_{ijk}^{(2)}(\boldsymbol{M}) &= \chi_{ijk}^{(2)}(\boldsymbol{M}) \\ m_y^- o\chi_{ijk}^{(2)}(\boldsymbol{M}) &= \chi_{ijk}^{(2)}(\boldsymbol{M}) \end{aligned}\right\} . \tag{7.8}$$

For example, this gives

$$\left.\begin{aligned} m_y^- o\chi_{zyy}^{(2)}(M,0,0) &= \chi_{zyy}^{(2)}(M,0,0) \\ \Leftrightarrow +\chi_{zyy}^{(2)}(-M,0,0) &= \chi_{zyy}^{(2)}(M,0,0) \end{aligned}\right\} \tag{7.9}$$

and

$$\left.\begin{aligned} m_y^- o\chi_{yyy}^{(2)}(M,0,0) &= \chi_{yyy}^{(2)}(M,0,0) \\ \Leftrightarrow -\chi_{yyy}^{(2)}(-M,0,0) &= \chi_{yyy}^{(2)}(M,0,0) \end{aligned}\right\} . \tag{7.10}$$

Table 7.1. Nonzero tensor elements of the nonlinear susceptibility tensor $\chi^{(2)}(\boldsymbol{M})$ for the (001) surface of a cubic crystal for the magnetization parallel to the x axis (longitudinal configuration), y axis (transverse configuration) and parallel to the z axes (polar configuration)

	Even in $\boldsymbol{M}$	Odd in $\boldsymbol{M}$
Longitudinal	$yzy = yyz$	$xyx = xxy$
$\boldsymbol{M}\Vert x$	$xzx = xxz$	$zyz = zzy$
	zzz	yzz
	zyy	yyy
	zxx	yxx
Transverse	$xxz = xzx$	$yxy = yyx$
$\boldsymbol{M}\Vert y$	$yyz = yzy$	$zxz = zzx$
	zxx	xxx
	zyy	xyy
	zzz	xzz
Polar	$xxz = xzx = yyz = yzy$	$xyz = xzy = -yxz = -yzx$
$\boldsymbol{M}\Vert z$	$zxx = zyy$	
	zzz	

As a result, we can separate the nonlinear susceptibility into two sets of tensor elements: those which are odd χ^{-} and even χ^{+} in the magnetization. These tensor elements fulfill the relation

$$\chi_{ijk}^{(2),\pm}(-\boldsymbol{M}) = \pm\chi_{ijk}^{(2),\pm}(\boldsymbol{M}) . \tag{7.11}$$

The tensor elements for the transversal configuration can be derived in an analogous way. In this case, $\boldsymbol{M} = (0, M, 0)$ and the corresponding symmetry operations are

$$\left.\begin{aligned} m_x^- o\chi_{ijk}^{(2)}(0, M, 0) &= \chi_{ijk}^{(2)}(0, M, 0) \\ m_y^+ o\chi_{ijk}^{(2)}(-M, 0, 0) &= \chi_{ijk}^{(2)}(M, 0, 0) \end{aligned}\right\} . \tag{7.12}$$

Finally, we consider the case of a polar Kerr effect, i.e., with the magnetization normal to the surface plane: $\boldsymbol{M} = (0, 0, M)$. For this case, the symmetry operations are

$$\left.\begin{aligned} m_x^- o\chi_{ijk}^{(2)}(0, 0, M) &= \chi_{ijk}^{(2)}(0, 0, M) \\ m_y^+ o\chi_{ijk}^{(2)}(0, 0, M) &= \chi_{ijk}^{(2)}(0, 0, M,) \\ R_{z,90^0}^+ o\chi_{ijk}^{(2)}(0, 0, M) &= \chi_{ijk}^{(2)}(0, 0, M) \end{aligned}\right\} . \tag{7.13}$$

Table 7.1 lists the so derived odd and even tensor components for all the Kerr configurations [32].

From Table 7.1, the effect of the magnetization on the SHG response can directly be understood qualitatively. Take, for example, a s-polarized input beam for the case of a longitudinal configuration. In that situation, there is only one even term χ_{zyy} and one odd term χ_{yyy} that will contribute to the Magnetization-induced Second-Harmonic Generation (MSHG) response. Because the even term will give rise to a p-polarized output and the odd term will give an s-polarized SHG signal, the total output polarization can be varied by changing the direction of the magnetization. This will lead to a nonlinear Kerr rotation, similarly to the linear case. However, as we will see later, the effects can be much larger than in the linear case. One of the reasons is that the odd component is of the same order of magnitude as the even one (there are also other, intrinsic, differences between the two cases that will be discussed in Sect. 7.3.2).

For the transverse geometry, it follows from Table 7.1, that the generated SHG signal will always be p-polarized, both for s- and p-polarized input beams. Again take, for example, an s-polarized input. The MSHG signal assumes the form

$$I_{2\omega}(\pm M) \approx |\alpha\chi^{(2)}_{xyy} \pm \beta\chi^{(2)}_{xyy}|^2 \,, \tag{7.14}$$

where the constants α and β include the Fresnel factors for the incoming and generated optical beams. Equation (7.14) shows that the presence of a magnetization does not change the SHG polarization but will only affect the output intensity. So similarly to the linear case, there is no nonlinear Kerr rotation in the transversal geometry. However, in contrast to the linear case, these intensity changes can be very large. Experimentally one then often measures the magnetic contrast ρ defined by

$$\rho = \frac{I_{2\omega}(+\boldsymbol{M}) - I_{2\omega}(-\boldsymbol{M})}{I_{2\omega}(+\boldsymbol{M}) + I_{2\omega}(-\boldsymbol{M})} \,. \tag{7.15}$$

The tensor components tabulated in Table 7.1 fulfill another important relationship, that can be derived from the breaking of the time-reversal symmetry due to a magnetic field. Including the explicit time and phase dependence, the induced polarization can be written as

$$\boldsymbol{P}^{(2)}_{2\omega}(\boldsymbol{r},t) = [\boldsymbol{\chi}^{+}(\boldsymbol{M}) + \boldsymbol{\chi}^{-}(\boldsymbol{M})]\boldsymbol{E}_{\omega}(\boldsymbol{r},t)\boldsymbol{E}_{\omega}(\boldsymbol{r},t) \,. \tag{7.16}$$

In the absence of dissipation (far away from resonances), time reversal implies

$$\left.\begin{array}{rcl} \boldsymbol{P}_{2\omega}(\boldsymbol{r},t) & \Rightarrow & \boldsymbol{P}^{*}_{2\omega}(\boldsymbol{r},t) \\ \boldsymbol{E}_{\omega}(\boldsymbol{r},t) & \Rightarrow & \boldsymbol{E}^{*}_{\omega}(\boldsymbol{r},t) \\ \boldsymbol{M} & \Rightarrow & -\boldsymbol{M} \end{array}\right\} , \tag{7.17}$$

leading to

$$\boldsymbol{P}^*_{2\omega}(\boldsymbol{r},t) = [\boldsymbol{\chi}^+(\boldsymbol{M}) - \boldsymbol{\chi}^-(\boldsymbol{M})]\boldsymbol{E}^*_\omega(\boldsymbol{r},t)\boldsymbol{E}^*_\omega(\boldsymbol{r},t)\,. \tag{7.18}$$

Adding (7.16) and (7.18) yields

$$\begin{aligned}&\boldsymbol{P}_{2\omega}(\boldsymbol{r},t) + \text{c.c.} = \\ &\quad\boldsymbol{\chi}^+(\boldsymbol{M})[\boldsymbol{E}_\omega(\boldsymbol{r},t)\boldsymbol{E}_\omega(\boldsymbol{r},t) + \text{c.c.}] + \boldsymbol{\chi}^-(\boldsymbol{M})[\boldsymbol{E}_\omega(\boldsymbol{r},t)\boldsymbol{E}_\omega(\boldsymbol{r},t) - \text{c.c.}]\,.\end{aligned} \tag{7.19}$$

The left-hand side of (7.19) is real, and so must be the right-hand side, for all $\boldsymbol{r}$ and t. From this it immediately follows that, in the absence of dissipation, $\boldsymbol{\chi}^+(\boldsymbol{M})$ is real and $\boldsymbol{\chi}^-(\boldsymbol{M})$ is purely imaginairy. This result also follows from a microscopic treatment, as was recently shown by Hübner and Bennemann [33].

7.3.2 The Nonlinear Magnetooptical Kerr Effect

The Nonlinear Magneto-Optical Kerr Effect (NOMOKE) or rotation describes the change of the nonlinear optical (SHG) response due to the presence of a magnetization (and not, as it may suggest, a nonlinear dependence on the magnetization of the Kerr rotation). The theoretical expressions in Sect. 7.5 will show that $\boldsymbol{\chi}^{(2)}(\boldsymbol{M})$ depends linearly on the magnetization. Because the nonlinear Kerr angle $\Phi_\mathrm{K}^{(2)}$ is related to a ratio of magnetic and nonmagnetic $\boldsymbol{\chi}^{(2)}(\boldsymbol{M})$ tensor components (see below), $\Phi_\mathrm{K}^{(2)}$ will also depend linearly on $\boldsymbol{M}$ (just as the linear Kerr rotation $\Phi_\mathrm{K}^{(1)}$). For this reason, it would be more correct to speak of a Kerr rotation of the nonlinear optical response.

The nonlinear optical effects are caused by the nonlinear surface polarization $\boldsymbol{P}(2\omega)$ that acts as a source term in the wave equation. The latter can now be written as

$$\nabla \times \nabla \times \boldsymbol{E}(2\omega) + \frac{\overleftrightarrow{\varepsilon}(2\omega)}{c^2}\frac{\delta^2}{\delta t^2}\boldsymbol{E}(2\omega) = -\frac{1}{\varepsilon_0 c^2}\frac{\delta^2}{\delta t^2}\boldsymbol{P}(2\omega)\,. \tag{7.20}$$

The solution of this wave equation includes the solution of the homogenous equation plus one particular solution of the inhomogenous equation. Similarly to the linear case, the homogeneous part yields complex refraction indices for the transmitted second-harmonic wave:

$$N_\mathrm{t}^2 = \varepsilon_0(2\omega) \pm \mathrm{i}\varepsilon_1(2\omega)\sin\theta_{2\mathrm{t}}\,. \tag{7.21}$$

The inhomogeneous part has solutions which no longer relate to indices of refraction but directly to the surface susceptibility $\boldsymbol{\chi}^{(2)}$.

For the calculation of the Kerr rotation the amplitude of the reflected electric field should be known. For this we have to make use of the Fresnel formulae for the nonlinear case. These formulae can be derived by using the continuity of the electric field parallel to the surface and of the magnetic field normal to the surface (p-polarized incident light is assumed) [4,9]. This leads to the following two equations:

$$-\cos\theta_{\mathrm{i}}E_{\mathrm{r}}(2\omega)=\cos\theta_{\mathrm{t}}E_{\mathrm{t}}(2\omega)+\frac{\cos\theta_{\mathrm{s}}N_{\mathrm{s}}^{2}\sin\theta_{\mathrm{s}}}{\varepsilon_{0}(2\omega)[\varepsilon_{0}(2\omega)-N_{\mathrm{s}}^{2}]}P(2\omega) \tag{7.22}$$

and

$$k_{\mathrm{r}}E_{\mathrm{r}}=k_{\mathrm{t}}E_{\mathrm{t}}+\frac{k_{\mathrm{s}}\sin\theta_{\mathrm{s}}}{\varepsilon_{0}(2\omega)-N_{\mathrm{t}}^{2}}P(2\omega)\,, \tag{7.23}$$

where E_{r}, k_{r} and E_{t}, k_{t} are the electric field and wave vectors of the reflected and transmitted waves and θ_{t} and θ_{i} are the refraction and incident angle (Fig. 7.3). N_{s} is the refraction index of the source term:

$$N_{\mathrm{s}}^{2}=\varepsilon_{0}(\omega)\pm\mathrm{i}\varepsilon_{1}(\omega)\sin\theta_{\mathrm{s}}\,. \tag{7.24}$$

The angle θ_{s} determines the orientation of the source-wave vector k_{s} and is equal to the refraction angle of the fundamental ray. The direction of the source term $\boldsymbol{P}(2\omega)$ is chosen perpendicular to the surface. The amplitude of the reflected electric field follows from the combination of these two equations:

$$E_{\mathrm{r}}\pm(2\omega)=\frac{1}{\varepsilon_{0}c^{2}}\frac{P(2\omega)\sin\theta_{\mathrm{s}}}{k_{\mathrm{t}}\cos\theta_{\mathrm{i}}+k_{\mathrm{r}}\cos\theta_{\mathrm{t}}}\frac{\varepsilon_{0}(2\omega)k_{\mathrm{s}}\cos\theta_{\mathrm{t}}-N_{\mathrm{s}}^{2}\sin\theta_{\mathrm{s}}}{\varepsilon_{0}(2\omega)[\varepsilon_{0}(2\omega)-N_{\mathrm{s}}^{2}]}\,. \tag{7.25}$$

Magnetism is included in the indices of refraction. With this formula for E_{r} (7.3) will lead to a nonlinear Kerr angle of [20]

$$\tan\Psi_{\mathrm{K}}^{(2)}=\mathrm{i}\left[\frac{\chi^{(2),-}}{\chi^{(2),+}}+(\mathrm{h.o.})\right]\,, \tag{7.26}$$

in which $\chi^{(2),+}$ and $\chi^{(2),-}$ are the nonmagnetic and magnetic contributions to $\boldsymbol{\chi}^{(2)}$. The big difference between the expressions for $\Psi_{\mathrm{K}}^{(2)}$ and $\Psi_{\mathrm{K}}^{(1)}$ is the factor $1/\sqrt{((1+\chi_{0})\cos^{2}\theta_{\mathrm{i}}+\chi_{0}(\omega))}$ in (7.5). This factor causes a small value of $\Psi_{\mathrm{K}}^{(1)}$ in (7.5), but it is not present in the nonlinear case. Therefore, it is clear that the nonlinear Kerr rotation for all θ_{i} is always enhanced by the factor $\sqrt{((1+\chi_{0})(\cos^{2}\theta_{\mathrm{i}}+\chi_{0}(\omega))}$. For example, for Fe at $\theta_{\mathrm{i}}=45°$ and $\lambda=830\,\mathrm{nm}$, this gives a reduction factor of 0.15 for $\Psi_{\mathrm{K}}^{(1)}$. Thus, one of the reasons for the difference in Kerr rotation is the fact that $\boldsymbol{\chi}^{(2)}$ depends sensitively on the magnetic properties of the surface because in the nonlinear case only the surface contributes. The destructive contributions to the Kerr rotation by bulk inter- and intraband transitions, that occur in the linear case, are now avoided.

Above this effect, in the nonlinear case one can select large magnetic $\boldsymbol{\chi}^{(2)}$ contributions by a proper choice of input polarization and angle of incidence. This is in contrast to the linear case where one always deals with small off-diagonal magnetic terms $\chi_{1}(\omega)$ relative to large diagonal nonmagnetic terms $\chi_{0}(\omega)$.

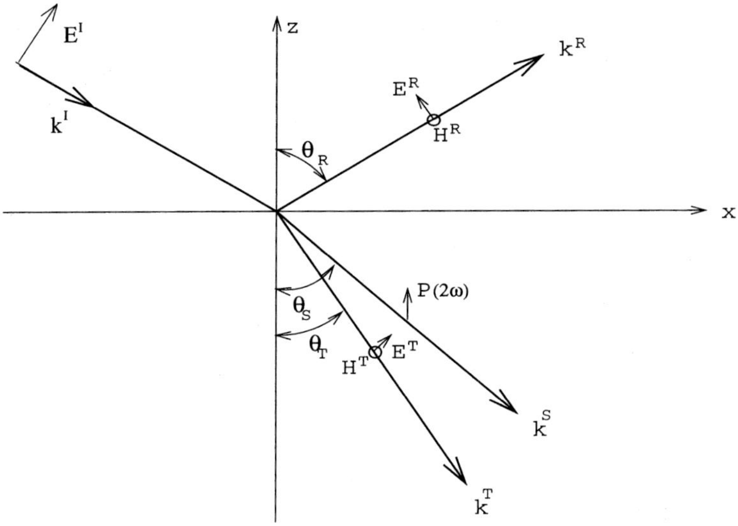

Fig. 7.3. The harmonic wave at the boundary of a nonlinear medium, with the electric field vector in the plane of reflection

This can easily be demonstrated with a simple example. Consider the case of a longitudinal configuration, with $\boldsymbol{M}$ parallel to the x axis. Inspection of Table 7.1 yields that for a nonmagnetic surface, the only nonzero tensor component that will be excited is $\chi^{(2)}_{zyy}$, which will lead to a purely p-polarized output for an s-input polarization. However, in the case of a magnetic surface, also the odd tensor component $\chi^{(2)}_{yyy}$ will be nonzero, leading to a turning of the SHG polarization away from the p-direction. In the limit of a very small angle of incidence, the z-component of the reflected SHG signal will be vanishingly small. Therefore, the $\chi^{(2)}_{zyy}$ contribution and thus the p component of the SHG signal, will be very small. This means that in the limit of near-normal incidence, the SHG signal from the magnetic surface will be s-polarized, instead of p-polarized for the nonmagnetic case, i.e., a nonlinear Kerr rotation of 90°. A similar enhancement of the magnetic (off-diagonal) term relative to the nonmagnetic (diagonal) one is however not possible for the linear case. The same configuration only leads to a small Kerr rotation, given by $\Psi^{(1)}_{\mathrm{K}} = \arctan(\chi^{(1)}_{yx}/\chi^{(1)}_{yy})$, because $\chi^{(1)}_{yx} \ll \chi^{(1)}_{yy}$.

For a p-polarized input and for the same configuration and angle of incidence, the nonlinear Kerr rotation will generally be much smaller. This is due to the fact that for this polarization, there are three nonzero even tensor elements χ_{zzz}, χ_{xzx} and χ_{zxx}, and two nonzero odd tensor components χ_{yzz} and χ_{yxx}. In this situation an increasing contribution from one tensor component is often (partly) compensated by a decrease in another term, resulting in less pronounced changes in the MSHG response and in smaller nonlinear Kerr rotations.

The nonlinear Kerr effect can also be used to determine the easy axis of magnetization. This is particularly useful to distinguish between in-plane and out-of-plane magnetization (as was also pointed out by Hübner and Bennemann [33]). Again, consider the case of near-normal incidence. As we have just shown, an s-polarized excitation beam will lead to a very large nonlinear Kerr rotation, whereas the p-polarized input will give rise to a much smaller rotation. In contrast, for a polar configuration, Table 7.1 suggests that for s-polarized input, the output will always be p-polarized, with or without a magnetization, i.e., for s-polarized input there will be no nonlinear Kerr rotation. Thus by looking at both the s- and p-polarization input responses, an immediate distinction beween in-plane and out-of plane magnetization can be obtained. To obtain the in-plane direction, one can do anisotropy measurements. By measuring the nonlinear Kerr rotation as a function of the azimuthal orientation, the longitudinal direction can be obtained from the maximum nonlinear Kerr rotation (actually, experimentally it will be more accurate to determine the orientation for which $\Phi_{\mathrm{K}}^{(2)}$ will vanish, which gives the transversal direction).

7.3.3 Nonlinear Magnetic Circular Dichroism

As already mentioned before, the origin of the linear magnetooptical Kerr effect is the difference in response for left- and right-handed circularly polarized light, leading to a (albeit small) rotation of the orientation of the polarization ellipse of a reflected light beam.

If one directly studies this difference, one can define a circular dichroism as

$$\delta = \frac{I^+ - I^-}{I^+ + I^-}, \tag{7.27}$$

where $I^{+(-)}$ indicates the reflected intensity of the right- (left)- handed circularly polarized input beam. For visible light, this leads to very small magnetic dichroism effects, of the order of less than one per cent. In the x-ray regime, magnetic dichroism effects can be as large as a 100%, by tuning to the proper core levels [34]. This has the added advantage of being element-specific, which is of enormous value for the study of multi-component magnetic systems. For this reason a special part of the new European synchrotron facility in Grenoble (the ESRF) will be dedicated to these studies. The latter indicates a disadvantage of this approach, namely the need of a synchrotron facility in order to apply it.

In the nonlinear optical case, one can also expect large circular dichroism effects, similar to the large nonlinear Kerr rotations. The first observations of this nonlinear circular dichroism were reported by Reif et al. [22]. Analogously to (7.27), one can define

$$\delta = \frac{I_{2\omega}^+ - I_{2\omega}^-}{I_{2\omega}^+ + I_{2\omega}^-}, \tag{7.28}$$

where $I_{2\omega}^{+(-)}$ indicates the generated second-harmonic intensity for right-handed and left-handed input polarizations, respectively. For the Heussler alloy PbMnSb, $\delta = 3.5\%$ was oberved, using 532 nm excitation. The real potential of this approach was recently demonstrated by Fiebig et al. [35] in a spectroscopic study of this nonlinear magnetic dichroism effect. By tuning to resonances, δ of a 100% can be obtained as well.

Though optical transitions are always much broader than core levels and also less specific (most relevant optical transitions are in the same small frequency range corresponding to 0.2 to 5 eV), the simplicity of this optical technique relative to a synchroton study certainly opens perspectives (not even mentioning the several orders of magnitude difference in costs).

7.4 The Nonlinar Magnetooptical Susceptibility

7.4.1 Quantum-Mechanical Expressions for $\chi^{(2)}(M)$

A quantum-mechanical derivation of the optical second-order nonlinar susceptibility can be found in Shen's book [4]. For the second-order dipole-tensor elements one has for the off-resonance case

$$\chi_{jkl}^{(2)}(2\omega) = -\frac{e^3}{\hbar^2}\sum_{\boldsymbol{q}}\sum_{g,n,n'}\left(\frac{(r_\mathrm{j})_{gn}(r_\mathrm{k})_{nn'}(r_\mathrm{l})_{n'g}}{(2\omega-\omega_{ng})(2\omega-\omega_{n'g})} + 5 \text{ similar terms}\right) f_g(\boldsymbol{q})\,, \tag{7.29}$$

where the summation runs over all possible states g, n, n' and wave vector $\boldsymbol{q}$, $\hbar\omega_{ng} = E_n - E_g$ and $f_g(\boldsymbol{q})$ is the Fermi distribution function for the state $|g, \boldsymbol{q}\rangle$.

There exist two ways in which the magnetization could influence the susceptibility:

1. The energy eigenvalues shift, which would result in different values of ω_{ng} for different spin states.
2. The electronic eigenfunctions change, because they are perturbed, which would lead to differences in the transition matrix elements $(r_\mathrm{i})_{ng}$ for the different spin states.

Both are a result of the combination of spin-orbit coupling and exchange interaction. Kittel has shown that the second phenomenon has the largest influence on the optical susceptibility [14]. We shall treat the spin-orbit interaction energy as being the perturbing Hamiltonian [15,16]:

$$H = H_0 + H_{\mathrm{s.o.}}\,, \tag{7.30}$$

with [36]

$$H_{\mathrm{s.o.}} = \frac{1}{2m^2c^2}[\nabla V(\boldsymbol{r}) \times \boldsymbol{p}] \cdot \boldsymbol{S} = h(\boldsymbol{r})\boldsymbol{L} \cdot \boldsymbol{S}\,. \tag{7.31}$$

Spin-orbit coupling changes the orbital momentum $\boldsymbol{p}$ of the electrons and thus the polarization of the generated light, much in the same way as an external magnetic field would do [16], so

$$-\mathrm{i}\hbar\dot{\boldsymbol{p}} = [H_{\mathrm{s.o.}}, \boldsymbol{p}] = [h(\boldsymbol{r})\boldsymbol{L}\cdot\boldsymbol{S}, \boldsymbol{r}] = -h(\boldsymbol{r})\boldsymbol{p}\times\boldsymbol{S}\,. \tag{7.32}$$

In nonferromagnetic materials both spin states are equally occupied so no net effect occurs. However, in ferromagnetic materials, exchange interaction splits the band structure and spin states are not equally occupied, causing a net rotation of the polarization [16].

Let $\psi(\boldsymbol{k},\boldsymbol{r})\alpha(s)$ be the unperturbed eigenfunctions of H_0, where $\alpha(s)$ are the eigenfunctions of the spin operator. In the mean-field approximation for the exchange forces the quantities $E_n(\boldsymbol{k}) \pm \delta$ are taken as the energy eigenvalues of the eigenfunctions $\psi(\boldsymbol{k},\boldsymbol{r})\alpha(\pm 1)$ [14,37]. Time-independent perturbation theory gives the eigenfunctions of the Hamiltonian in (7.30) [14]:

$$\phi_{\lambda,\pm 1}(\boldsymbol{r}) = [\psi_n(\boldsymbol{k},\boldsymbol{r}) \pm \xi_n(\boldsymbol{k},\boldsymbol{r})]\alpha(\pm 1)\,. \tag{7.33}$$

Here λ denotes all the necessary quantum numbers, and

$$\xi_n(\boldsymbol{k},\boldsymbol{r}) = \sum_{m\neq n} b_{nm}(\boldsymbol{k})\psi_m(\boldsymbol{k},\boldsymbol{r})\,, \tag{7.34}$$

with

$$b_{nm}(\boldsymbol{k}) = \frac{\mathrm{i}\hbar^2/4m^2c^2}{E_n(\boldsymbol{k}) - E_m(\boldsymbol{k})}\int \psi_m^*(\boldsymbol{k})(\nabla V\times\nabla)_{z_0}\psi_n(\boldsymbol{k})\mathrm{d}r\,, \tag{7.35}$$

where z_0 is the quantization direction of the spin. Within a first-order approximation the energies stay the same, so

$$(H_0 + H_{\mathrm{s.o.}})\phi_{\lambda,\pm 1} = [E_{\mathrm{n}}(\boldsymbol{k}) \pm \delta]\phi_{\lambda,\pm 1}\,. \tag{7.36}$$

We now obtain the tensor elements describing SHG from the magnetized system by using the wave functions $\phi_{\lambda,\pm 1}$ of (7.33) in (7.29). Considering only terms up to first order in b_{nm} one may rewrite the susceptibility in even (+) and odd (−) terms as

$$\chi_{jkl}^{(2)} = \chi_{jkl}^{(2),+} + \chi_{jkl}^{(2),-} \tag{7.37}$$

where $\chi_{jkl}^{(2),+}$ is the part of the nonlinear susceptibility tensor that is independent of the spin orientation. The elements of $\chi_{jkl}^{(2),+}$ are of zeroth order in b_{nm}:

$$\chi_{jkl}^{(2),+} = -\frac{e^3}{\hbar^2}\sum_{\boldsymbol{k}}\sum_{g,n,n'\sigma} \left(\frac{(r_j)_{gn}(r_k)_{nn'}(r_l)_{n'g}}{(2\omega-\omega_{ng})(2\omega-\omega_{n'g})} + 5\text{ similar terms}\right) f_g(\boldsymbol{k})\,. \tag{7.38}$$

The result is, of course, the same as for the nonmagnetic system, see (7.29). $\chi^{(2),-}$ is the tensor that does depend on the spin orientation of the electron. The elements of $\chi^{(2),-}$ are of first order in b_{nm}

$$\chi^{(2),-}_{jkl} = -\frac{e^3}{\hbar^2}\sum_{\boldsymbol{k}}\sum_{g,n,n'\sigma}$$
$$\left(\sum_m \pm b^*_{gm}\frac{(r_j)_{gn}(r_k)_{nn'}(r_l)_{n'g}}{(2\omega-\omega_{ng})(2\omega-\omega_{n'g})} + 35 \text{ similar terms}\right) f_g(\boldsymbol{k},\sigma)\,, \tag{7.39}$$

where σ is a spin index. Whether there is a plus or a minus sign in (7.39) depends on the spin of the electron. If both spin states are filled, all contributions to $\chi^{(2),-}_{jkl}$ will cancel, because of the summation over σ. If, however, a magnetization is present, both spin states are not equally occupied, and (7.39) will give a contribution to the susceptibility.

In accordance with the symmetry arguments given before, (7.38) shows that there are no contributions of first order in $\chi^{(2),+}_{jkl}$, i.e., the lowest-order effect of the magnetization on the even elements is of second order. This implies that the even elements are less dependent on the size of the magnetization than the odd elements (that are of first order in b_{nm}).

7.4.2 Theoretical Estimates for $\chi^{(2)}(M)$

Some theoretical values for $\chi^{(2),-}_{jkl}$ have been estimated for nickel and iron. Ru-Pin Pan et al. made a rather crude approximation for Ni(100) by assuming that both all odd and all even tensor elements are equal, and obtained [15]

$$|\omega^2\chi^{(2),-}|_{\mathrm{Ni}} \approx 10^{12}\,\mathrm{V^{-1}s^{-2}m} \approx 10^{-16}\,\mathrm{esu}\,. \tag{7.40}$$

The same surface has been more thoroughly analysed by Hübner et al. After including the complete band structure and dipole-transition matrix elements they obtained [39,40]

$$|\omega^2\chi^{(2)}_{xzz}|_{\mathrm{Ni}} \approx 10^{17}\,\mathrm{V^{-1}s^{-2}m} \approx 10^{-11}\,\mathrm{esu}\,. \tag{7.41}$$

Pustogowa et al. used a similar approach for the surface of iron and found [41,42]

$$|\omega^2\chi^{(2)}_{xzz}|_{\mathrm{Fe}} \approx 10^{16}\,\mathrm{V^{-1}s^{-2}m} \approx 10^{-12}\,\mathrm{esu}\,. \tag{7.42}$$

All these values are well above the detection limit of SHG, that is $10^9 - 10^{10}\,\mathrm{V^{-1}s^{-2}m}$ or 10^{-19}–10^{-18} esu [38,43] implying that magnetization-induced tensor elements should be obervable.

Furthermore, the estimated ratios of odd and even elements are high enough to induce obervable changes of the Second-Harmonic (SH) intensity on inverting the magnetization. For nickel both Ru-Pin Pan et al. [15] and Hübner et al. [40] estimated

$$|\chi^{(2),-}/\chi^{(2),+}|_{\mathrm{Ni}} = 0.07\,, \tag{7.43}$$

and for iron Pustogowa et al. deduced [41,42]

$$|\chi^{(2),-}/\chi^{(2),+}|_{\mathrm{Fe}} = 0.18\,. \tag{7.44}$$

Reif et al. deduced the latter ratio by simply scaling the nickel ratio of (7.43) by the ratio of the bulk magnetic moments for iron an nickel, $\mu_{\mathrm{Fe}}/\mu_{\mathrm{Ni}} = 3.8$, and obtained [17]

$$|\chi^{(2),-}/\chi^{(2),+}|_{\mathrm{Fe}} = 0.27\,. \tag{7.45}$$

The value is in close agreement with the result of a simple analysis of their experiments on the Fe(110) surface, and also close to the value given by Pustogowa et al., see (7.44). Scaling by the surface magnetic moments instead of the bulk magnetic moments does not significantly alter the estimate. Employing a similar approach for cobalt gives, with $\mu_{\mathrm{Co}}/\mu_{\mathrm{Ni}} = 3.1$ [44]

$$|\chi^{(2),-}/\chi^{(2),+}|_{\mathrm{Co}} = 0.22\,. \tag{7.46}$$

Experimentally (Sect. 7.7.1) we find $|\chi^{(2),-}/\chi^{(2),+}|_{\mathrm{Ni}} = 0.66$ for an excitation wavelength of 800 nm (the estimates were done for 532 nm). For the Co/Cu interface we find a very similar ratio $|\chi^{(2),-}/\chi^{(2),+}|_{\mathrm{Co/Cu}} = 0.89$.

All these estimates have been made for a particular frequency. In a number of recent papers, Hübner et al. presented ab initio calculations of the linear and nonlinear magnetooptical spectra for Ni and Fe. Nonlinear spectra show a much richer structure, that is connected with the differences in selection rules and the surface sensitivity of MSHG. Spectroscopic MSHG experiments will therefore be an excellent and sensitive way to probe magnetic surface states. So far, this has been a rather unexplored area of research.

7.5 SHG from Multilayers

To find the nonlinear optical response of a metallic multilayer we must know the nonlinear source polarization at the interfaces, and calculate the generated SH fields. Nonlinear optical generation from simple multilayer systems has been treated by Dick et al. [45] for layers of organic molecules, by Yeganeh et al. [46] for semi-conductors and by Koopmans et al. for thin C_{60} films [47]. Their results were either analyzed by simultaneously solving the boundary conditions at all interfaces, or by describing the multiple reflections of both the fundamental and the SH beams in terms of reflection and transmission coefficients. A well-known concept in optics is to treat the boundary conditions at an interface and the propagation of light through a homogeneous slab in terms of matrices on the basis of forward and backward travelling waves. The matrices relate the field components on both sides of the interface and the layer, respectively. Describing the full multilayer is thus reduced to a simple matrix multiplication [48]. Although this approach is completely analogous to simultanuously solving the boundary conditions at all interfaces, it has the strong advantage of being much more flexible, when it comes to solving systems with different numbers of layers.

For our MSHG calculations we assume that the system contains only electric dipole sources of SH radiation at the interfaces [24]. The model is, in fact, related to work by Sipe [49]. He developed a Green-function formalism for calculating fields generated by sources in the presence of a multilayer geometry. We use the nonlinear boundary conditions, as derived by Heinz [50], giving the discontinuity of the second-harmonic fields at the nonlinear interface (i.e., at the source). Our treatment of the emitted fields is similar to his approach. After introducing infinitesimal vacuum sheets between all the layers we come to an equivalent description of both the propagation of the fundamental and the second harmonic, which is very convenient for converting the model into a computer program.

The influence of the total multilayer on the transmission of the fundamental wave can be described as

$$\begin{pmatrix} E^{+}_{\omega,p,0} \\ E^{-}_{\omega,p,0} \end{pmatrix} = \prod_{j=1}^{m} T_j^{\omega,p} \begin{pmatrix} E^{+}_{\omega,p,m+1} \\ 0 \end{pmatrix}, \tag{7.47}$$

where the system contains m layers, and $E^{+}_{\omega,p,0}$ and $E^{-}_{\omega,p,0}$ represent, respectively, the forward and backward fields at the front side, and $E^{+}_{\omega,p,m+1}$ the forward field at the backside of the sample (notice that there can be only a forward field at the backside of the sample). $T_j^{\omega,p}$ is obtained from

$$T_j^{\omega,p} = M_{0,j}^{\omega,p} \otimes C_j^{\omega} \otimes M_{0,j}^{\omega,p}, \tag{7.48}$$

where $M_{0,j}^{\omega,p}$ describes the interface between media 0 and j (the index 0 refers to a vacuum sheet), and C_j^{ω} describes layer j:

$$\left.\begin{aligned} M_{j-1,j}^{\omega,p} &= \frac{1}{f^{\omega,+}_{j-1,j-1}} \begin{pmatrix} f^{\omega,+}_{j-1,j} & -f^{\omega,-}_{j-1,j} \\ -f^{\omega,-}_{j-1,j} & f^{\omega,+}_{j-1,j} \end{pmatrix} \\ f^{\omega,\pm}_{j',j} &= \hat{n}_{j'}(\omega)\cos\hat{\theta}_{\omega,j} \pm \hat{n}_j(\omega)\cos\hat{\theta}_{\omega,j'} \\ C_j^{\omega} &= \begin{pmatrix} e^{-\mathrm{i}\hat{k}_{j,z}(\omega)t_j} & 0 \\ 0 & e^{-\mathrm{i}\hat{k}_{j,z}(\omega)t_j} \end{pmatrix} \\ \hat{k}_{j,z}(\omega) &= \frac{\hat{n}_j(\omega)\omega}{c}\cos\hat{\theta}_{\omega,j} \\ \hat{n}_j(\omega)\sin\hat{\theta}_{\omega,j} &= \sin\theta_{\omega,0} \end{aligned}\right\}, \tag{7.49}$$

where $\theta_{\omega,0}$ is the (real) angle of incidence in the vacuum. With (7.47) we relate the forward and backward fields at all interfaces in the multilayer to

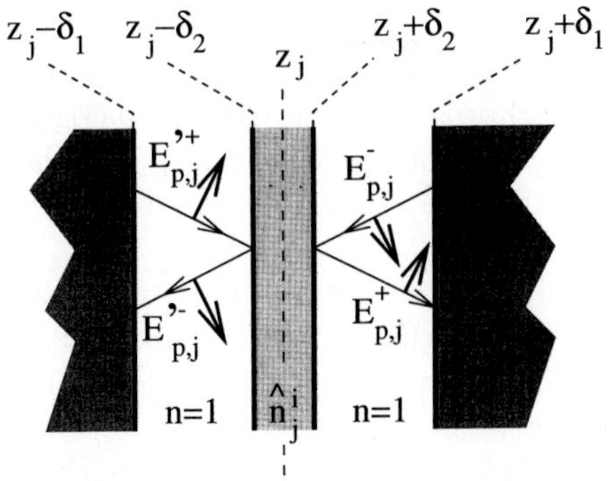

Fig. 7.4. Infinitesimal nonlinear sheet of thickness $2\delta_2$, inside an infinitesimal vacuum sheet of thickness $2\delta_1$, at interface j at $z = z_j$. $\boldsymbol{E}'^{\pm}_{p,j-1}$ represent the p-polarized second-harmonic fields in the vacuum sheet with $z_j - \delta_1 < z_j - \delta_2$ ('just to the left of the nonlinear sheet') and $\boldsymbol{E}^{\pm}_{p,j-1}$ the polarized SH-fields in the vacuum sheet with $z_j + \delta_2 < z < z_j + \delta_1$ ('just to the right of the nonlinear sheet'), $\hat{n}^i_j$ is the refractive index of the interface. The label 2ω has been omitted, for clarity

the incoming fundamental field. Using the ideas of Bloembergen and Pershan [9] we introduce an infinitesimal nonlinear sheet of thickness $2\delta_2$, inside an infinitesimal vacuum sheet of thickness $2\delta_1$ (Fig. 7.4). This sheet has a refractive index $\hat{n}^i_j$, the refractive index of the interface. If we take the $\boldsymbol{x}, \boldsymbol{z}$ plane as the plane of incidence, $k_y = 0$, and we find that the fundamental fields induce a 2ω polarization in the sheet

$$P^{\mathrm{sh}}_{2\omega,k}(\boldsymbol{r}_j, t) = P^{\mathrm{sh}}_{2\omega,k,j} e^{\mathrm{i}[2k_x(\omega)x - 2\omega t]}\,, \tag{7.50}$$

where $\boldsymbol{r}_j$ is in the plane of interface j and

$$P^{\mathrm{sh}}_{2\omega,k,j} = \chi^{(2)}_{klm,j}(2\omega)\varepsilon_{\omega,l,j}\varepsilon_{\omega,m,j} \tag{7.51}$$

with $k, l, m = x, y, z$, and $\chi^{(2)}_{klm,j}(2\omega)$ represents the nonlinear susceptibility tensor elements of interface j. $\varepsilon_{\omega,j}$ is the total fundamental field in the nonlinear sheet, and autosummation is implied. The total fundamental field in the nonlinear sheet is, in general, a sum of forward and backward fields, that are derived from the fields in the vacuum sheet

$$\begin{pmatrix} \varepsilon^{+}_{\omega,p,j} \\ \varepsilon^{-}_{\omega,p,j} \end{pmatrix} = \frac{1}{u^{\omega,+}_j} \begin{pmatrix} u^{\omega,+}_j & -u^{\omega,-}_j \\ -u^{\omega,-}_j & u^{\omega,+}_j \end{pmatrix} \begin{pmatrix} \varepsilon^{+}_{\omega,p,j} \\ \varepsilon^{-}_{\omega,p,j} \end{pmatrix}, \tag{7.52}$$

with $u^{\omega,+}_j = \hat{n}^i_j(\omega)\cos\theta_{\omega,0} \pm \hat{\theta}^i_{\omega,j}$, where $\hat{\theta}^i_{\omega,j}$ is the complex angle of propagation in the nonlinear sheet, and p indicates the polarization. The total fundamental field in the nonlinear sheet j follows from

$$\left.\begin{aligned} \varepsilon_{\omega,p,j} &= (-\varepsilon^{+}_{\omega,p,j} + \varepsilon^{-}_{\omega,p,j})\cos\hat{\theta}^i_{\omega,j} \\ \varepsilon_{\omega,z,j} &= \ (\varepsilon^{+}_{\omega,p,j} + \varepsilon^{-}_{\omega,p,j})\sin\hat{\theta}^i_{\omega,j} \end{aligned}\right\}. \tag{7.53}$$

Substituting these equations into (7.51) gives the total polarization of the sheet. The presence of the source $\boldsymbol{P}^{\mathrm{sh}}_{2\omega,j}$ causes a discontinuity of the SH fields at interface j. This discontinuity is derived from the boundary conditions, accounting for the presence of a (nonlinear) source polarization [4,50]. The result is

$$\left.\begin{aligned} \Delta E_{2\omega,x,j} &= -\tfrac{ik_x(2\omega)}{[\hat{n}^i_j(2\omega)]^2} 4\pi P^{\mathrm{sh}}_{2\omega,z,j} \\ \Delta E_{2\omega,y,j} &= -\tfrac{i2\omega}{c} 4\pi P^{\mathrm{sh}}_{2\omega,x,j} \end{aligned}\right\}, \tag{7.54}$$

where the equations express the difference between the complex amplitudes of the field components in the vacuum sheets on both sides of nonlinear sheet j

$$\left.\begin{aligned} \Delta E_{2\omega,x,j} &= E_{2\omega,x,j} - E'_{2\omega,x,j} \\ \Delta H_{2\omega,y,j} &= H_{2\omega,y,j} - E'_{2\omega,y,j} \end{aligned}\right\}, \tag{7.55}$$

where $\boldsymbol{E}'_{2\omega,j}$ and $\boldsymbol{H}'_{2\omega,j}$ are the electric, respectively, magnetic fields in the vacuum sheet with $z_j - \delta_1 < z < z_j - \delta_2$, and $\boldsymbol{E}_{2\omega,j}$ and $\boldsymbol{H}_{2\omega,j}$ are the fields in the vacuum sheet with $z_j + \delta_2 < z < z_j + \delta_1$ (Fig. 7.4). Separating forward and backward field components gives

$$\left.\begin{aligned} \Delta E_{2\omega,x,j} &= -\cos\theta_{\omega,0}(E^+_{2\omega,p,j} - E^-_{2\omega,p,j}) + \cos\theta_{\omega,0}(E'^+_{2\omega,p,j} - E'^-_{2\omega,p,j}) \\ \Delta H_{2\omega,y,j} &= -(E^+_{2\omega,p,j} + E^-_{2\omega,p,j}) + (E'^+_{2\omega,p,j} - E'^-_{2\omega,p,j}) \end{aligned}\right\}, \tag{7.56}$$

Here, we have used the nonlinear equivalent of Snell's law [4]. Because there is no dispersion in vacuum, we obtain $\theta_{2\omega,0} = \theta_{\omega,0}$. After introducing

$$\kappa_{2\omega,p,j} \equiv \frac{E^+_{2\omega,p,j} - E^-_{2\omega,p,j}}{E^+_{2\omega,p,j} + E^-_{2\omega,p,j}}, \tag{7.57}$$

and an analogous definition for $\kappa'_{2\omega,p,j}$, we arrive at

$$E'^+_{2\omega,p,j} + E'^-_{2\omega,p,j} = \frac{\Delta E_{2\omega,x,j} - \kappa_{2\omega,p,j}\cos\theta_{\omega,0}\cdot\Delta H_{2\omega,y,j}}{(\kappa'_{2\omega,p,j} - \kappa_{2\omega,p,j})\cos\theta_{\omega,0}}. \tag{7.58}$$

The values of κ are easily derived from (7.47)–(7.49) (now for 2ω), once we divide the multilayer into two parts at the interface j, and realize that the system has only outgoing SH fields at both ends. For the second-harmonic wave outgoing at the backside, we find

$$\begin{pmatrix} E^+_{2\omega,p,j} \\ E^-_{2\omega,p,j} \end{pmatrix} = \prod_{q=j}^{m} T^{2\omega,p}_q \begin{pmatrix} E^+_{2\omega,p,m+1} \\ 0 \end{pmatrix}, \tag{7.59}$$

with $T^{2\omega,p}_q$ as in (7.48). From this equation we directly calculate the value of $\kappa_{2\omega,p,j}$. The value of $\kappa'_{2\omega,p,j}$ is obtained after studying the SH wave outgoing at the front side. In the latter case the multilayer is treated in the backward direction, which implies exchanging $E^{+}_{2\omega,p,j}$ and $E^{-}_{2\omega,p,j}$ and replacing z by $-z$ in $C^{2\omega}_{j}$. This analysis also yields the ratio

$$\frac{E^{-}_{2\omega,p,0}}{E'^{+}_{2\omega,p,j} + E'^{-}_{2\omega,p,j}} . \qquad (7.60)$$

As (7.58) gives the denominator of this ratio, we can calculate $E^{-}_{2\omega,p,j}$, the outgoing SH field at the front side, which is exactly the field that we detect in a standard SH reflection geometry. Similar expressions can, of course, be deduced for the s-polarized fields.

7.6 Experimental Considerations

The study of nonlinear-optical effects requires excitation by strong electromagnetic fields. Therefore nonlinear optics has strongly developed since the discovery of the laser, the more because the first lasers happened to be pulsed ones. Considering the fact that for surface nonlinear optics, input intensities are of the order of 1 MW/cm^2, it is obvious that almost any material and, in particular, a thin metallic film, would immediately be destroyed if it would be subjected to such a laser power in a continous mode. Instead, pulsed lasers are utilized, through which one can reach such peak powers, whereas the average power is usually far below 1 W.

The second-harmonic signal (in photons per pulse) in reflection from a surface is given by [4,51]

$$S(2\omega) = \frac{32\pi^3\omega}{\hbar c^3 \epsilon(\omega)\epsilon(2\omega)^{1/2}} |\chi_{\mathrm{s}}^{(2)}|^2 I^2(\omega) A T , \qquad (7.61)$$

for an input laser pulse with intensity $I(\omega)$, cross section A and pulse width T. Here $\chi_{\mathrm{s}}^{(2)}$ indicates the effective surface nonlinear susceptibility, including the Fresnel factors for the transmission and reflection of the fundamental and second-harmonic fields at the interface. For a metal surface, $\chi_{\mathrm{s}}^{(2)}$ is of the order of 10^{-15} esu. A pulsed Nd:YAG laser at 1.06 µm with $I(\omega) = 10$ MW/cm^2, $A = 0.1$ cm^2 and $T = 10$ ns would yield 10^2 photons/pulse or 3000 photons/s for a typical pulse-repetition rate of 30 Hz. Including a typical overall detection efficiency of $\approx 5\%$ (resulting from the several colour filters, polarizers, the monochromator and quantum efficiency of the photomultiplier), this yields a signal of 150 photon counts/s. It is a rather low, but easily detectable signal with modern (gated) photon counting techniques. Note that this means that the nonlinear reflectivity $I(2\omega)/I(\omega) \approx 10^{-15}$. This requires an extremely strong reduction of the fundamental laser intensity on the detection side, which is obtained by employing several colour filters and a monochromator (Fig. 7.5). Though peak powers of 100 MW are readily available with standard YAG lasers, such powers cannot be used, as they would damage

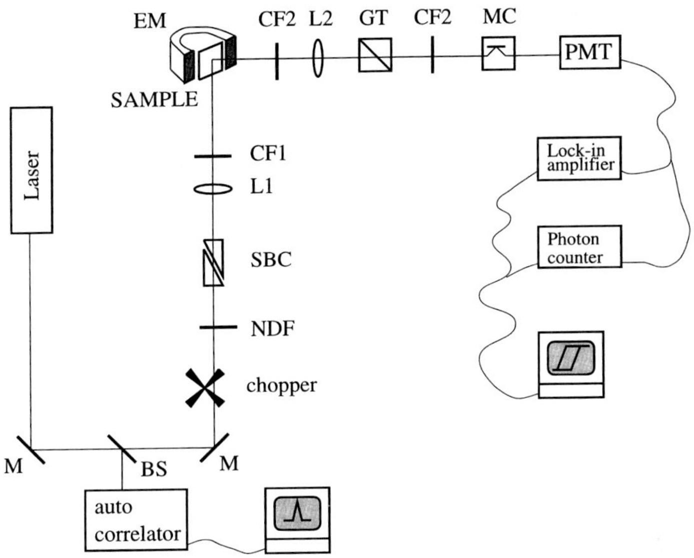

Fig. 7.5. Schematic representation of a MSHG/NOMOKE set-up (*M* mirror, *B* beam splitter, *NDF* neutral density filter, *SBC* Soleil-Babinet compensator, *L* lens, *CF* colour filter, *EM* electromagnet, *GT* Glan Taylor polarizer, *MC* monochromator, *PMT* photomultiplier tube)

the sample under investigation. Recent developments in high repetition rate, mode-locked lasers, have changed and dramatically improved the experimental possibilities [52]. A mode-locked Ti:sapphire laser, with an average output power of 1 W, delivers pulses of about 100 fs at a repetition rate of 82 MHz. For a typical experimental condition (see the examples in Sect. 7.8), one utilizes an average input power of 100 mW focused on an area of 10^{-4} cm^2. This corresponds to a peak intensity of $I = 1.2 \times 10^8$ W/cm^2, or a laser fluence of 12 µJ/cm^2. For the same susceptibility this yields only 1.4×10^{-4} photons per pulse. However, due to the high repetition rate it corresponds to 1.2×10^4 photons per second, that is a *100 times increase* of the signal.

Figure 7.5 exhibits a schematic experimental set-up for the NOMOKE experiments, with a Ti:sapphire laser pumped by an Argon-Ion laser. The output has a Gaussian mode and is continuously tunable from 720 to 1100 nm with two- to three-mirror sets. At 835 nm the laser has an average output power of approximately 1.3 W. The laser fluence is regulated by a set of neutral-density filters. The polarization of the incoming fundamental beam is adjustable with a Soleil-Babinet compensator, that is used as a tunable half-wave plate. A lens is employed to focus the beam on the sample to a spot diameter of about 100 µm. Before reaching the sample, the light is passed through a colour filter (CF1). It removes all light with frequencies below 570 nm, thus eliminating any possible signals at the second-harmonic wave-

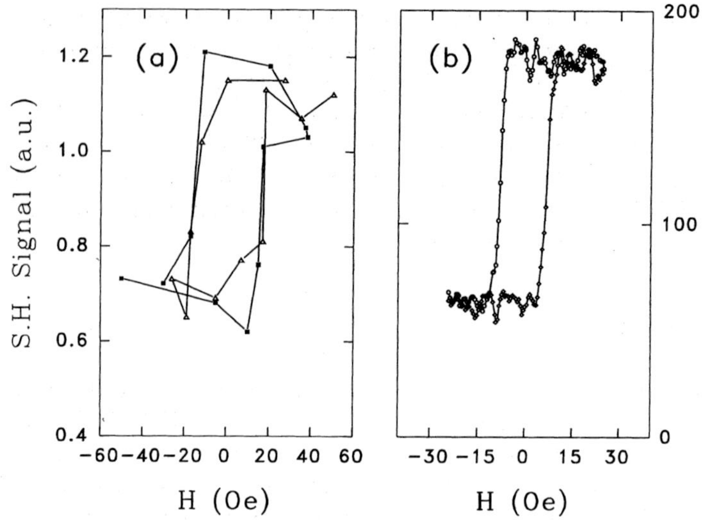

Fig. 7.6. MSHG hysteresis: (**a**) from a CoAu film as measured with a Nd:YAG laser, (**b**) from a CoCu film as measured with a Ti:sapphire laser

length that have not been generated at the sample. The sample is mounted between the poles of an electromagnet. The magnetization $\boldsymbol{M}$ is either parallel to the surface and in the plane of incidence (longitudinal configuration) or parallel to the surface and perpendicular to the plane of incidence (transversal configuration). In case of perpendicular magnetization, the set up can also be modified to a polar geometry. Because of the enormous difference between the intensities of the reflected fundamental and the generated second harmonic, and the fact that, in general, the beams are not spatially separated, proper filtering is of vital importance. Therefore two colour filters (CF2) are used to select the second harmonic light. A UV-transparent lens (L2) focuses the second-harmonic light into the photomultiplier tube (PMT: Thorn EMI, 9789B). A chopper is applied in combination with a lock-in amplifier to improve the signal-to-noise ratio. In the longitudinal configuration an analyser (GT) is used to measure the plane of polarization of the second-harmonic signal. In the transversal configuration the signal is always p-polarized (Sect. 7.4), and thus no analyzer is needed. Due to its poor optical transmission at shorter wavelengths, omitting the analyzer significantly improves the signal-to-noise ratio. For small signals, a (gated) photon counting system can be used.

The advantage of a high repetition rate mode-locked laser versus a low repetition Q-switched laser is clearly demonstrated in Fig. 7.6. In Fig. 7.6 a, an MSHG hysteresis curve is plotted, measured on a 500 Å Co film, covered by 50 Å Au using the frequency-doubled output of a Nd:YAG laser at 532 nm as the fundamental incident radiation [18]. The data taking for such a figure required about two hours in total. In contrast, the hysteresis curve in Fig. 7.6 b was measured within 5 min using a Ti:sapphire laser [19]. A further advantage of the use of a Ti:sapphire laser is its high stability and the fact that it can easily be used for the linear MOKE experiments as well. In addition, this type of laser is easily tunable. Though the standard tunability range of a pulsed Ti:sapphire laser is limited to the range of about 720–1100 nm, this tuning range can be extended by means of different nonlinear optical techniques, like parametric oscillators and amplifiers [4].

7.7 Examples

Though studies of clean surfaces in an ultra-high vacuum environment are of interest and importance for the development of MSHG and NOMOKE, one of the most attractive features of a nonlinear optical technique is the possibility to probe buried interfaces of multilayer systems. A substantial effort is thus devoted to the study of such buried magnetic interfaces. However, to experimentally prove the surface sensitivity of MSHG, in situ studies of clean surfaces with the possibility of surface modification are indispensable. The first of such experiments was reported by Reif et al. [17] who showed a MSHG effect on Fe(001) of about 25%, that disappeared in time due to surface contamination. Though these first results were still rather noisy they stimulated a lot of theoretical and experimental work. We will therefore begin this part with the description of some MSHG results on Fe, starting with the very early results and concluding with some recent ones on very clean Fe surfaces and thin films. The first results on buried magnetic interfaces were obtained for Co/Au films. We will show how interface sensitivity was demonstrated here by varying the number of interfaces. Though both these types of early experiments were very important, clear demonstration of sensitivity for surface magnetism was obtained by simultaneously performed in situ MOKE and MSHG studies on fcc Co films on Cu(001) in UHV [19,54,55]. Comparison of MSHG with linear MOKE results in combination with CO and O_2 dosage experiments unambiguously proved the surface sensitivity of MSHG. It also revealed some very interesting results for extremely thin films (below 5 ML). After this, we will proceed with discussing the Cu/Co/Cu(001) multilayer system. Here, both the Co and the Cu overlayer thickness will be varied, which will give rise to very different results. In particular, we will see how the Cu overlayer-thickness variation on a 10 ML Co film will give rise to an extremely strong oscillatory-like behavior of the MSHG response. This appears to be the first indication of the extreme sensitivity of MSHG for the presence of quantum-well states in thin metallic films. Thereafter nonlinear magnetooptical Kerr rotation experiments will be discussed on a number of Fe and Co systems. Enormous enhancements (up to a factor of 10^3) of the nonlinear Kerr rotation relative to the linear equivalent are oberved and can be explained by the fundamental differences between the two techniques (Sect. 7.4). After this, we will briefly discuss some first results on polycrystalline Ni surfaces and some very interesting applications of MSHG on antiferromagnetic materials. Finally we will discuss very recent MSHG and NOMOKE studies of magnetic quantum-well systems. It will appear that in the presence of quantum wells, the latter totally dominate the nonlinear optical response. The reason for this extreme sensitivity seems to originate in the spatial distribution of the electronic density of states of these quantum-well states, though a real quantitative understanding is still lacking.

7.7.1 MSHG Studies of Fe Surfaces

The first experimental evidence of the influence of surface magnetism on second-harmonic generation were reported on Fe(110) by Reif et al. [17].

The experiment was done on a clean Fe(110) surface in UHV, in the transversal geometry. For the excitation, the frequency-doubled output of a Nd:YAG laser at 532 nm was used, which provided 6 ns pulses at a repetition rate of 20 Hz. Figure 7.7 exhibits the experimental results for the magnetization up and down, as a function of time. For both, input and output, p-polarization was used. Though the signals are quite noisy, a clear difference in the SHG response for magnetization up and down can be observed.

From an analysis of the data an initial magnetic effect of $\chi^{(2),-}/\chi^{(2),+} \approx 0.25$ was found. The decay of this ratio (with a time constant $\tau \approx 30$ min) was attributed to the disappearance of the surface magnetic moments due to the absorption of background CO gas (the base pressure was $\approx 10^{-7}$ Pa).

Though these results also raised some questions (e.g., as to why the nonmagnetic contribution did not seem to change due to CO adsorption, in contrast to other observations [52]), they were very important as they not only demonstrated the magnetic effects in SHG, but also the potential of MSHG to probe (changes in) surface magnetism.

Recently, Vollmer et al. performed a very detailed study of Fe(110), with a mode-locked Ti:sapphire laser [53]. In the wavelength range of 740–840 nm they found that the surface sensitivity for MSHG strongly depended on the input polarization used for the experiment. For s-polarized input, the relative magnetic contribution decreased sharply with a few Langmuirs of oxygen dosage. The relative magnetic effect ρ became 0 at an exposure of 10 L, though the average SHG intensity remained nearly constant. The effect on the p-polarized input was a lot smaller, suggesting a sizeable bulk contribution in this configuration. Interestingly, our recent experiments with p-polarized excitation on an oxidized Fe(100) single-crystal surface in air also revealed a strong effect of $\rho(\mathrm{pp}) \approx 40\%$ [56]. The latter thus may include bulk effects as well. For very clean surfaces Vollmer et al. observed a nonlinear Kerr rotation close to 90°, that strongly decreased due to very small amounts of oxygen $\leq 1\%$. This strong decrease of $\boldsymbol{\Phi}_{\mathrm{K}}^{(2)}$ seems to be in contradiction with our observation of a very large rotation, even for oxygen-covered surface (Sect. 7.7.5). Those results were obtained on a (100) surface and for near

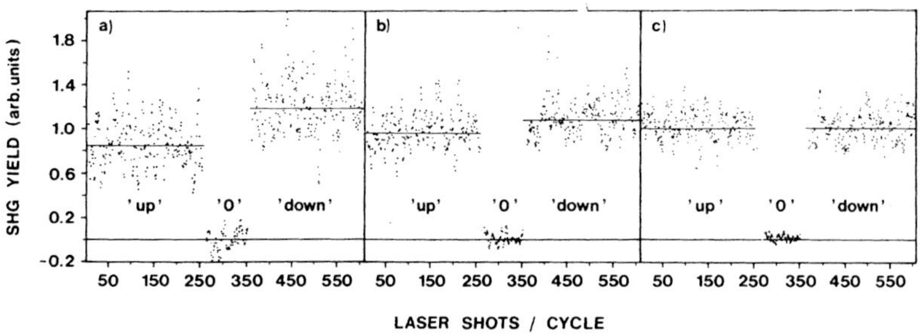

Fig. 7.7. Relative magnetization dependence of the SH signal for three different times elapsed since sample preparation [(**a**)≈ 45 min, (**b**) ≈ 60 min, (**c**) ≤ 180 min]. Shown is, in each panel, an averaged experimental cycle, consisting of 250 pulses with magnetization "up", 100 pulses with no SHG signal, and 250 pulses with magnetization "down". The solid lines represent the average of the respective regions of interest [17]

normal incidence. As Fe also forms magnetic oxides, this interesting system clearly requires more study. For example, preliminary results from a single-crystal $Mn_{0.6}Zn_{0.35}Fe_{2.05}O_4(100)$ surface showed $\rho(\mathrm{sp}) \approx 10\%$ [57]. At this point it may be useful to repeat an earlier remark: "interface sensitivity needs to be verified for any given system" (Sect. 7.3).

Wierenga et al. studied the MSHG response for thin Fe films, epitaxially grown on a Cu(001) single-crystal substrate. Using the 800 nm output of the Ti:sapphire laser, a magnetic effect $\rho(\mathrm{pp}) \approx 10\%$ was observed for a 20 ML bcc Fe film [52].

From a very crude analysis (ignoring phase factors), this would correspond to a ratio $\chi^{(2),-}/\chi^{(2),+} \approx 5\%$. Here one has to realize that this value describes the effective result from two contributing interfaces (the Fe/vacuum and the Fe/Cu interface) that partly compensate each other. For a 2 ML fcc Fe film on Cu(001), it was found that $\rho(\mathrm{pp}) = \rho(\mathrm{sp}) = 0$. This is consistent with the prediction that one can make based on the tensor elements listed in Table 7.1. For a polar configuration, a p-polarized input only gives an s-polarized SHG, so $I_{\mathrm{pp}}(2\omega) = \rho(\mathrm{pp}) = 0$. In the case of an s-polarized output, only the odd (magnetic) component χ_{xyz} is excited. Apart from the fact that its absolute value is small, the absence of an even element prevents the possibility of nonlinear interference, and thus $\rho(\mathrm{sp}) = 0$. Recently, Vollmer et al. have measured the nonlinear polar Kerr rotation of a 2 ML Fe film. At an angle of incidence of 38° they found a nonlinear polar Kerr rotation of $4° \pm 1°$ [58].

7.7.2 MSHG Study of Co/Au Multilayers

Some of the pilot experiments demonstrating the feasibility of MSHG were performed on Co/Au multilayers [54]. In contrast to the Co/Cu systems which will be discussed later, Co/Au multilayers can be studied ex situ because of

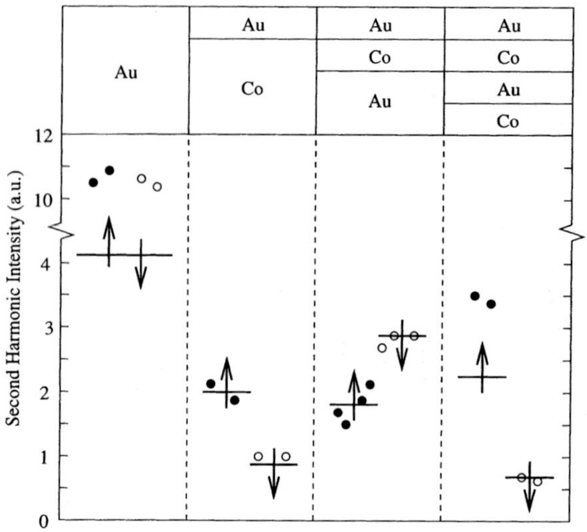

Fig. 7.8. The SH signals from the pure Au film and samples. A: 1 Co/Au interface. B: 2 Co/Au interfaces. C: 3 Co/Au interfaces. ↑: film saturated upwards; ↓: film saturated downwards

the protective inert Au layer. In this sense, they more resemble systems of practical interest. In the following we will discuss some of the earliest results on Co/Au multilayers, in which interface sensitivity was demonstrated by varying the number of interfaces.

The samples consisted of thin films of Co and Au, evaporated at a rate of about 2 Å/s, while the substrate was kept at room temperature. The pressure was 5×10^{-7} Torr while evaporating. The substrates were optical-quality glass plates, cleaned in ethanol and blow-dried with nitrogen. Four systems were studied: Sample A has one Co/Au interface: glass+500 Å Co+50 Å Au, sample B has two Co/Au interfaces: glass+500 Å Au+50 ÅCo+50 Å Au, sample C has three Co/Au-interfaces: glass+50 Å Co+50 Å Au+50 Å Co+50 Å Au, the fourth sample is a 1500 Å thick Au film that served as a reference. Although the films are polycrystalline, they are isotropic on the scale of the laser-beam diameter $\approx 60\,\mathrm{mm}^2$. For the experiments the 532 nm output of a frequency-doubled, Q-switched seeded Nd:YAG laser was used with 8 ns pulse width. The pulse intensity was kept below $7\,\mathrm{mJ/cm}^2$. The samples were mounted in-between the poles of an electromagnet. The applied magnetic field was in the plane of the sample and was perpendicular to the plane of incidence (transversal geometry).

Figure 7.8 presents the results of experiments on the pure Au film, and samples A, B and C at near-normal incidence and pp-like polarization. For each sample the second-harmonic intensities were measured for upward ($\boldsymbol{M}\uparrow$) and downward ($\boldsymbol{M}\downarrow$) saturation. As is to be expected, the signal from the Au film is not dependent on the magnetization. The inversion of the ($\boldsymbol{M}\uparrow$)

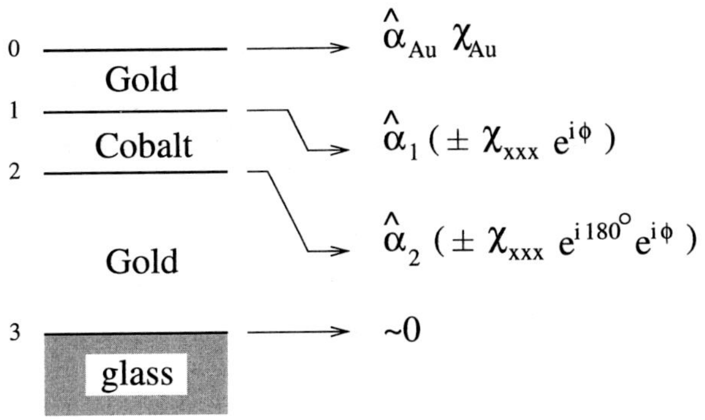

Fig. 7.9. A schematic example of the samples under study. The relative contributions of the interfaces are indicated

and ($\boldsymbol{M}\downarrow$) levels between samples A and B and between samples B and C is most striking. This is an extremely strong indication that the interfaces play a significant role.

Figure 7.9 depicts a typical example of the samples under study. The complex coefficients $\hat{\alpha}_n$ that describe the local electromagnetic fields are calculated with

$$\hat{\alpha}_n = \frac{\hat{A}_n(\omega)^2}{\hat{S}_n(2\omega)}, \tag{7.62}$$

where $\hat{A}_n(\omega)$ is the ratio of amplitudes of the fundamental field at the interface n and the incoming field outside the sample, and $\hat{S}_n(2\omega)$ is the ratio of the generated second-harmonic field at the interface n and the outgoing field. Both $\hat{A}_n(\omega)$ and $\hat{S}_n(2\omega)$ are calculated taking account of the multiple reflection theory (Sect. 7.5).

For the air/Au interface there ought to be no second-harmonic generation for normal incidence and pp-like polarization, since χ_{xxx} is zero for an isotropic nonmagnetic interface. However, experiments on the pure-Au film show significant SH signal in this configuration. This can be explained by the experimental limitations allowing a minimum angle of incidence of 4°. To first-order approximation χ_{xxx}, χ_{xxz} and χ_{zxx} are excited. We analyse the results as though χ_{zxx} were the strongest tensor element. (Note that for their calculations of the nonlinear Kerr rotation, Hübner and Bennemann assumed that $\chi_{zxx} < \chi_{xzx}$). However, similar results can be obtained from any other combination. Sample B contains two interfaces between Co and Au that only differ by a mirror plane parallel to the interface. It was shown that the tensor elements of such interfaces have opposite phases. This relation reduces the number of free parameters in the systems to three: an even tensor element at the air/Au interface: $\chi^{(2)}_{zxx,\mathrm{top}}$, an odd tensor element at each Au/Co interface: $\chi^{(2)}_{xxx,\mathrm{bur}}(\boldsymbol{M})$, and $\phi-$ the phase difference between these

Table 7.2. The α's as calculated from multiple reflection theory. The phase of $\hat{\alpha}_{\mathrm{Au}}$ was chosen to be zero

	$\hat{\alpha}_{\mathrm{Au}}$	$\hat{\alpha}_{\mathrm{t}}$
pure Au	0.543	
sample A	0.223	$0.142\mathrm{e}^{-\mathrm{i}21^0}$
sample B	0.370	$0.112\mathrm{e}^{+\mathrm{i}18^0}$
sample C	0.257	$0.123\mathrm{e}^{-\mathrm{i}32^0}$

two tensor elements. The values of these parameters are found by analyzing the system with the model of Sect. 7.5. The contributions from the Co/glass and the Au/glass interfaces are negligible for samples A and B because of the thick metal layer between the air/Au and metal/glass interfaces. Furthermore the contributions of the comparable Co/quartz interface have been measured exciting from the quartz side. It is concluded that the contributions of the Co/glass interface in sample C are negligible.

The total second-harmonic signal of the whole system is now described by

$$I(\pm M) = |\hat{\alpha}_{Au}\chi_{\mathrm{Au}} \pm \sum_n (-)^{n-1}\hat{\alpha}\chi_{xxx}(\boldsymbol{M})\mathrm{e}^{i\phi}|^2 , \tag{7.63}$$

where ϕ is the phase difference between χ_{Au} and $\chi_{xxx}(\boldsymbol{M})$. The '+' sign refers to $\boldsymbol{M}\uparrow$, the '−' sign to $\boldsymbol{M}\downarrow$.

The values of $\hat{\alpha}_{\mathrm{Au}}$ and $\hat{\alpha}_{\mathrm{t}} = \sum n(-)^{n-1}\hat{\alpha}_n$ for all samples are listed in Table 7.2. We have used the indices of refraction measured by Johnson and Christy [59]. It is found that the calculated values for $\hat{\alpha}_{\mathrm{Au}}$ and $\hat{\alpha}_{\mathrm{t}}$ are not critically depending on the exact thicknesses of the layers. Changing the size of both Co layers in sample C by as much as 10% caused variations of only a few % in the α's.

Results of the calculations for samples A, B and C are indicated by the solid lines in Fig. 7.8. The values for the fit parameters are:

$$\chi^{(2)}_{xxx,\mathrm{bur}}(\boldsymbol{M})/\chi^{(2)}_{xxx,\mathrm{top}}(\boldsymbol{M}) = 0.27 \quad \text{and} \quad \phi = 40^\circ$$

where $\chi^{(2)}_{xxx,\mathrm{top}}$ is used as a scaling parameter. It is noted that the calculated values are also not critically depending on the exact thicknesses of the layers.

The model clearly describes the inversion of $I(\boldsymbol{M}\uparrow)$ and $I(\boldsymbol{M}\downarrow)$ between samples A and B, and between samples B and C. It does explain that sample C gives a larger magnetic field-induced SH effect than sample A. Although surprising and counterintuitive, as C involves deeper interfaces, it is purely the result of multiple reflections in these multilayer systems. The calculated values of the pure-Au film and sample C are in better agreement with the experiment than in the analysis of [54].

In an earlier publication we found $\phi = 88^\circ$, and argued that this value was to be expected, because the time-inversion properties of $\chi(\boldsymbol{M})$ would predict a phase difference of $\phi = 90°$ (Sect. 7.4.1). However, this argument is only valid if dissipation is negligible. A brief look at the dielectric constants shows that this is probably not the case, as the energy of the second-harmonic photons is lying near the interband transition of Au. This implies that the phase difference between odd and even elements is no longer a priori determined, and a deviation from 90° seems likely.

Obviously, the approach of using only one even and one odd nonlinear tensor component is a crude approximation. The Hübner and Bennemann results indicate that even contributions of relatively small tensor elements can be of major importance to the proper description of the nonlinear magnetic effect [33].

Nevertheless, the agreement with the experimental results is quite good. This is due to the fact that for a fixed angle of incidence (as in the experiments described above), the contributions of the various $\chi^{(2)}$ components scale in the same way. The most important conclusion of these multilayer experiments is the fact that for the analysis of such a multilayer system, a proper description of the linear and nonlinear electromagnetic response of the complete system, including multiple reflections, is an essential ingredient. The latter is corroborated by more recent work, where a correlation between preparation conditions, interface roughness and magnetic properties was clearly established by measuring a complete angle of incidence dependence of the MSHG response of a Pt/CoNi/Pt system [94].

7.7.3 MSHG and MOKE Study of Co/Cu(001) Films

Because the Co/Cu(001) system is one of the most thoroughly studied and well-known examples of epitaxially grown thin magnetic films, it can serve as a perfect example to demonstrate the merits and limitations of a new technique such a MSHG. MSHG in combination with MOKE was used to investigate thin Co films of thicknesses between 1 and 20 ML grown on a Cu(001) substrate [19]. The samples were prepared in a UHV system with a base pressure of 5×10^{-11} Torr. Cleaning the Cu(001) substrate consisted of serveral cycles of Ar^+ sputtering followed by annealing at 600°C. The Co films were grown at a rate of approximately 1 ML/min, while the Cu(001) substrate was kept at a temperature of approximately 100°C. It has been shown that the growth of Co on Cu(001) is pseudomorphic [60]. Epitaxial growth was verified for every film by monitoring the (0,0) Medium Energy Electron Diffraction (MEED) spot intensity while depositing (Fig. 7.10). After preparation the film quality was checked by Auger Electron Spectroscopy (AES); all contaminations were below 1 atomic %, except carbon, which was typically 2–3%.

The magnetization was parallel to the [110] direction of the Co film, the easy axis of the film, and perpendicular to the optical plane of the SHG experiments (transversal configuration). The MOKE measurements were done in the longitudinal configuration. For the SHG experiments we used the output

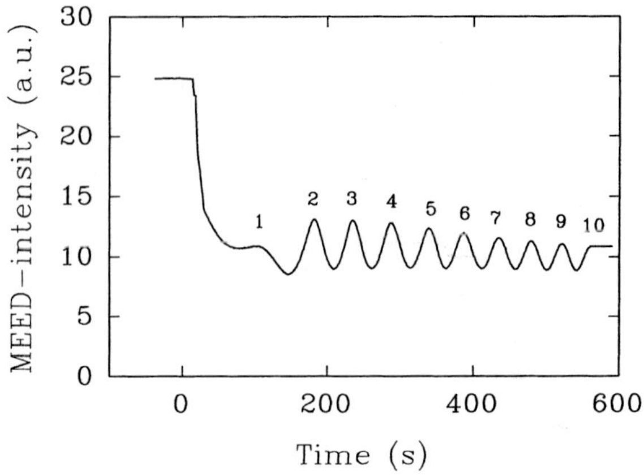

Fig. 7.10. Typical MEED pattern while growing 10 ML of cobalt on Cu(001). A peak in the MEED intensity corresponds to a full monolayer

of a Ti:sapphire laser at 800 nm. The pulse intensity of the incoming beam was about 16 μJ/cm^2. At an angle of incidence of 35°, the pp polarization combination was studied (i.e., both fundamental and second-harmonic beams are polarized in the plane of incidence). Because for the nonlinear case, we are in the transversal geometry, a p-polarized input will only give rise to a p-polarized SHG signal, as directly follows from inspection of Table 7.1. So for the experiments no analyzer was needed, because the s-polarized SH output was indeed negligible.

Figure 7.11 exhibits the relative magnetic effect ρ and the MOKE amplitude M_r as a function of the Co thickness for Co on Cu(001). The difference between the two sets of data is striking: M_r increases almost linearly with Co thickness, whereas ρ hardly changes after 3 ML of Co. The origin of these different behaviors lies in the probing depth of the two techniques. MOKE is a bulk probe, and the total Kerr rotation is proportional to the amount of material, taking absorption losses into account. On the other hand, MSHG is an interface-sensitive probe that is independent of the bulk film thickness. The deviation of the linear thickness dependence of M_r is a result of the absorption in the thin films. Usually this is taken into account by a Lambert-Beer's type of analysis. A more accurate approach is to use a multiple reflection model, similar to Moog et al. [61] and Lissberger et al. [62] For thicknesses below 20 ML, both approaches give an equally good fit. Above 20 ML, the effects of multiple reflections start to play a noticable role, as can be seen in the inset of Fig. 7.11. For the calculation, the bulk indices of refraction as listed in [48] were used. From the close agreement between experiment and calculation, one can conclude that the MOKE results can accurately be described by bulk refractive indices for Co thicknesses above 3 ML.

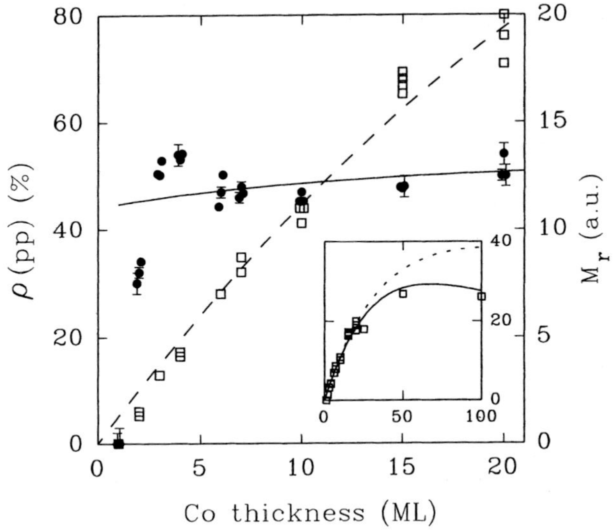

Fig. 7.11. Co film thickness dependence of the relative nonlinear magnetic effect ρ(pp) data and theoretical fit; *open symbols / dashed line* MOKE data and fit. The inset shows the MOKE data and fits for thicker Co films; *dashed line*: Lambert Beer's law, *solid line*: multiple reflection model

To analyze the MSHG results one should realize that there are two interfaces that contribute to the SH signal: the Co/Cu and the vacuum/Co interface. To determine the relative strength of the SH signal from these two interfaces the SH signal from a Co film on Cu(001) as a function of carbon monoxide exposure was measured, as gas adsorption is known to strongly reduce the SHG from metal surfaces [7,10].

Figure 7.12 displays the pp SH intensities of a 7 ML Co film on Cu(001) for the positive and negative magnetizations as a function of CO dosage. One can observe that the signals change until a dosage of 1 Langmuir (1 L= 10^{-6} Torr/s), whereafter they become constant until at least 40 L. The original value $\rho \approx 45\%$ has increased to $\rho \approx 70\%$. Comparable effects have been observed on adsorbing O_2 and for different Co film thicknesses. It was noted that a dosage of a few Langmuirs of O_2 to magnetized Ni(110) and Fe(110) crystals reduces the pp SH signal, generated from a 532 nm Nd:YAG beam, by up to a factor of 20 depending on initial cleanness. It is therefore assumed that the CO eliminates all SH contributions by the Co/vacuum interface. The total response can then be calculated using the multiple reflection model that includes the relevant nonlinear tensor elements and the boundary conditions for nonlinear sources at the interfaces [24]. The line in Fig. 7.11 presents the result of such a fit, that is seen to give a good description of the experimental results above 5 ML. For the analysis only one odd and one even term for each interface was used, like it was done for the Co/Au example. This is, of

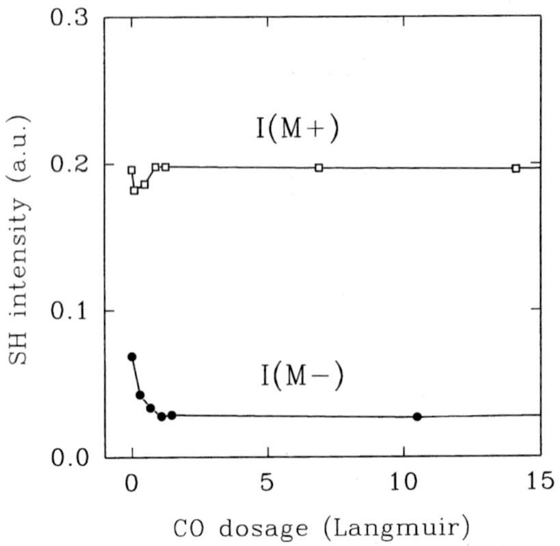

Fig. 7.12. The pp SH intensities of a 7 ML Co film on Cu(001) for positive and negative magnetic saturation as a function of the CO dosage. The signals for 15 L CO remain constant until, at least, 40 L

course, a simplification, as Table 7.1 suggests that for the pp configuration, there are 3 even and 3 odd terms. Including all these terms, however, gives too many fit parameters, without yielding any more physical insight. Since for a fixed angle of incidence these various components can be scaled to each other, using only one even and one odd component is more realistic. For the Co/Cu interface we found the following ratio for these components:

$$\chi^{(2),-}_{\mathrm{Co/Cu}}/\chi^{(2),+}_{\mathrm{Co/Cu}} = 0.89\,. \tag{7.64}$$

This means that the magnetic contributions are of the same order of magnitude as the nonmagnetic ones. This accounts for the large relative effects ($\rho > 40\%$). Comparing the even and odd elements at the vacuum/Co and Co/Cu interfaces gives

$$\chi^{(2),-}_{\mathrm{vac/Co}}/\chi^{(2),-}_{\mathrm{Co/Cu}} = -1.4\,, \tag{7.65}$$

and

$$\chi^{(2),+}_{\mathrm{vac/Co}}/\chi^{(2),+}_{\mathrm{Co/Cu}} = -1.9\,. \tag{7.66}$$

This indicates that at the vacuum/Co and Co/Cu interfaces the amplitudes of both even and odd tensor elements are comparable. These ratios are found for any combination of one even and one odd tensor element. Recently, these experiments were repeated [55]. The results are in perfect agreement with those reported here, except for the fact that the background CO contamination was somewhat lower. This resulted in a larger Co/vacuum contribution

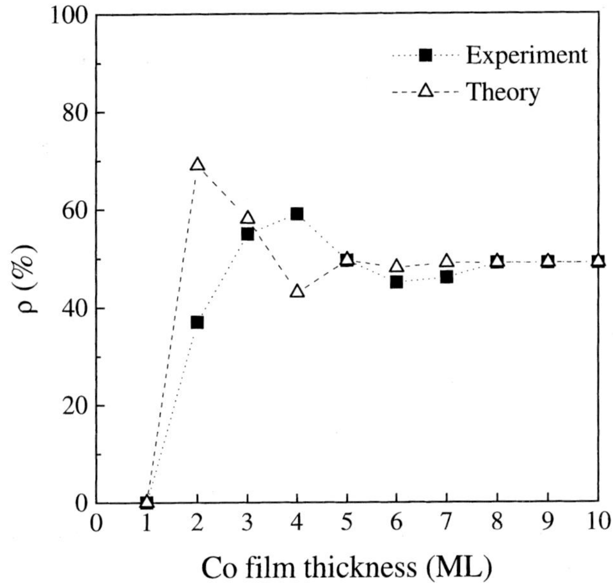

Fig. 7.13. Comparison between measured (*downward triangle*) and calculated (*upward triangle*) relative magnetic effect ρ for Co/Cu(001) as a function of Co thickness

and consequently a smaller total magnetic effect, in accordance with the analysis given here that shows that the two interface contributions are out of phase.

The results described above show that above 5 ML, both the MOKE and MSHG response can accurately be described by a multiple reflection model with the bulk indices of refraction. The strong thickness dependence of the MSHG results below 5 ML can, of course, not be described by such an approach, as it takes at least 2 ML to define an interface.

In order to understand the microscopic origin of those results, electronic structure calculations, using a linear Korringa-Kohn-Rostoker (LKKR) method, have been performed. With this method the magnetic moment per Co layer can be calculated as a function of the Co layer thickness. Preliminary results indicate that the magnetic moments of such a Co over-layer indeed changes for the first few monolayers [63]. If we make the very crude assumption that the multiple reflection model works even between 1 and 5 ML, and adopting the same bulk optical constants as before, one can then calculate the MSHG response from our Co/Cu(001) system, using the calculated magnetic moment distribution and assuming $\chi^{(2)}(M)$ to be proportional to the magnetic moment in the layer. The results of such a calculation are displayed in Fig. 7.13. Despite the crude approximations, Fig. 7.13 shows a very good qualitative agreement with the experimental results plotted in Fig. 7.11. These results reveal that, in accordance with the theory of Sect. 7.4, MSHG can indeed be used to probe the magnetic moments at interfaces.

In a more recent experiment, the Kirschner group has shown that MSHG is even capable to detect the small ($\sim$ 10%) changes of the Co moments at step edges [95] for epitaxially grown Co on Cu(001).

7.7.4 MSHG and MOKE Studies of Cu/Co/Cu(001) Films

An important application of MSHG is the study of magnetic multilayers. For such a study to be successful, one has to be able to distinguish the different contributions of the various interfaces. To investigate these possibilities Cu/Co/Cu(001) structures have been used as model system [64]. MSHG and MOKE experiments were done on a 10 ML Cu/Co/Cu(001) system, with varying Co thicknesses. The samples were prepared in the same way as before (Sect. 7.7.3). The Cu overlayers were grown at 1 ML/min, while keeping the substrate at 70°C. It has been shown that the growth of Cu on Co(001) is pseudomorphic as well [60]. For the MSHG and MOKE experiments the same system as described in Sect. 7.7.3 has been used.

Figure 7.14 shows the experimental Co thickness dependence of ρ for Cu/Co/Cu(001) plus the theoretical prediction based on the results of Fig. 7.11. For these calculations, the solutions for the Co/Cu interface were utilized in the Cu/Co/Cu(001) trilayers to calculate ρ(pp) of these systems. The only fitting parameter was the nonmagnetic nonlinear contribution of the Cu/vacuum interface, that was not present before. The agreement between experimental and predicted dependence is quite good. More importantly,

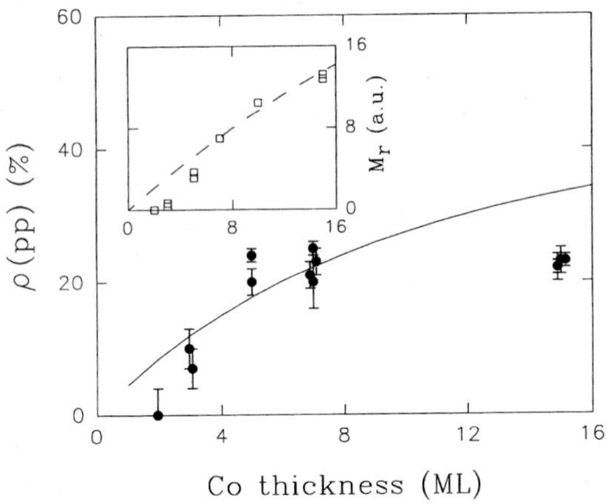

Fig. 7.14. Experimental and calculated Co film-thickness dependence of ρ(pp) in Cu/Co/Cu(001); *dots* experimental data, *solid line* calculation based on fitting parameters of Fig. 7.11. The inset shows the experimental and calculated MOKE amplitude

Fig. 7.14 suggests that such a multiple reflection model is *required* in order to understand the MSHG response of such a multilayer at all. The Cu/Co/Cu trilayer is very symmetric and because of mirror symmetry $\chi^{(2)}_{\mathrm{Co/Cu}} = -\chi^{(2)}_{\mathrm{Cu/Co}}$. Thus, without taking the multiple reflections into account, which yield different local fields at the two interfaces, their total response and thus ρ would have been zero. (Here we ignore possible differences that may exist between the absolute values of the two interface contributions, due to differences in the local structure). Figure 7.14 also presents the MOKE data for this system, with a theoretical fit including multiple reflections directly calculated from the Co/Cu results, using no adjustable parameters.

These results reveal that, in principle, the MSHG response of a multilayer system can be understood, in which several interfaces do contribute to the nonlinear optical signal. The next step will be to increase the number of layers up towards systems of practical interest, consisting of 30 to 50 layers. Another interesting road to explore is whether it would be possible to "tune" to a specific interface, using either the wavelength or the angle of incidence as the controlling parameter. For the bulk response using MOKE, Penissard et al. have demonstrated this beautifully. By changing the wavelength, they could select a specific Co layer of a Co/Au multilayer sample [65]. In principle, this must also be possible for the interface contributions, but so far this is an unexplored area.

7.7.5 MSHG Study of Cu/Co/Cu(001): The Appearance of QW Oscillations

One of the exciting recent discoveries in magnetic multilayer systems is the obervation of the oscillatory exchange coupling between two ferromagnetic films as a function of a nonmagnetic spacer layer [80,81]. The subsequent discovery of the existence of quantum-well states in such very thin metallic films, which can be responsible for this observed coupling, has increased the interest in these systems [68–70].

In contrast to semiconductor multilayers, where quantum wells are the results of differences in the bandgaps of alternating materials, for metallic thin films, quantum wells arise as a result of the differences in the density of states (DOS) between neighboring layers. In a simple picture, these differences in the density of states give rise to a potential step at the interfaces. Electrons impinging on such an interface will partly be reflected, and depending on the thickness of the film, a standing electronic wave can be obtained, i.e., the system acts as a kind of Fabry-Perot interferometer for electrons [66]. This is, in principle, true for any combination of metallic films, or for a metallic film on a nonmetallic substrate [67]. In the case of a magnetic-nonmagnetic interface, the differences in DOS, and thus the potential step, will be spin-dependent, due to the spin dependence of the band structure of the magnetic film. This will lead to a spin-dependent scattering at the interfaces, and thus to a spin-dependent quantization condition. As a result, even in a nonmagnetic spacer

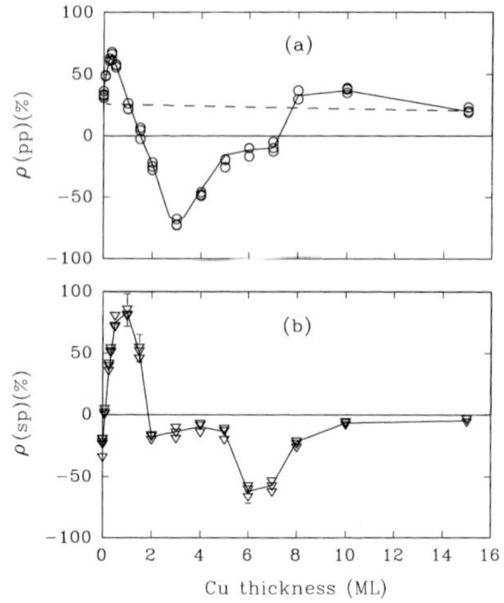

Fig. 7.15. a, b The relative magnetic effects ρ(pp) and ρ(sp) as a function of Cu coverage on a 10 ML Co film on Cu(001); (**a**) *circles* ρ(pp), *dashed line* result of model calculation including optical interference but neglecting quantum-well oscillations; (**b**) *triangles* ρ(sp). The *solid lines* are guides to the eye

layer, the quantum-well states may become spin-polarized. A good example is the Co/Cu/Co system. The DOS for Cu is very similar to the minority spin DOS of Co. However, the majority DOS of Co is very different. Therefore, the minority spin will hardly "see" the interface whereas the majority spins will be more strongly scattered resulting in a spin-dependent confinement in the Cu layer.

Recent photoemission studies of thin Cu layers on a Co(001) substrate have indeed been able to identify spin-polarized quantum-well states in these Cu films. As such confinement will lead to an oscillating density of states, optical transitions and, in particular, the MSHG response should also be affected by the presence of such quantum-well states. To further investigate this hypothesis, the Cu/Co(001) system has been studied, where quantum-well states have been clearly identified [68–70]. As a substrate for the experiments 10 ML Co grown on Cu(001) was used.

Figure 7.15 exhibits the MSHG results for this Cu/Co/Cu system, as a function of Cu coverage. One oberves strong oscillatory-like variations in ρ(pp) with amplitudes up to $\rho = 75\%$. From the simultaneously measured MOKE amplitude it is concluded that above 2 ML of Cu, the MOKE results are consistent with a normal multiple reflection analysis, using bulk refractive indices and a *constant* Kerr rotation. In the MSHG data of Fig. 7.15, two interfering periods can be distinguished: one of about 5 ML modulated by a substructure of 2–3 ML. Interestingly, the 5 ML periodicity has been observed by photoemission [68–70], whereas such a short-period oscillation has been observed by Johnson et al. [71] in MOKE experiments on the Co/Cu/Co system. The extreme sensitivity of MSHG for these oscillations may be understood from its surface and interface specificity, in combination with results of recent calculations that indicate that for very thin films these quantum-well states primarily affect the DOS at the interfaces [72]. Furthermore, Carbone et al. [70] have shown that the induced spin polarization in the nonmagnetic material is primarily located at the interfaces.

Assuming a bulk-like electronic structure of the Cu films, the Cu thickness dependence of $\rho(\mathrm{pp})$ can be calculated from the multiple reflection model using the values of the various tensor elements, as derived from the experiments. The dashed line in Fig. 7.15 a represents the result of such a calculation, clearly indicating that classical interference by no means explains the oberved strong oscillations. The calculations of the local magnetic moments (like was done for Co/Cu(001)) showed a similarly negative result: the induced moments in Cu and their variation with thickness are absolutely too small to be able to explain even qualitatively the observed oscillations [63]. The conclusion must be that the observed oscillatory response is due to the effect that the quantum-well states have on the nonmagnetic $\chi^{(2)}_{\mathrm{Cu}}$ response. This last conclusion is supported by recent results from R. Vollmer on the same system [98]. Detailed investigations of Cu/Fe and Au/Co systems indicate that quantum well effects in the MSHG response are not only very pronounced but require detailed modeling for a correct interpretation [98–100].

7.7.6 NOMOKE Rotation from Fe

These very first experiments on the nonlinear Kerr rotation from thin, centrosymmetric magnetic films, were stimulated by the prediction of large enhancements of $\boldsymbol{\Phi}^{(2)}_{\mathrm{K}}$ by Pustogowa and Hübner [20]. To enable a direct comparison with their theoretical results, experiments were performed on Fe thin films and single crystals. All experiments were done ex situ, and therefore required protective cover layers. The first sample consisted of a thin Fe film (thickness: 2 nm), covered with a 2 nm Cr film deposited by rf diode and dc magnetron sputtering, respectively. As a substrate a (100) silicon wafer was employed, with a thermal oxide layer of about 525 nm. The substrate was on a rotating table, which moved with a velocity of 0.96 m/min and 3.97 m/min underneath the Fe and Cr targets, respectively. Both targets were equiped with screens for getting uniformity of the layer thickness better than 1%. The base pressure was 2×10^{-7} Torr and the Ar pressure was 5 mTorr. For the second-harmonic experiments the 770 nm output of a mode-locked (80 MHz) Ti:sapphire laser was applied. The pulse width was 70 fs and the input power was 100 mW focused on a spot diameter of 100 μm. The experiments were done in the longitudinal configuration, i.e., the magnetization $\boldsymbol{M}$ was in the plane of the sample and in the optical plane of incidence (see inset of Fig. 7.16). Figure 7.16 plots the polarization dependence of the SH signal for an s-polarized input at an angle of incidence of 45° and for $\boldsymbol{M}$ along $\hat{x}$ and

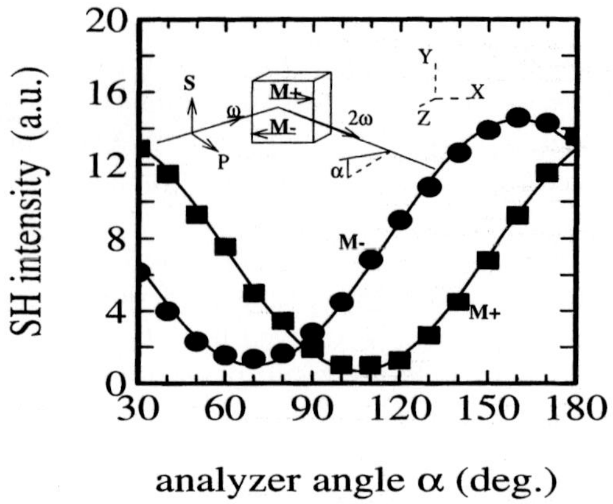

Fig. 7.16. Output-polarization dependence of SHG reflection from an Fe/Cr multilayer, for s-polarized input. *Filled squares*: $\boldsymbol{M}\|\hat{x}$. *Crosses*: $\boldsymbol{M}\|-\hat{x}$. The inset shows the experimental configuration

$-\hat{x}$, respectively. The difference between the two minima of the curves corresponds to $2\Phi_{\mathrm{K}}^{(2)} \approx 34°$, i.e., a nonlinear Kerr rotation $2\Phi_{\mathrm{K}}^{(2)}$ of 17°. In this geometry a linear Kerr angle of 0.03° was measured for the same Ti:sapphire input beam. These observations correspond to an enhancement of almost *three orders* of magnitude for the nonlinear Kerr rotation! The small linear rotation can be compared with the bulk Fe value of 0.1°. These observations can be understood by an analysis based on the theoretical considerations of Sect. 7.4.2, including the multiple reflections, as discussed in Sect. 7.6.

The nonlinear Kerr angle can be expressed in the s and p components of the reflected SH field, denoted by $E_{\mathrm{s}}(2\omega)$ and $E_{\mathrm{p}}(2\omega)$, respectively. The relevant tensor components and their appearance for s- and p-polarized input have been compiled in Table 7.3. Defining $R \equiv \mathrm{Re}\{E_{\mathrm{s}}(2\omega)/E_{\mathrm{p}}(2\omega)\}$, $I \equiv \mathrm{Im}\{E_{\mathrm{s}}(2\omega)/E_{\mathrm{p}}(2\omega)\}$, and $A^2 = R^2 + I^2$ we get

$$\boldsymbol{\Phi}_{\mathrm{K}}^{(2)} = \frac{1}{2}\arctan\left(\frac{2R}{(1-A^2)}\right) + \phi_0\,, \tag{7.67}$$

with $\phi_0 = 0$ for $A^2 \leq 1$, $\phi_0 = 90°$ for $A^2 > 1$ and $R \geq 0$, and $\phi_0 = -90°$ for $A^2 > 1$ and $R < 0$. (7.67) is completely analogous to (7.26), as derived by Pustogowa et al. It is easily verified that in the limit $A \ll 1$, (7.63) reduces to $\boldsymbol{\Phi}_{\mathrm{K}}^{(2)} = R$. However, the nonlinear case generally is far from this limit, since $\boldsymbol{\Phi}_{\mathrm{K}}^{(2)}$ can become as large as 90°.

Inspection of Table 7.3 reveals that the s-input configuration is particularly simple, with only one even, χ_{zyy}^{+} and one odd, χ_{yyy}^{-}, contributing element per interface. For normal incidence, only the χ_{yyy}^{-} contribution survives. From $E_{\mathrm{p}}(2\omega) = 0$ we then find $\boldsymbol{\Phi}_{\mathrm{K}}^{(2)} = \pm 90°$. Moving away from normal incidence,

Table 7.3. The nonzero elements of the SH susceptibility tensor for an isotropic surface in the longitudinal configuration ($\boldsymbol{M}//\hat{x}$). The two columns list the elements that are even and odd in the magnetization, respectively. The occurence of the elements in the p- and s-input configuration is indicated within brackets

χ^+		χ^-	
zxx	(p)	yxx	(p)
zyy	(s)	yyy	(s)
zzz	(p)	yzz	(p)
$xzx = xxz$	(p)	$zyz = zzy$	
$yzy = yyz$		$xyx = xxy$	

the ratio $|E_\mathrm{s}(2\omega)/E_\mathrm{p}(2\omega)|$, and as a consequence $\boldsymbol{\Phi}_\mathrm{K}^{(2)}$, decrease so that $\boldsymbol{\Phi}_\mathrm{K}^{(2)}$ is tunable over a wide range while scanning the angle of incidence.

A similar argument for a large $\boldsymbol{\Phi}_\mathrm{K}^{(2)}$ close to normal incidence is found for a p-polarized input configuration. Table 7.3 suggests that close to normal incidence, the dominant even tensor components have one z index, and thus vanish at normal incidence, whereas the dominant odd tensor element, χ^-_{yxx}, is finite at normal incidence. (This seems in contradiction with [Ref. [9], Fig. 3]. However, there one has only considered χ^-_{yxx} that also vanishes at normal incidence).

Figure 7.17 displays the observed nonlinear Kerr rotation for both s- and p-polarized input, as a function of the angle of incidence. A small in-plane anisotropy was observed in the experimental results, that is possibly induced by the sputtering process. The data plotted in Fig. 7.17 have been averaged over this azimuthal anisotropy. The solid curves are theoretical fits, based on the multiple reflection model, with the unknown interface-tensor elements of Table 7.3 as parameters and the following assumptions. Only Fe is expected to contribute to the magnetic (odd) nonlinear susceptibility, because the Cr film is antiferromagnetic. Furthermore, the top Cr layer will be oxidized so that the major contribution to the nonmagnetic, nonlinear susceptibility is also expected to originate from the Fe film. Therefore, effective SH susceptibilities are assigned to the Fe and Cr layers together. It was verified that the actual position within the Cr/Fe top layer did not significantly change the results of the fits. In addition to these SH sources at the top layer, a nonmagnetic SH source at the Si/oxide interface was incorporated. The bulk optical constants of the metals and silicon were obtained from [59] and [73], respectively. Because of the limited number of parameters involved in the s-polarization configuration (χ^-_{yyy} and χ^+_{zyy} at the top layer, and χ^+_{zyy} at the Si/oxide interface) one finds a unique fit to these experimental data points. The fit in Fig. 7.17 includes a relative maximum of $|\boldsymbol{\Phi}_\mathrm{K}^{(2)}|$ near $\boldsymbol{\Theta}_\mathrm{i} \approx 65°$, that is due to an enhancement effect through multiple reflections in the thick

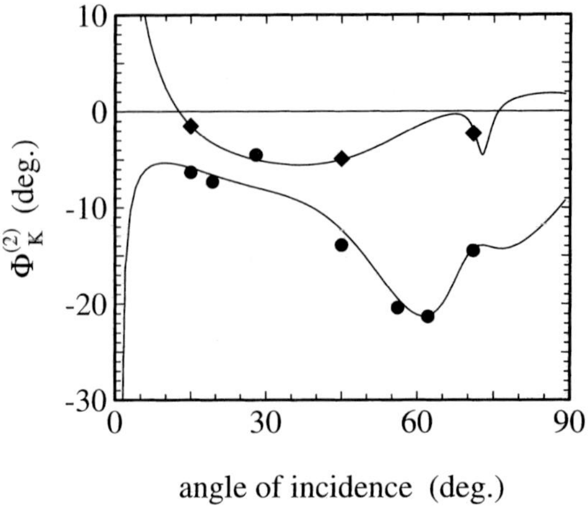

Fig. 7.17. Nonlinear Kerr rotation $\Phi_K^{(2)}$ for an Fe/Cr multilayer as a function of the angle of incidence. *Dots*, s-input polarization; *diamonds*, p-input polarization. The curves are theoretical fits

silicon oxide layer. Similar, but smaller, enhancement factors due to the substrate are also known for the linear Kerr angle [74].

For p-polarization, several combinations of tensor elements give satisfying fits. Fig. 7.17 offers one of such solutions, obtained by choosing fixed values for the relative phase factors and fitting the absolute values of the tensor components. At 45° angle of incidence, one experimentally finds for the p input $\boldsymbol{\Phi}_K^{(2)} = 4.9°$.

These results show once more that for a proper understanding and analysis of the nonlinear response, the total electromagnetic response, including linear and nonlinear multiple reflections, has to be considered. The importance of such substrate effects can be illustrated in the following way. From the thinfilm analysis one can simply calculate the response of an Fe surface, by letting the thickness of the Fe film tend to infinity. Similarly, one can vary the thickness of the silicon oxide layer for the thinfilm structure.

Figure 7.18 presents the results of such simulations. For the thin oxide layer, the enhancement effect around $\theta_i = 60°$ has totally vanished, which demonstrates the importance of the multiple reflections and of the local-field effects and thus the role of the substrate. Without going through a complete calculation, one can simply get an intuitive idea about this role. Let us compare two systems: (i) the surface of a bulk magnetic medium with refractive index $n = n_B$, and (ii) a thin film of the same material on a substrate with $n = n_S$, and neglecting the SH contribution from the film-substrate interface. In the limit of vanishing film thickness and $\sin\theta \ll n$ and using the standard Fresnel reflection theory, one can easily show that for

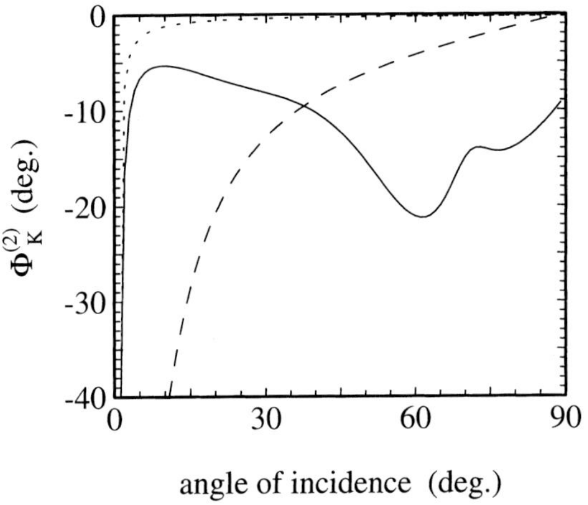

Fig. 7.18. Nonlinear Kerr rotation $\Phi_{\mathrm{K}}^{(2)}$ for an Fe/Cr multilayer and a bulk Fe surface for s-polarized input as a function of the angle of incidence. The *bold solid line* is the theoretical fit of Fig. 7.17. The dashed line is a simulation for $\Phi_{\mathrm{K}}^{(2)}$ of a clean Fe surface, based on the multilayer fit. The *dotted line* is a simulation for $\Phi_{\mathrm{K}}^{(2)}$ for a Fe/Cr multilayer on a silicon wafer with a 10 nm oxide

s-polarized input the ratio of the reflected SH fields due to the odd (E_{s}) and even (E_{p}) terms is given by

$$\frac{E_{\mathrm{s}}(2\omega)}{E_{\mathrm{p}}(2\omega)}\alpha\frac{1+R_{\mathrm{s}}(2\omega)}{1+R_{\mathrm{p}}(2\omega)}=\frac{1}{n(2\omega)}, \tag{7.68}$$

with n equal to n_{B} or n_{S}. Similar relations hold for p-polarized input. Equation (7.68) shows that the MSHG effects will be more pronounced for a thin film on a low-refractive-index substrate ($n_{\mathrm{S}} \ll n_{\mathrm{B}}$) than for a bulk material.

For the bulk surface, Fig. 7.18 indicates a smooth variation of $\boldsymbol{\Phi}_{\mathrm{K}}^{(2)}$ as a function of the angle of incidence. In accordance with (7.67), Fig. 7.18 shows that $\boldsymbol{\Phi}_{\mathrm{K}}^{(2)}$ can be tuned at will between 0° and 90°, by varying the angle of incidence. This is a direct result of the fact that at near-normal incidence the even contribution $\chi_{xyy}^{(2),+}$ vanishes, whereas the odd magnetic term $\chi_{xyy}^{(2),-}$ gives a finite contribution.

The prediction of this large tunable nonlinear Kerr rotation was confirmed by experiments on single-crystalline Fe whiskers [75]. The (100) surfaces of whiskers were capped by MBE-grown Au and Cr layers, while some experiments were also performed on uncapped, oxidized Fe whiskers. The experiments were done using the 833 mm output of the Ti:sapphire laser. Figure 7.19 illustrates the measured angular dependence of the nonlinear Kerr rotation in the longitudinal configuration and for s-polarised excitation.

In this figure the experimental results can be seen for an Fe sample with a Cr top layer, a Fe sample with an Au top layer and for an uncapped

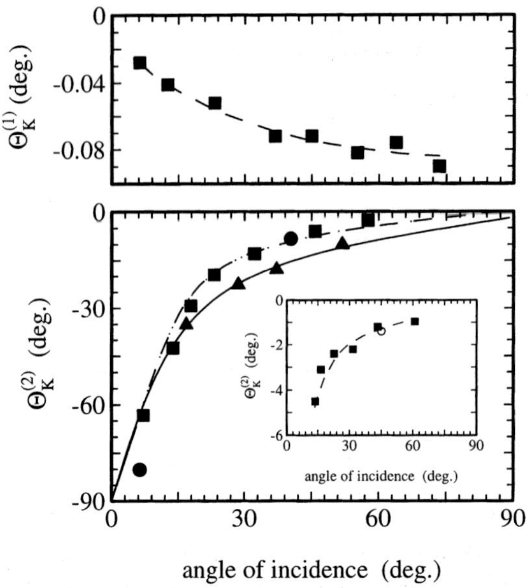

Fig. 7.19. Linear and nonlinear Kerr rotation on a Fe surface for s-polarized input in the longitudinal configuration as a function of the angle of incidence. *Triangles*, Fe/Cr; *dots*, Fe/Au; *squares*, uncapped Fe. The *solid line* is a theoretical fit for the Fe/Cr sample. The *dotted line* is the theoretical prediction for a clean Fe surface from [23]. The inset shows $\Phi_K^{(2)}$ for p-polarized input. The *open circle* is the calculation from [20]

oxidyzed Fe sample. One finds a maximum Kerr angle of 80° at an angle of incidence of 6°. This corresponds to an enhancement of more than a factor of 10^3, compared with the value for the linear Kerr rotation of 0.03° as obtained in the same experimental set-up, using the fundamental incident beam. The solid line in Fig. 7.19 is a theoretical fit for the Fe/Cr sample based on (7.67) and a model for SHG from interfaces [24]. Furthermore, the experimental results are in quite good agreement with the predicted behavior from Fig. 7.18.

The inset of Fig. 7.19 depicts the experimental results for the longitudinal configuration with p-polarized incident light. In Table 7.3 it can be seen that now one is dealing with two odd tensor components, both of which give rise to s-polarized SHG and three even components which produce p-polarized SHG. Therefore, it is to be expected that for the same incidence angles the ratio $|E_S(2\omega)/E_p(2\omega)|$ and thus $\boldsymbol{\Phi}_K^{(2)}$ will be smaller compared to the s-input configuration. Near normal incidence the influence of the components χ_{ijk} with "z" for i, j or k will again be small. Since all even tensor components have at least one z index, the p-polarized SHG will then be small. The s-polarized SHG, which is now dominated by χ_{yxx}^{-}, is finite. The result is that for small angles of incidence the Kerr rotation will again increase. This is clearly con-

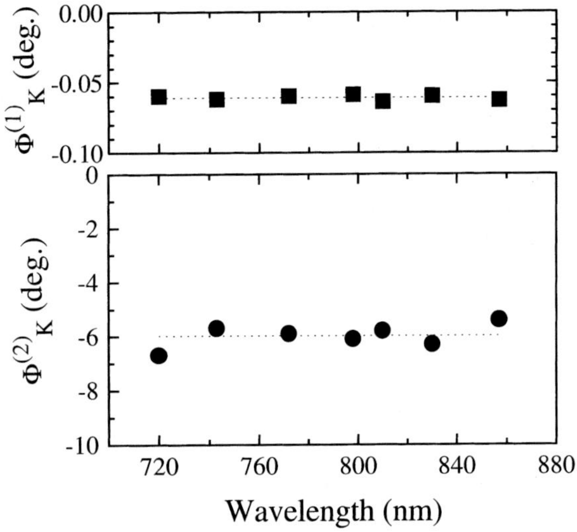

Fig. 7.20. Experimental wavelength dependence of the nonlinear (*dots*) and linear (*squares*) Kerr angle for the s-input longitudinal configuration at an incident angle of 45°

firmed by the experiments, but seems in contradiction with the theoretical θ_i dependence [20]. However, there one has only considered the odd component χ^-_{yzz} that also vanishes near normal incidence. At an angle of incidence of 45° the experimental and theoretical configurations should be equivalent. Here one finds $\boldsymbol{\Phi}_K^{(2)} = 1.2°$ in excellent agreement with the theoretical prediction of 1.4° from Pustogowa and Hübner [20]. In a recent paper, these researchers have included the proper Fresnel dependence in their expression for $\boldsymbol{\Phi}_K^{(2)}$, showing a good agrement with the experimental findings [76].

In the longitudinal configuration the wavelength dependence of both $\boldsymbol{\Phi}_K^{(1)}$ and $\boldsymbol{\Phi}_K^{(2)}$ at an angle of incidence of 45° have also been measured (Fig. 7.20). The average value for $\boldsymbol{\Phi}_K^{(1)} = 0.062°$ is in excellent agreement with the bulk Fe value of 0.061° [77]. Figure 7.20 shows that in the tuning range of the Ti:sapphire laser the Kerr rotations are constant, in agreement with the theoretical prediction of Pustogowa et al. [20]. From their calculations it follows that for $\boldsymbol{\Phi}_K^{(2)}$ a strong wavelength dependence is expected above 900 nm and below 600 nm. Future experiments in these wavelength ranges will be interesting.

7.7.7 MSHG Study of Ni

Though the first theoretical calculations for the magnetic $\chi^{(2)}$ components were done for Ni [16], so far there has been little experimental MSHG work on this material. Böhmer et al. published the results of an MSHG study of polycrystalline nickel surfaces [78]. The experiments were done in air, using picosecond laser pulses at 600 nm for the excitation. In the longitudinal configuration, they observed a nonlinear Kerr rotation of 4°. In the transversal configuration they found anisotropies of 13 and 19%, for s $\rightarrow$ p and p $\rightarrow$ p polarization combinations, respectively, by measuring the azimuthal magnetization dependence. From this azimuthal dependence, Böhmer et al. concluded that the (antiferromagnetic) nickel-oxide layer does not affect the results. From a comparison of their measured MOKE and MSHG hysteresis loops, there may be an indication of a slightly different behavior for the two cases. Though the accuracy was not good enough to draw definite conclusions from these results, this is an interesting point for future studies. Expecially for the surface of a bulk crystal, the surface-domain structures may be different from that of the bulk, due to differences in required magnetic energy to order these domains. Finally, it should be mentioned that preliminary MSHG studies on a clean Ni surface by Wierenga et al. were unsuccessful [52]. In contrast to the results discussed above, these experiments were performed using the frequency-doubled output of a Nd:YAG laser at 532 nm as the fundamental excitation. Comparing the two wavelengths used (600 and 532 nm, respectively) with the theoretical work of Hübner et al. [16], one finds that the magnetic component $\chi^{(2)}_{xzz}$ shows a maximum near 600 nm, and a zero near 532 nm. As this may be accidental, a more complete spectroscopic MSHG study is clearly desirable. A first attempt in this direction has recently been made [96]. A spectroscopic MSHG study (around 2.7 eV) of a Ni(100) surface in UHV shows strong resonant effects that are extremely sensitive to oxygen contamination. The results can be described by a simple model that takes into account a Ni surface state and spin-split bulk derived states.

7.7.8 MSHG Study of Cr_2O_3

The study of Cr_2O_3, performed by Fiebig et al. [35], suggested a very new and powerful application of the nonlinear Kerr effect, by visualizing the domain structure in an antiferromagnetic material. Contrary to the other examples, bulk effects play a dominating role here. The MSHG spectrum of Cr_2O_3 was studied in the region of 1.7–2.9 eV as a function of temperature. Cr_2O_3 is antiferromagnetic below the Néel temperature $T_N = 307.5$ K. The magnetic point-group symmetry is $\underline{\bar{3}m}$, that gives rise to a bulk electric dipole contribution $\chi^{(2)}$, that changes sign under the time-reversal operation. Above T_N, only (much weaker) magnetic-dipole contributions are allowed. The different contributions can be separated spectroscopically and by their temperature dependence. In the spectral region of spin-forbidden transitions, there was no polarization dependence observed. However, in the region that includes spin-allowed transitions (between 2 and 3 eV), a strong dependence on the circular polarization of the fundamental beam was oberved. This polarization dependence is nicely demonstrated in Fig. 7.21, which depicts the temperature

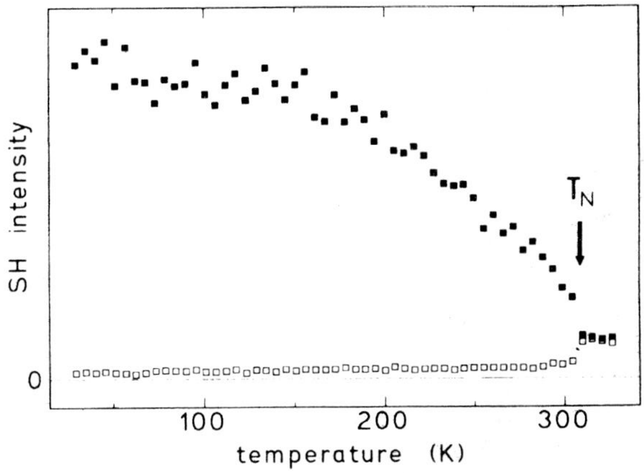

Fig. 7.21. Temperature dependence of the SH intensity of Cr_2O_3 at 2.00 eV in one domain. The Néel temperature $T_N = 307.5$ K is marked by an *arrow*. *Full* and *open* squares refer to right-hand and left-hand circularly polarization of the incoming laser beam, respectively

dependence of the SHG intensity for both right- and left-handed circularly polarized excitation beams.

The contrast between the two polarizations is striking. This polarization dependence could also be used to distinguish between different domains with opposite orientation of the antiferromagnetic vector, as the effect exactly reverses going from one domain to another. As such antiferromagnetic domains can not be distinguished in a linear magnetooptical experiment, this gives a unique possibility for future applications of nonlinear magnetooptics. In a recent publication, the same researchers have already demonstrated this, by visualizing the antiferromagnetic domain structure in Cr_2O_3 (Fig. 7.22).

7.7.9 MSHG Study of Co/Au: Observations of Quantum-Well States

As already discussed in Sect. 7.7.5, the discovery that magnetic films seperated by a nonmagnetic spacer layer could be coupled antiferromagnetically [80], and the subsequent discovery that this coupling could oscillate between ferromagnetic and antiferromagnetic [81], has triggered intense efforts to understand these phenomena [82]. Magneto-Optical Kerr Effect (MOKE) and direct and inverse photoemission experiments have shown direct evidence of such Quantum-Well (QW) states in thin, noble metal films on magnetic substrates (Fe and Co) [68,83], and in a bcc Fe(100) layer on Au(100) [84]. More recently, oscillations of the Kerr angle with changes of the overlayer-film thickness have been observed for a Au/Co/Au system [85]. These last results suggested the existence of quantum-size effects in a thin Au film and also gave an indication of possible interface contributions. As the latter could

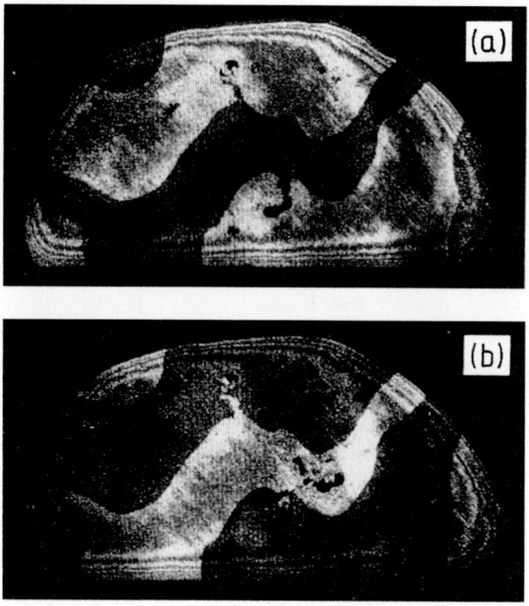

Fig. 7.22. MSHG visualization of antiferromagnetic domain structure in Cr_2O_3 [79]

only be obtained in an indirect way, a more detailed study of the interface effects can better be done using nonlinear magnetooptics. The study of the Cu/Co/Co(001) system (Sect. 7.7.5) already indicated the extreme sensitivity of MSHG for the presence of such QW states.

The investigation of possible QW states in thin Au film was initiated with the polar NOMOKE effect and the direct (nonmagnetic) SHG response. The experiments were done with the Ti:sapphire laser set-up, slightly modified to study the polar configuration.

The sample consisted of a 6 ML Co(0001) film on a 30 nm thick Au buffer layer, covered with a stepped Au wedge, consisting of 13 terraces of thickness t_{Au} = 6–18 monolayer. The terraces have a width of 1.5 mm, so the laser beam can easily be focused on each separate step. Details about the sample preparation can be found in [85]. Note that this sample has a perpendicular easy magnetization axis.

Table 7.1 infers that in order to observe the polar Kerr effect, p-polarized incident light has to be used, as for s-polarization, only even elements can be excited.

For p-polarized incident light the Co/Au interfaces have one odd component that gives rise to s-polarized SHG and three even components which produce p-polarized SHG. Experimentally, the nonlinear Kerr angle $\Phi_{\mathrm{K}}^{(2)}$ is found by fitting a $\cos^2 \alpha$ dependence to the measured intensity as a function

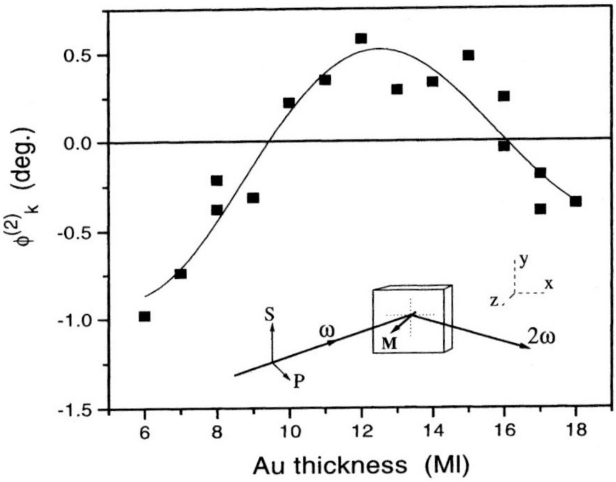

Fig. 7.23. Nonlinear polar Kerr rotation as a function of the Au overlayer thickness. The *solid line* is a fit to (7.69). The inset shows the experimental configuration

of the analyzer angle α, for opposite directions of the magnetic field. This leads to a precision in the nonlinear Kerr angle of $\pm 0.12°$.

The experimental variation of $\Phi_K^{(2)}$ versus t_{Au} plotted in Fig. 7.23 reveals a clear oscillatory behavior. These measurements have been done with p-polarized incident light at a wavelength of 740 nm and at an angle of incidence of 45°. The observed amplitude of the oscillation is about two orders of magnitude larger than its linear counterpart, as measured on the same sample [85]. The solid line in the figure is a least-square fit to the expression

$$\Phi_K^{(2)} = A \exp(t_{Au}/\delta) \cos\left(2\pi \frac{t_{Au}}{\Lambda} + \gamma\right) + \beta \,. \tag{7.69}$$

From this fit the period Λ of the oscillation is found to be 14 ± 3 ML with an attenuation length δ of 18 ± 4 ML. The precision of this value is not very large because of the limited size of the sample, and because of the uncertainty in the measured Kerr angles. The maximum measured Kerr angle is about 1.0°. Although this value is a factor of 10 larger than its linear counterpart, the rotation is not as large as one would expect from the nonlinear Kerr experiments discussed earlier. One reason for this is the phase difference of 180° between the tensor components of the mirrored interfaces Au/Co and Co/Au. Because of the small thickness of the Co film (6 ML), the SH signals from these interfaces will partly cancel each other, which reduces the magnetic effects.

The observed oscillation in the Kerr angle suggests that the quantum-size effects have a strong influence on the tensor components. Since these effects will not only appear in the magnetic (odd), but particularly in the nonmagnetic tensor elements (from the Au response), the QW states should also be observable in the total SH intensity.

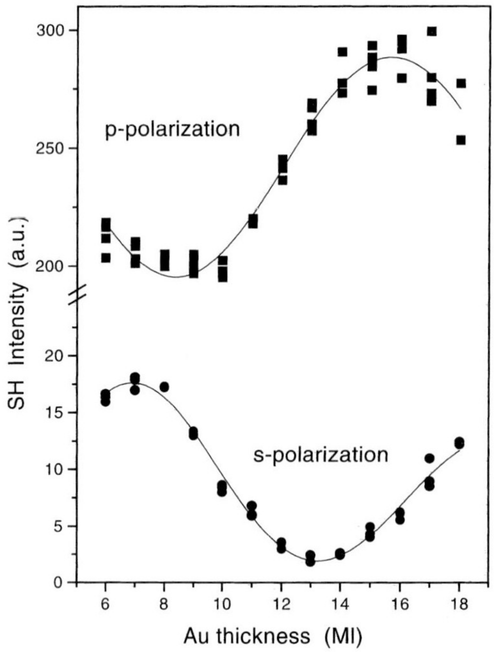

Fig. 7.24. Total generated SH intensity as a function of the Au overlayer. *Dots*, s-polarized incoming light; *squares*, p-polarized incoming light. The solid lines are fits to (7.68)

In Fig. 7.24 the SHG response of the Au film for p-polarized input is plotted as a function of the Au thickness showing a very clear oscillation. The *solid line* is a fit to (7.69) (with a period of 13.9 ML), but now for the intensity instead of the Kerr angle. However, now no damping is observed.

For comparison the linear reflection as a function of the Au thickness was measured as well. Within the experimental uncertainty of ±2% a constant value was found. The observed period of about 13.5 ML is quite remarkable, as the linear Kerr measurements (at 632 nm) on the same sample gave a period of 7.8 ML [85]. Experiments on the oscillatory interlayer coupling between Co films across Au(111) gave a period of 4.5 ML [86]. The discrepancy between the periods in the coupling and the Kerr measurements is probably related to the fact that for the coupling, only the electrons at the Fermi surface are relevant [66]. The difference between the oscillation period in the linear and nonlinear Kerr rotation could possibly be due to the differences in the selection rules that are involved in the linear and nonlinear optical response. This point, however, needs more theoretical investigations (see e.g. [91,92,99,100]. The question remains why (M)SHG is so extremely sensitive for the presence of QW states in these thin metallic films. The answer can probably be found in recent theoretical work [63], showing that these QW states have their density of states primarily at the interfaces. Oscillations in DOS, therefore, couple most strongly to the nonlinear optical surface response, and much more weakly to the linear optical bulk response. A similar oscillatory behavior was recently observed in the SHG response of a thin Au film, deposited on a Si substrate [87].

Together, these experimental and theoretical findings clearly show that (M)SHG will be one of the best techniques to study the properties of the interesting QW states. Clearly, future experiments should be done as a function of incident wavelength as well, to enable real spectroscopic studies of these states. Recent developments in laser technology and parametric oscillators have brought this within experimental possibilities. For example, a recently developed fs parametric oscillator using a KTP crystal, extends the tunability range of a Ti:sapphire laser to above 2 μm [88]. Preliminary spectroscopic studies on the same Au/Co system indicate that the oscillatory part of the signals originates from the intraband transitions in Au [93].

7.8 Summary and Future Prospects

The presence of a magnetization drastically alters the nonlinear optical response of metallic surfaces and interfaces. Though the origin of the nonlinear magnetooptical effects is the same as for the linear one, namely the spin-orbit coupling, the phenomena are quite different. The induced (*surface*) nonlinear polarization acts as a source term in the wave equation and gives rise to an inhomogeneous solution of the wave equation. This leads to a nonlinear Kerr rotation that is about one order of magnitude larger than its linear equivalent. The differences in the symmetry properties of the nonlinear and linear optical tensors can be exploited to obtain further enhancements of up to three orders of magnitude.

Apart from these strong magnetic effects, magnetization-induced second-harmonic generation is particularly attractive because of its ability to study the magnetic properties of buried interfaces. The large nonlinear Kerr rotations and large intensity changes as well as the interface specificity have been very clearly demonstrated by a number of studies.

The rather quick development of nonlinear magnetooptics in the last few years has been made possible through the mutual stimulation of experimental and theoretical work and the development of modelocked fs lasers. Right now, there is a fairly basic understanding of the origin and the strength of the nonlinear magnetooptical effects, and soon we expect the same for the experimental observation that (M)SHG is extremely sensitive to the presence of quantized electronic states in ultrathin metallic films (see e.g. [91]). What is clearly lacking at the moment are more extended spectroscopic nonlinear measurements. These should be able to reveal the nature and the energies of the relevant electronic surface and interface states, with much higher resolution than possible by electron spectroscopies. Another important aspect for

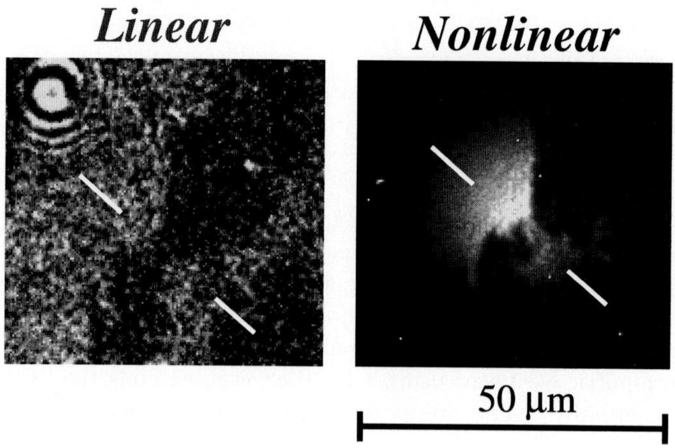

Fig. 7.25. Linear and second harmonic images of the same domain pattern in a 9 nm thick CoNi film sandwiched by Pt, showing much more details in the MSHG image. For example, the change of the contrast along the cross section between the white lines cannot be inferred from the linear picture

future applications is the sensitivity to interface roughness. SHG is known to be strongly affected by surface roughness. Qualitatively this has recently been demonstrated in a MSHG study of Au/Fe and CoNi/Pt interfaces. However, still much work needs to be done to turn this into a quantitative tool. An almost unexplored area so far is the application of nonlinear magnetooptics for the study of magnetic domain structures (see Fig. 7.25). Particularly for multilayered systems, MSHG can yield unique information about domain structures at deeper interfaces. In addition, nonlinear imaging techniques can yield other unique information, for instance regarding the domain structure in antiferromagnetic materials, which is altogether impossible with linear techniques. Finally, a very exciting new area is the study of spin dynamics, making use of pump probe techniques with fs mode locked lasers [97].

Acknowledgements. I would like to thank all the (former and present) members of our research group in Nijmegen that have contributed to the work described here. In particular, I want to mention H.A. Wierenga, W. de Jong, A.F. van Etteger, M. Groot Koerkamp and A. Kirilyuk. I owe a special thanks to the last two for their help in finalizing the manuscript. I thank Profs. J. Reif, D. Fröhlich and R.V. Pisarev for permission to use some of their figures, and Prof. J. Kirschner for the very fruitful collaboration. This work was part of the research program of the Stichting voor Fundamenteel Onderzoek der Materie (FOM) and financially supported by the Nederlandse Organisatie voor Wetenschappelijk Onderzoek (NWO) and by European Research Programmes Bright Euram and Human Capital and Mobility (ERBCHBGCT 930444) and TMR (NOMOKE network).

References

1. M. Campagna, D.T. Pierce, F. Meier, K. Sattler, H.C. Siegmann: Adv. Electron, and Electron. Phys. **41**, 113 (1976)
2. D.L. Abraham, H.H. Hopster: Phys. Rev. B **62**, 1157 (1989)
3. R. Feder (ed.): Polarized Electrons in Surface Physics (World Scientific, Singapore 1985). J. Kessler: Polaized Elechan, F, 2nd edn. Spinger Ser. on Atoms and Plasmas, Vol. 1 Springer, Berlin, Heidelberg 1985
4. Y.R. Shen: The principles of nonlinear optics (Wiley, New York, 1984).
5. Th. Rasing: Appl. Phys. A **59**, 531 (19994)
6. T.F. Heinz; in Nonlinear Surface Electromagnetic Phenomena, ed. By H.E. Ponath, G.I. Stegeman (North Holland, Amsterdam 1991) p. 335
7. G.L. Richmond, J.M. Robinson, V.L. Shannon: Prog. Surf. Sci. **28**, 1 (1988)
8. P.A. Franken, A.E. Hill, C.W. Peters, G. Weinreich: Phys. Rev. Lett. **7**, 118 (1961)
9. N. Bloembergen, P.S. Pershan: Phys. Rev. **128**, 606 (1962)
10. S. Janz, H.M. van Driel: Int. J. Nonlinear Opt. Phys. **2**, 1 (1992)
11. X.D. Zhu, Th. Rasing, Y.R, Shen: Phys. Rev. Lett. **61**, 2883 (1988): Chem. Phys. Lett. **155**, 459, (1989)
12. W. de Jong, A.F. van Etteger, C.A. van 't Hof, P.J. van Hall, Th. Rasing: Surf. Sci. **331-333**, 1372 (1995)
13. O.A. Aktsipetrov, E.D. Mishina: Sov. Phys. Dokl. **29**, 37 (1984)
14. P.N. Argyres: Phys. Rev. **97**, 334 (1995)
15. Ru-Pin Pan, H.D. Wie, Y.R. Shen: Phys. Rev. B **39**, 1229 (1989); Ru-Pin Pan, Y.R. Shen: Chin. J. Phys. (Tapei), **25**, 175 (1987)
16. W. Hübner, K.H. Bennemann: Phys. Rev. B **40**, 5973 (1989)
17. J. Reif, J.C. Zink, C.M. Schneider, J. Kirschner: Phys. Rev. Lett. **67**, 2878 (1991)
18. G. Spierings, V. Koutsos, H.A. Wierenga, M.W.J. Prins, D. Abraham, Th. Rasing: Surf. Sci. **287**, 747 (1993); J. Magn. Magn. Mat. **121**, 109 (1993)
19. H.A. Wierenga, W. de Jong, M.W.J. Prins, Th. Rasing, R. Vollmer, A. Kirilyuk, H. Schwabe J. Kirschner: Phys. Rev. Lett. **74**, 1462 (1995)
20. U. Pustogowa, W. Hübner, K.H. Bennemann; Phys. Rev. B **49**, 10031 (1994); Appl. Phys. A **59**, 611 (1994)
21. O.A. Aktsipetrov, O.V. Braginskii, D.A. Esikow: Sov. J. Quantum Electron. **20**, 259 (1990)
22. J. Reif, C. Rau E. Matthias: Phys. Rev. Lett. **71**, 1931 (1993)
23. B. Koopmans, M. Groot Koerkamp, Th. Rasing, H.v.d. Berg: Phys. Rev. Lett. **74**, 3692 (1995)
24. H.W. Wierenga, M.W.J. Prins, Th. Rasing: Physica B **204**, 281 (1995)
25. R.P. Hunt: J. Appl. Phys. **38**, 1652 (1967)
26. M.J. Freiser: IEEE Trans. **4**, 152 (1968)
27. M. Weber, A. Liebsch: Phys. Rev. B **35**, 7411 (1987) (A. Liebsch: Phys. Rev. B **36**, 7378 (1987)
28. Recent work by R. Vollmer, M. Straub, J. Kirschner. Surf. Sci. 352–352, **684** (1996) shows that bulk contributions can be sizeable even for metallic surfaces, depending on the polarization and frequency of the incident fundamental radiation.
29. H.W.K. Tom, G.D. Aumiller: Phys. Rev. B **33**, 8818 (1986)

30. P. Guyot-Sionnest, Y.R. Shen: Phys. Rev. B **35**, 4420 (1987); P. Guyot-Sionnest, W. Chen, Y.R. Shen: Phys. Rev. B **33**, 8254 (1986); D.L. Mills: Nonlinear Optics, 2nd edn. (Springer, Berlin, Heidelberg 1998)
31. N.N. Akkmediev, S.B. Borisov, A.K. Zvezdin, I.L. Lyubchanskii, Y.V. Melikhov: Sov. Phys.-Sol. State **27**, 650 (1985); S. Kielich, R. Zawadny: Acta Physica Polonica A **43**, 579 (1973); Optica Acta **20**, 867 (1973)
32. Note that Table 7.1 is slightly different from that presented by Ru-Pin Pan in [15]; this is probably due to overlooking of additional symmetry elements of the 100 surface.
33. W. Hübner, K.H. Bennemann: Phys. Rev. B **52**, 13411 (1995)
34. G. van der Laan, B.T. Thole, G.A. Sawatzky, J.B. Goedkoop, J.C. Fuggle, J.M. Esteva, R.Karnatak, J.P. Remeika, H.A. Dabkowska: Phys. Rev. B **34**, 6529 (1985)
35. M. Fiebig, D. Fröhlich, B.B. Krichevtov, R.V. Pisarev: Phys. Rev. Lett. **73**, 2127 (1994)
36. Quantum Physics S. Gasiorowicz (Wiley, New York, 1974)
37. Solid State Physics, N.W. Ashcroft, N.D. Mermin (Saunders College, Philadelphia, 1978)
38. W. Hübner, K.-H. Bennemann: Vacuum **41**, 514 (1990)
39. W. Hübner, K.-H. Bennemann: Surf. Sci. **242**, 299 (1991)
40. W. Hübner: Phys. Rev. B **42**, 11553 (1990)
41. U. Pustogowa, W. Hübner, K.-H. Bennemann: Phys. Rev. B **48**, 8607 (1993)
42. U. Pustogowa, W. Hübner, K.-H. Bennemann: Surf. Sci. **307-309**, 1129 (1994)
43. Y.R. Shen: Ann. Rev. Mater. Sci. **16**, 69 (1986)
44. A.J. Freeman, C.L. Fu: J. Appl. Phys. **61**, 3356 (1987)
45. B. Dick, A. Gierulski, G. Marowsky, G.A. Reider: Appl. Phys. B **38**, 107 (1985)
46. M.S. Yeganeh, J. Qi, J.P. Culver, A.G. Yodh, M.C. Tamargo: Phys. Rev. B **46**, 1603 (1992)
47. B. Koopmans, A. Anema, H.T. Jonkman, G.A. Savatzky, F. van der Woude: Phys. Rev. B **48**, 2759 (1993)
48. Handbook of optical constants of solids, ed. By E.D. Palik (Academic Orlando 1985)
49. J.E. Sipe: J. Opt. Soc. Am. B **4**, 481 (1987)
50. T.F. Heinz: Nonlinear optics of surfaces and adsorbates (PhD Thesis, Lawrence Berkeley Laboratory, Berkeley, USA (1982) unpublished)
51. Th. Rasing, Y.R. Shen: MRS Bull. 28-30 July 1988
52. W. Koechner: Solid-State Laser Engineering 5th edn., Springer Ser. Opt. Sci, Vol. 1 (Springer, Berlin, Heidelberg 1999)
53. H.A. Wierenga: Magnetization induced optical second harmonic generatino on magnetic multilayers, PhD Thesis, University of Mijmegen, the Netherlands (1995)
54. H.A. Wierenga, M.W.J. Prins, D.L. Abraham, Th. Rasing: Phys. Rev. B **50**, 1282 (1994)
55. R. Vollmer, A. Kirilyuk, H. Schwabe, J. Kirschner, H.A. Wierenga, W.de Jong, Th. Rasing: J. Magn. Magn. Mat. **148**, 295 (1995)
56. M. Groot Koerkamp, Th. Rasing: J. Magn. Magn. Mat. **156**, 213 (1996)
57. Th. Rasing, H.A. Wierenga: Ferroelectrics **162**, 217 (1994)
58. R. Vollmer, M. Straub, J. Kirschner: Surf. Sci. **352**, 937 (1996)
59. P.B. Johnson, R.W. Christy: Phys. Rev. B **9**, 5056 (1974): Phys. Rev. B **6**, 4370 (1972)

60. A. Cebollada, R. Miranda, C.M. Schneider, P. Schuster, J. Kirschner: J. Magn. Magn. Mater. **102**, 25 (1991)
61. E.R. Moog, C. Liu, S.D. Bader, J. Zak: Phys. Rev. B **39**, 6949 (1989)
62. R. Gamble, P.H. Lissberger: J. Opt. Soc. Am. A **5**, 1533 (1988)
63. P. van Gelderen, S. Crampin, Th. Rasing, J. Inglesfield: Phys. Rev. B **54**, R2343 (1996)
64. H.A. Wierenga, W. de Jong, M.W.J. Prins, Th. Rasing, R. Vollmer, A. Kirilyuk, H. Schwabe, J. Kirschner: Surf. Sc. **331-333**, 1294 (1995)
65. G. Penissard, P. Meyer, J. Ferré, D. Renard: J. Magn. Magn Mat. **146**, 55 (1995)
66. P. Bruno: J. Appl. Phys. **76**, 6972 (1994)
67. O. Keller, A. Zayats, A. Liu, K. Pedersen, F.A. Pudomin, E.A. Vinogradov: Opt. Commun. **115**, 137 (1995)
68. J.E. Ortega, F.J. Himpsel, G.J. Mankey, R.F. Willis: Phys. Rev. B **47**, 1540 (1993): Phys. Rev. Lett. **69**, 844 (1992)
69. N.B. Brookes, Y. Chang, P.D. Johnson: Phys. Rev. Lett. **67**, 354 (1991)
70. C. Carbone, E. Vescovo, O. Rader, W. Gudat, W. Eberhardt: Phys. Rev. Lett. **71**, 2805 (1993)
71. M.T. Johnson, S.T. Purcell, N.W.E. McGee, R. Goehoorn, J.aan de Stegge, W. Hoving: Phys. Rev. Lett. **68**, 2688 (1992)
72. See for example, S. Krompiewski, F. Süss, B. Zellermann, U. Kray: J. Magn. Magn. Mater. **148**, 198 (1995) and other contributions to 14th Int. Coll. Magn. Films and Surf., E-MRS Symp. On magnetic ultrathin films, multilayers and surfaces, Düsseldorf 1994, J. of Magn. Magn. Mater **148** (1995)
73. D.E. Aspnes, A.A. Studna: Phys. Rev. B. **27**, 985 (1983)
74. D.O. Smith; J. Appl. Phys. **36**, 1120 (1965)
75. M. Groot Koerkamp, Th. Rasing: Surf. Sci. **933**, 352–354 (1996)
76. W. Hübner, K. H. Bennemann: Phys. Rev. B **52**, 13411 (1995)
77. A.J. Kolk, M. Orlovic; J. Appl. Phys. **34**, 1060 (1963)
78. K. Böhmer, J. Hohfeld, E. Matthias: Appl. Phys. A **60**, 203 (1993)
79. M. Fiebig, D. Fröhlich, G. Sluyterman, R.V. Pisarev; Appl. Phys. Lett **66**, 2906 (1995)
80. P. Grünberg, R. Schreiber, Y. Pang, M.B. Brodsky, H. Sowers; Phys. Rev. Lett. **57**, 2442 (1986)
81. S.S.P. Parkin, N. More, K.P. Roche; Phys. Rev. Lett. **64**, 2304 (1990)
82. P. Bruno, C. Chappert; Phys. Rev. Lett. **67**, 1602 (1991); **67**, 2592(E) (1991); Phys. Rev. B **46**, 261 (1992)
83. D. Hartmann, W. Weber, A. Rampe, S. Popovic, G. Güntherodt: Phys. Rev. B **48**, 16837 (1993)
84. F.J. Himpsel: Phys. Rev. B **44**, 5966 (1991)
85. R. Mégy, A. Bounouh, Y. Suzuki, P. Beauvillain, P. Bruno, C. Chappert, B. Lecuyer, P. Veillet: Phys. Rev. B **51**, 5586 (1995)
86. V. Grolier, D. Renard, B. Bartenlian, P. Beauvillain, C. Chappert, C. Dupas, J. Ferré, M. Galtier, E. Kolb, M. Mulloy, J.P. Renard, P. Veillet: Phys. Rev. Lett. **71**, 3023 (1993)
87. K. Pedersen, P. Morgen: Physica Scripta T **54**, 238 (1994)
88. A. Anema. C.A. van 't Hof, A.F. van Etteger, Th. Rasing: unpublished
89. A. Kirilyuk, V. Kirilyuk, Th. Rasing, V.V. Pavlov, R.V. Pisarev: J. Magn. Soc. Jpn. **20**, 361–364 (1996)

90. V. Kirilyuk, A. Kiriluk, Th. Rasing: Appl. Phys. Lett. **70**, 2306–2308 (1997)
91. T. Luce, W. Hübner, A. Kirilyuk, Th. Rasing, K.H. Bennemann: Phys. Rev. B. **57**, 7377–7384 (1998)
92. A. Kirilyuk, Th. Rasing, R. Mégy, P. Beauvillain: Phys. Rev. Lett. **77**, 4608–4611 (1996)
93. A. Kirilyuk, A.V. Petukhov, Th. Rasing, R. Megy, P. Beauvillain: Surf. Sci. **409**, 377–379 (1997)
94. A. Kirilyuk, Th. Rasing, M.A.M. Haast, J.C. Lodder: Appl. Phys. Lett. **72**, 2331 (1998)
95. Q.Y. Jin, H. Regensburger, R. Vollmer, J. Kirschner: Phys. Rev. Lett. **80**, 4056 (1998)
96. K.J. Veenstra, E. Jurdik, A. Petukhov, Th. Rasing, to be published
97. F. Bentivegna, P. Yvernault, A.V. Petukhov, Th. Rasing: J. Magn. Soc. Jpn. S **1**, 131 (1999)
98. R. Vollmer, in *Nonlinear Optic in Metals*, ed. by K.H. Bennemann, Oxford Science Publ. (1998), pp. 42–131
99. Th. Rasing, in *Nonlinear Optics in Metals* (etc., see [98]), pp. 132–218
100. T. Luce, K.H. Bennemann, in *Nonlinear Optics in Metals* (see [98]), pp. 437–477

Index

Springer Series in Surface Sciences

Editors: G. Ertl, R. Gomer and D. L. Mills

Founding Editor: H.K.V. Lotsch

1 **Physisorption Kinetics**
By H. J. Kreuzer, Z. W. Gortel

2 **The Structure of Surfaces**
Editors: M. A. Van Hove, S. Y. Tong

3 **Dynamical Phenomena at Surfaces, Interfaces and Superlattices**
Editors: F. Nizzoli, K.-H. Rieder, R. F. Willis

4 **Desorption Induced by Electronic Transitions, DIET II**
Editors: W. Brenig, D. Menzel

5 **Chemistry and Physics of Solid Surfaces VI**
Editors: R. Vanselow, R. Howe

6 **Low-Energy Electron Diffraction**
Experiment, Theory and Surface Structure Determination
By M. A. Van Hove, W. H. Weinberg, C.-M. Chan

7 **Electronic Phenomena in Adsorption and Catalysis**
By V. F. Kiselev, O. V. Krylov

8 **Kinetics of Interface Reactions**
Editors: M. Grunze, H. J. Kreuzer

9 **Adsorption and Catalysis on Transition Metals and Their Oxides**
By V. F. Kiselev, O. V. Krylov

10 **Chemistry and Physics of Solid Surfaces VII**
Editors: R. Vanselow, R. Howe

11 **The Structure of Surfaces II**
Editors: J. F. van der Veen, M. A. Van Hove

12 **Diffusion at Interfaces: Microscopic Concepts**
Editors: M. Grunze, H. J. Kreuzer, J. J. Weimer

13 **Desorption Induced by Electronic Transitions, DIET III**
Editors: R. H. Stulen, M. L. Knotek

14 **Solvay Conference on Surface Science**
Editor: F. W. de Wette

15 **Surfaces and Interfaces of Solids**
By H. Lüth*)

16 **Atomic and Electronic Structure of Surfaces**
Theoretical Foundations
By M. Lannoo, P. Friedel

17 **Adhesion and Friction**
Editors: M. Grunze, H. J. Kreuzer

18 **Auger Spectroscopy and Electronic Structure**
Editors: G. Cubiotti, G. Mondio, K. Wandelt

19 **Desorption Induced by Electronic Transitions, DIET IV**
Editors: G. Betz, P. Varga

20 **Scanning Tunneling Microscopy I**
General Principles and Applications to Clean and Adsorbate-Covered Surfaces
Editors: H.-J. Güntherodt, R. Wiesendanger
2nd Edition

21 **Surface Phonons**
Editors: W. Kress, F. W. de Wette

22 **Chemistry and Physics of Solid Surfaces VIII**
Editors: R. Vanselow, R. Howe

23 **Surface Analysis Methods in Materials Science**
Editors: D. J. O'Connor, B. A. Sexton, R. St. C. Smart

24 **The Structure of Surfaces III**
Editors: S. Y. Tong, M. A. Van Hove, K. Takayanagi, X. D. Xie

25 **NEXAFS Spectroscopy**
By J. Stöhr

26 **Semiconductor Surfaces and Interfaces**
By W. Mönch
2nd Edition

27 **Helium Atom Scattering from Surfaces**
Editor: E. Hulpke

28 **Scanning Tunneling Microscopy II**
Further Applications and Related Scanning Techniques
Editors: R. Wiesendanger, H.-J. Güntherodt
2nd Edition

29 **Scanning Tunneling Microscopy III**
Theory of STM and Related Scanning Probe Methods
Editors: R. Wiesendanger, H.-J. Güntherodt
2nd Edition

30 **Concepts in Surface Physics**
By M. C. Desjonquères, D. Spanjaard*)

31 **Desorption Induced by Electronic Transitions, DIET V**
Editors: A. R. Burns, E. B. Stechel, D. R. Jennison

32 **Scanning Tunneling Microscopy and its Application**
By C. Bai

33 **Adsorption on Ordered Surfaces of Ionic Solids and Thin Films**
Editors: H.-J. Freund, E. Umbach

34 **Surface Reactions**
Editor: R. J. Madix

35 **Applications of Synchrotron Radiation**
High-Resolution Studies of Molecules and Molecular Adsorbates on Surfaces
Editor: W. Eberhardt

36 **Kinetics of Metal-Gas Interactions at Low Temperatures: Hydriding, Oxidation, Poisoning**
By E. Fromm

*) Available as a textbook

Printing: Mercedes-Druck, Berlin
Binding: Stürtz AG, Würzburg